Peptide Hormone Action

A Practical Approach

The Practical Approach Series

SERIES EDITORS

D. RICKWOOD

Department of Biology, University of Essex
Wivenhoe Park, Colchester, Essex CO4 3SQ, UK

B. D. HAMES

Department of Biochemistry, University of Leeds
Leeds LS2 9JT, UK

Affinity Chromatography
Animal Cell Culture
Animal Virus Pathogenesis
Antibodies I and II
Biochemical Toxicology
Biological Membranes
Biosensors
Carbohydrate Analysis
Cell Growth and Division
Centrifugation (2nd edition)
Clinical Immunology
Computers in Microbiology
DNA Cloning I, II, and III
Drosophila
Electron Microscopy in Molecular Biology
Fermentation
Flow Cytometry
Gel Electrophoresis of Nucleic Acids (2nd edition)
Gel Electrophoresis of Proteins (2nd edition)
Genome Analysis

HPLC of Small Molecules
HPLC of Macromolecules
Human Cytogenetics
Human Genetic Diseases
Immobilised Cells and Enzymes
Iodinated Density Gradient Media
Light Microscopy in Biology
Liposomes
Lymphocytes
Lymphokines and Interferons
Mammalian Development
Medical Bacteriology
Medical Mycology
Microcomputers in Biology
Microcomputers in Physiology
Mitochondria
Mutagenicity Testing
Neurochemistry
Nucleic Acid and Protein Sequence Analysis
Nucleic Acids Hybridisation

Peptide Hormone Action

A Practical Approach

Edited by

K. SIDDLE

and

J. C. HUTTON

University of Cambridge
Department of Clinical Biochemistry
Addenbrooke's Hospital
Hills Road, Cambridge CB2 2QR, UK

OXFORD UNIVERSITY PRESS
Oxford New York Tokyo

Oxford University Press
Walton Street, Oxford OX2 6DP

Oxford is a trade mark of Oxford University Press

Published in the United States
by Oxford University Press, New York

British Library Cataloguing in Publication Data
Peptide hormone action.
1. Hormones: Polypeptides. Biochemistry
I. Siddle, K.   II. Hutton, J. C.   III. Series
574.1927
ISBN 0–19–963070–4
ISBN 0–19–963071–2 pbk

Library of Congress Cataloging in Publication Data
Peptide hormone action: a practical approach/edited by J. C. Hutton, K. Siddle.
(Practical approach series)
Includes bibliographical references.
Includes index.
1. Peptide hormones—Research—Methodology.   2. Peptides hormones—
Receptors—Research—Methodology.   3. Second messengers
(Biochemistry)—Research—Methodology.   I. Siddle, K.   II. Hutton, J.C.   III. Series.
[DNLM: 1. Hormones—analysis.   2. Peptides—analysis.   WK 185 P4233]
QP572.P4P43   1990   612.4'05—dc20   90–7347
ISBN 0–19–963070–4 (hard)
ISBN 0–19–963071–2 (pbk.)

Typeset by Cotswold Typesetting Ltd, Gloucester
Printed by Information Press Ltd, Eynsham, Oxford

Preface

A full understanding of all the cellular processes involved in the biosynthesis, secretion, and action of a polypeptide hormone requires familiarity with almost every branch of biochemistry. However it is possible here to delve into only a few topics specifically relevant to endocrinology. It is fortunate perhaps that secretion and action are frequently reflections of similar processes. After all, the stimuli for secretion are most often other hormones and neurotransmitters, which may themselves be polypeptides. In one instance diverse stimuli converge on the regulation of one process, exocytosis, whereas in the other a single molecular event, the binding of a hormone to a specific receptor, may result in a cascade of cellular responses.

Insulin features prominently, though not exclusively, in this the second volume on methods for Polypeptide Hormones. No doubt this in part reflects the research interests of the editors. However it is also an indication of the importance of this hormone in the development of ideas and techniques in hormone research. Insulin has figured prominently as a model for studies of hormone biosynthesis and processing and for receptor characterization, as well as being significant in the early days of peptide sequencing, crystallography, and radioimmunoassay. Paradoxically, for what is probably the most intensively studied hormone, insulin remains one of the least well understood in terms of both its stimulus–secretion coupling and its mechanism of action.

In compiling this volume we have been fortunate to secure contributions from laboratories which are internationally acknowledged in their fields and have made important contributions to the development and application of the techniques described. While the companion volume, *Peptide Hormone Secretion*, concerns itself with analytical techniques for isolation, localization, and quantitation of hormones and for the study of biosynthesis and secretion, this volume covers aspects of hormone action with a special emphasis on receptors and second messenger systems. Topics covered include: hormone binding and receptor characterization; cyclic nucleotides, calcium, inositol phosphates and diacylglycerol as second messengers; and receptor-associated tyrosine kinases. Once again the intention is to provide both underlying principles and specific experimental protocols. It is hoped that this volume thus provides both an introduction for the newcomer to the field and some incentive to established workers to venture into less familiar aspects of the discipline.

Cambridge   K. SIDDLE
1990   J. C. HUTTON

vii

Contents

Contents

Contents

Contents

Contributors

A. K. CAMPBELL
Department of Medical Biochemistry, University of Wales College of Medicine, Heath Park, Cardiff CF4 4XN, UK.

D. R. COOPER
Division of Endocrinology/Metabolism, Department of Medicine, University of South Florida College of Medicine and Veterans Adminstration Hospital, Tampa, FL 33612, USA.

R. L. DORMER
Department of Medical Biochemistry, University of Wales College of Medicine, Heath Park, Cardiff CF4 4XN, UK.

R. V. FARESE
Division of Endocrinology/Metabolism, Department of Internal Medicine, University of South Florida College of Medicine and Veterans Administration Hospital, Tampa, FL 33612, USA.

S. GAMMELTOFT
Department of Clinical Chemistry, Bispebjerg Hospital, DK-2400 Copenhagen NV, Denmark.

M. B. HALLETT
Department of Surgery, University of Wales College of Medicine, Heath Park, Cardiff CF4 4XN, UK.

C. J. KIRK
Department of Biochemistry, University of Birmingham, P.O. Box 363, Birmingham B15 2TT, UK.

T. LAKEY
Molecular Pharmacology Group, Department of Biochemistry, University of Glasgow G12 8QQ, UK.

A. J. MORRIS
Department of Pharmacology, CB #7365 FLOB, University of North Carolina, NC 27514, USA.

G. MURPHY
Molecular Pharmacology Group, Department of Biochemistry, University of Glasgow, Glasgow G12 8QQ, UK.

J. E. PESSIN
Department of Physiology and Biophysics, The University of Iowa, Iowa City, Iowa 52242, USA.

N. PYNE

Department of Physiology and Pharmacology, University of Strathclyde, 204 George Street, Glasgow, UK.

S. B. SHEARS

Department of Pharmacology, National Institute for Environmental Health Sciences, N.I.H., Research Triangle Park, NC 27709, USA.

J. L. TREADWAY

Department of Physiology and Biophysics, The University of Iowa, Iowa City, Iowa 52242, USA.

M. WHITE

Research Division, Joslin Diabetes Center, Harvard Medical School, 1 Joslin Place, Boston, MA 02215, USA.

Abbreviations

αIR	insulin receptor antibody
AMP	adenosine monophosphate
ANF	atrial natriuretic factor
αPY	phosphotyrosine antibody
ATP	adenosine triphosphate
β-OG	n-octyl-β,D-glucopyranoside
BAPTA	1,2,-bis(2-aminophenoxy)ethane-N,N,N',N'-tetraacetic acid
Bq	Becquerel
BSA	bovine serum albumin
c.p.m.	counts per minute
cAMP	adenosine 3':5, cyclic monophosphate
cDNA	complementary DNA
cGMP	guanosine 3':5, cyclic monophosphate
CHO	Chinese hamster ovary
Ci	Curie
cIMP	inosine 3':5, cyclic monophosphate
CNBr	cyanogen bromide
Con A	concanavalin A
DAG	diacylglycerol
DEAE	diethylaminoethyl
DETAPAC	diethylenetriamine-pentaacetic acid
DHFR	dihydrofolate reductase
DiC$_5$	1,2-dipentanoyl glycerol
DiC$_8$	1,2-dioctanoyl glycerol
DMEM	Dulbeccos modified Eagles medium
DMF	dimethylformamide
DMSO	dimethylsulfoxide
DSS	disuccinimidyl suberate
DTT	dithiothreitol
EDTA	ethylenediamine tetraacetate
EGF	epidermal growth factor
EGTA	ethyleneglycolbis(β-aminoethyl)ether N,N,N',N'-tetraacetate
FACS	fluorescence-activated cell sorter
FPLC	fast performance liquid chromatography
G-3-PO$_4$	glycerol-3-phosphate
G-protein	guanine nucleotide-binding regulatory protein
G$_i$	adenylate cyclase-inhibitory G-protein
Gpp[NH]p	guanylyl-imidodiphosphate
Gro$P$ ins	glycerophosphoinositol
G$_s$	adenylate cyclase-stimulatory G-protein
GTP	guanosine triphosphate
HBSS	Hanks balanced salt solution

Hepes	*N*-2-hydroxyethylpiperazine *N'*-2-ethanesulfonic acid
HIR	human insulin receptor
HPLC	high performance liquid chromatography
IBMX	3-isobutyl-1-methylxanthine
IGF	insulin-like growth factor
Ins	*myo*inositol
Ins*P*	inositol phosphate
KLH	keyhole limpet haemocyanin
KRH	Krebs–Ringer Hepes buffer
M	molarity
MES	2[*N*-morpholino] ethanesulfonic acid
Mops	3-(*N*-morpholino)-propanesulfonic acid
mRNA	messenger RNA
NAD	nicotinamide adenine dinucleotide
NADH	nicotinamide adenine dinucleotide (reduced)
NBF	nucleoside binding factor
NMP	nucleoside monophosphate
NMR	nuclear magnetic resonance
NP-40	Nonidet P-40
NSB	non-specific binding
OAG	1-oleoyl, 2-acetyl-glycerol
PAGE	polyacrylamide gel electrophoresis
PAK	proteolysis-activated kinase
PBS	phosphate buffered saline
PCA	perchloric acid
PDBu	phorbol-12,13-dibutyrate
PDD	phorbol 12,13-didecanoate
PDE	phosphodiesterase
PDGF	platelet-derived growth factor
PI	phosphatidylinositol
P_i	inorganic phosphate
PI-glycan	phosphatidylinositol glycan
PIC	phosphoinositidase C
PIP_2	phosphatidylinositol-4',5'-bisphosphate
PMA	phorbol 12-myristate 13-acetate (12-O-tetradecanoyl phorbol 13-acetate, TPA)
PMSF	phenylmethylsulphonyl fluoride
PNPP	*p*-nitrophenyl phosphate
POPOP	1,4-bis[5-phenyloxazol-2-yl] benzene
PPO	2,5-diphenyloxazole
PtdIns	phosphatidylinositol
PtdIns*P*	phosphatidylinositol phosphate
r.p.m.	revolutions per minute
RIA	radioimmunoassay
R_T	total receptor number
SDS	sodium dodecyl sulfate
$T_{1/2}$	half time of association or dissociation
TBS	Tris-buffered saline

TCA	trichloroacetic acid
TEMED	N,N,N',N'-tetramethylethylenediamine
TEN	Tris/EDTA/NaCl buffer
TFA	trifluoroacetic acid
TLC	thin layer chromatography
TPA	12-O-tetradecanoyl phorbol 13-acetate (phorbol 12-myristate 13-acetate, PMA)
TPEN	N,N,N',N'-tetrakis-(2pyridyl-methyl)-ethylenediamine
TREACL	triethanolamine hydrochloride
Tris	tris(hydroxymethyl) aminomethane
Tym(P)	phosphotyramine
Tyr	tyrosine
WGA	wheat germ agglutinin

1

Peptide hormone receptors

STEEN GAMMELTOFT

1. Introduction

Peptide hormones induce their cellular actions via interaction with specific cell surface receptors. The characterization of hormone receptors has primarily been based on their cellular functions: recognition of hormone, transmission of chemical signal, and activation of cellular effector systems such as transport carriers, metabolic enzymes, and secretory vesicles. In addition, receptors mediate also proteolytic degradation of their peptide ligand via endocytosis of the receptor–hormone complex. Peptide hormone receptors are integral membrane proteins consisting of an extracellular domain with ligand binding activity, a transmembrane domain, and a cytoplasmic domain involved in signal transduction. Recently, receptors for several peptide hormones, as well as polypeptide growth factors and neuropeptides have been purified and their amino-acid sequences resolved from the cDNA structures of their precursors. It should be emphasized that peptide hormone receptors show no principal differences in their structure and function from receptors for growth factors and neuropeptides, and they can be considered as one group of cellular proteins. Based on structural homologies between receptors, different classes have been recognized. In spite of the recent progress in resolving amino-acid sequences of receptors from the nucleotide sequence of their cDNAs, functional assays of receptor-binding, internalization and cellular actions remain fundamental in the identification of receptors.

The subject of this chapter is the competitive radioligand assay and its application in studies of receptor-binding and endocytosis of peptide hormones. This remarkably simple methodology involves non-covalent association of radioisotope-labelled hormone with receptor and its competitive inhibition by unlabelled hormone. The development of the assay has been critically dependent on the preparation of bioactive radioiodinated hormone, availability of isolated cells, plasma membrane fractions, or solubilized receptors, and application of rapid centrifugation or filtration procedures. Other techniques for detection of peptide hormone receptors: affinity labelling, immunochemical interaction, and biosynthetic or metabolic labelling, will be treated in a subsequent chapter.

2. Labelling of peptide hormones

Almost all receptor studies have been carried out with peptide hormones labelled with ^{125}I. There are several advantages to this technique: the radioiodination is fairly easy to perform, a product of high specific radioactivity $(0.5–5 \times 10^{16}$ Bq/mol or $0.15–1.5 \times 10^{16}$ Ci/mol) is obtained, and the half-life $(T_{\frac{1}{2}} = 60.2$ days) is reasonably long. Among other radioisotopes, ^{131}I was used in early work, and labelling with ^3H has been introduced only by few laboratories. The major drawbacks are the short half-life $(T_{\frac{1}{2}} = 8.05$ days) of ^{131}I and the low specific radioactivity of ^3H-labelled peptides (1–2.5 MBq/nmol or 30–80 μCi/nmol). Iodination of a peptide or protein is generally performed by chemical substitution of hydrogen in tyrosyl residues. Iodine substitution in the histidine residues can also occur under certain conditions. In the iodination reaction, iodide (e.g. $^{125}I^-$) has to be oxidized to iodine using, for example, chloramine-T, iodate (IO_3^-), lactoperoxidase and H_2O_2, or iodogen. Peptides which do not contain tyrosine or histidine may be iodinated by the Bolton–Hunter procedure where a chemically reactive group containing ^{125}I is covalently linked to the amino-terminus (1). The Bolton–Hunter procedure may also be used when a tyrosine residue is located in the receptor-binding or antibody-binding region of the peptide as in cholecystokinin or gastrin (2). Another approach has been to introduce a tyrosine residue by chemical synthesis of analogues of the peptide. These tyrosine–peptide analogues can then be iodinated by oxidation and used as tracers. Examples are shorter peptides like Tyr_1-somatostatin and Tyr_8-substance P which have been used as tracers in receptor-binding studies.

3. Iodination procedures

3.1 Chloramine-T method

The general principle was introduced by Hunter and Greenwood (3) for the iodination of growth hormone. The chloramine-T method has been widely used in various modifications for iodination of many peptide hormones. The nature of the iodinated peptide depends on the iodination conditions, in particular the pH, the molar ratio of iodine to peptide, the peptide concentration and the amount of chloramine-T. All these parameters may be varied in order to obtain an iodinated peptide preparation which is chemically homogenous, fully biologically active, immunoreactive, and structurally unaltered. In general, the following conditions are recommended: a physiological pH of 7.4, a molar ratio of iodine to peptide below 1.0, and peptide and chloramine-T concentrations of the same order of magnitude. In particular, the chloramine-T concentration should be as low as possible to avoid oxidation damage of the peptide. The result of the iodination reaction will often be a heterogeneous mixture of various monoiodinated, and diiodinated peptide isomers as well as native peptide. These molecules can be separated using chromatographic or electrophoretic methods. A procedure for iodination of insulin is given as example (*Protocol 1*) (4).

Protocol 1. Iodination of insulin, using chloramine-T

Materials
(a) Na ^{125}I either from New England Nuclear, Boston, Massachusetts, USA (sp. act. 6.3×10^{11} Bq/mg or from The Radiochemical Centre, Amersham, UK. (sp. act. 4.8–6.3×10^{11} Bq/mg). The ^{125}I is dissolved in NaOH, pH 7–11 at a concentration of 3.7–10×10^9 Bq/ml.

(b) Phosphate buffer 0.25 mol/litre, pH = 7.4 prepared by mixing 8 ml KH_2PO_4 (0.25 mol/litre) and 42 ml Na_2HPO_4 (0.25 mol/litre) and adjusting pH at 7.4.

(c) Phosphate buffer 0.04 mol/litre, pH = 7.4 prepared by diluting 10 ml 0.25 mol/litre phosphate buffer (b) with 52.5 ml H_2O and adjusting pH at 7.4.

(d) Chloramine-T 2.5 mg/ml in 0.04 mol/litre phosphate buffer (c).

(e) Sodium metabisulphite 5 mg/ml in 0.04 mol/litre phosphate buffer (c).

(f) Bovine serum albumin 25 mg/ml in 0.04 mol/litre phosphate buffer (c).

(g) Gel filtration buffer: Ammonium acetate 0.1 mol/litre, pH = 7.4 with bovine serum albumin 2 mg/ml.

(h) Sephadex G-50 medium column (30 cm × 1 cm) equilibrated with (g).

(i) Insulin 1 mg/ml in HCl 10 mmol/litre.

Procedure
1. In an iodination vial at room temperature add the reagents in the following order:
   (a) 60 μl 0.25 mol/litre phosphate buffer
   (b) 10 μl Na^{125}I(3.7×10^7 Bq $\sim$ 1 mCi)
   (c) 10 μl insulin ($\sim$ 10 μg)
   (d) 10 μl chloramine-T ($\sim$ 0.4 μg).

2. Incubate for 45 sec while mixing vigorously on a whirly mixer.

3. Stop the reaction by addition of 10 μl sodium metabisulphite ($\sim$ 50 μg).

4. Add 250 μl bovine serum albumin solution.

5. The incorporation of iodine in insulin is checked by precipitation in 10% trichloroacetic acid. It should be at least 80% and depends on the efficiency of the stirring. Another protocol adds significantly lower amounts of chloramine-T ($\sim$ 0.5–1 μg) in a step-wise manner and checks the incorporation after each step. At 80–90% incorporated iodine the reaction is stopped by addition of sodium metabisulphite.

6. ^{125}I-insulin can be separated by gel filtration from ^{125}I-insulin aggregates and free ^{125}I-iodide. The mixture is applied on a Sephadex G-50 column and

Protocol 1. *Continued*

eluted with ammonium acetate and bovine serum albumin at a flow-rate of 0.5 ml/min. Fractions of 1.0 ml are collected and aliquots are counted for radioactivity. The labelled insulin preparation is stored at $-20°C$.

3.2 Iodogen method

Iodogen was first described by Fraker and Speck (5) as a reagent for iodination of proteins and cell membranes. Virtually insoluble in water, iodogen (1,3,4,6-tetrachloro-3α, 6α-diphenylglycouril) can be dissolved in an organic solvent and the iodination tube coated with the solution. Iodogen mediates rapid iodination in the solid phase with aqueous solutions of iodine and peptides. The reaction is simply stopped by decanting the tube. Therefore no reducing agent is needed. The advantages of the method is its simplicity and the reduced exposure of peptide to the oxidant in the two-phase system. Consequently, oxidative damage of the peptide is minimized. A procedure for iodination of pancreatic polypeptide is given as example (*Protocol 2*) (6).

Protocol 2. Iodination of pancreatic polypeptide, using Iodogen

Materials
(a) $Na^{125}I$ (see 3.1.1)
(b) Phosphate buffer 0.25 M, pH = 7.4
(c) 1,3,4,6-tetrachloro-3α,6α-diphenylglycouril (iodogen)
(d) Dichloromethane
(e) Pancreatic polypeptide 1.5 nmol
(f) HPLC solvent: Acetonitrile 28% in trifluoroacetic acid in water
(g) Nucleosil 300–5 C18 column (25 cm × 0.4 cm) equilibrated with (f).

Procedure
1. Iodogen is dissolved in 2.0 ml dichloromethane and the inside of an Eppendorf test-tube is coated with 20 μl of this solution by gently turning the tube while the solvent evaporates.
2. Pancreatic polypeptide 1.5 nmol is reconstituted in 40 μl phosphate buffer and added to the tube.
3. $Na^{125}I$ (3.7×10^7 Bq ~ 1 mCi) is added and the tube is kept on ice for 5 min.
4. The reaction is stopped by adding 50 μl of the HPLC solvent.
5. The mixture is applied on a Nucleosil 300–5 C18 column and eluted isocratically at 50°C with a flow-rate of 1.0 ml/min. Fractions of 0.5 ml are collected and aliquots of 10 μl counted in a γ counter.

3.3 Bolton–Hunter method

Bolton and Hunter (1) described a method for labelling of proteins which involved iodination of *N*-succinimidyl 3-(4-hydroxy-phenyl) proprionate with ^{125}I and then using the derivative thus formed as an acylating agent to react with free amino groups in proteins. This method is applicable in three cases: peptides which are susceptible to oxidation damage and denatured following iodination; peptides where tyrosine residues are included in the receptor- or antibody-binding region; and peptides which do not contain tyrosine residues. In contrast, Bolton–Hunter labelling can not be used in peptides where free α or ε amino groups, i.e. amino terminus and lysine are involved in the biological activity. A procedure for iodination of cholecystokinin-33 is given as example (*Protocol 3*) (2).

Protocol 3. Iodination of cholecystokinin with Bolton–Hunter reagent

Materials

(a) ^{125}I-labelled Bolton–Hunter reagent either from New England Nuclear, Boston, Mass., USA (sp. act. 81.4 TBq $\sim$ 2200 Ci/mmol) or the Radiochemical Centre, Amersham, UK (sp. act. 74 TBq $\sim$ 2000 Ci/mmol) is dissolved in dry benzene.

(b) Borate buffer 0.1 M, pH = 8.5

(c) Borate buffer 0.1 M with glycine 0.2 M

(d) Gel filtration buffer: sodium phosphate 0.05 M, pH = 7.5

(e) Sephadex G-50 column (1 m × 1 cm) equilibrated with (d)

(f) Cholecystokinin-33 10 μg

Procedure

1. The ^{125}I-labelled Bolton–Hunter reagent is evaporated under nitrogen at room temperature for about 20 min.

2. Cholecystokinin-33 10 μg dissolved in 20 μl borate buffer is added to the Bolton–Hunter reagent, and kept refrigerated for 20 h.

3. The reaction is finished by adding 250 μl borate–glycine buffer and applying the sample on a Sephadex G-50 column. The column is eluted with the sodium phosphate buffer at a flow rate of 3 ml/h. Fractions of 1 ml are collected and aliquots of 10 μl are counted for radioactivity. The fractions containing ^{125}I-labelled cholecystokinin-33 are diluted with acetic acid 0.2 M to the appropriate concentration of radioactivity and stored at $-20°C$.

3.4 Other iodination methods

Two other methods, the lactoperoxidase and iodate methods, have been described for iodination of peptides or proteins by oxidation. The principle of the lactoperoxidase method is that iodide (e.g. $^{125}I^-$) is oxidized to iodine by H_2O_2 and lactoperoxidase (7). The oxidation by this reaction has been considered to cause less oxidative damage of the peptide compared with the chloramine-T reaction. A careful adjustment of the amount of chloramine-T to the amount of peptide (see Section 3.1) may, however, reduce the oxidative damage. The reaction scheme and details are described in references (7) and (8).

In the iodate method, IO_3^- is used as oxidative reagent, and the reaction scheme and the details have been described by Jørgensen and Larsen (9). The theoretical maximal specific activity of an iodinated peptide prepared by this method is only two-thirds of that of the ^{125}I, because one ^{127}I atom (from IO_3^-) is incorporated for every two ^{125}I atoms.

3.5 Heterogeneity of iodinated peptides

Oxidative iodination of peptides may result in heterogenous preparations containing various monoiodinated and diiodinated peptide isomers. The presence of monoiodinated peptide isomers depends on the number of tyrosine residues in the particular peptide as well as the iodination method. The percentage radioactivity in diiodinated peptides depends on the degree of iodination and the method. Two forms of diiodinated peptides exist: those with diiodotyrosine and those with two monoiodotyrosines.

The content and properties of mono- and diiodinated isomers have been studied in great detail with iodinated insulin preparations (8, 10, 11). Insulin contains four tyrosines, two in the A-chain (residues A14 and A19) and two in the B-chain (residues B16 and B26). After the iodination the reaction mixture will consist of insulin, the four monoiodoinsulin isomers and various diiodoinsulin isomers. The ratio diiodoinsulin/monoiodoinsulin/insulin will increase with the degree of iodination, i.e. mol iodide/mol insulin in the iodination mixture. The distribution of the iodine among the four tyrosine residues will depend on their availability during the reaction. This availability is dependent on several factors such as pH, ionic strength, and protein denaturation.

Insulin is a globular protein with a hydrophobic core and hydrophilic surface and all four tyrosines are on the outside of the monomeric molecule. At the concentration of insulin in the iodination mixture ($\sim 5 \times 10^{-5}$ mol/litre) insulin forms dimers ($Kd \sim 10^{-6}$ mol/litre), and the tyrosines B16 and B26 are buried in the non-polar core of the dimer, whereas tyrosines A14 and A19 are on the surface of the dimer. This explains that 80–90% of the iodine is located in the A chain and only 10–20% in the B chain. The fractions of unreacted insulin, monoiodoinsulin, and diiodoinsulin in the iodination mixture using the chloramine-T method and the procedure described in *Protocol 1* have been estimated as 70% uniodinated and 30% iodinated insulin. About one-third of the radioactivity will be in diiodoinsulins, and the rest will be distributed between the four monoiodoinsulin isomers with an overweight ($\sim 70\%$) in $[^{125}I]$monoiodo[Tyr A14]insulin. Iodination of insulin with the lactoperoxidase and iodogen methods gives a similar distribution, whereas the iodate method results in predominant incorporation of the iodine in tyrosine A19. This is probably because the iodination is carried out under acid conditions (pH ~ 3).

In the iodination reaction two other forms of radioactivity are present: free $^{125}I^-$ and labelled polymers. Both molecules are removed by gel filtration on Sephadex G-50, whereas native, monoiodinated, and diiodinated insulin elutes as approximately one peak. Free $^{125}I^-$ and labelled polymers are also formed as decay products primarily from [^{125}I]diiodoinsulins, and it is therefore recommended to rechromatograph the labelled insulin preparation 2–4 weeks after iodination. Alternatively, the monoiodoinsulin isomers can be separated from the mixture using polyacrylamide gel electrophoresis, ion-exchange chromatography, or HPLC. This is highly advisable for three reasons. First, the four monoiodoinsulin isomers show different biological properties. In isolated fat cells, the binding affinities and biological potencies of the A14, A19, B16, and B26[^{125}I]monoiodoinsulin isomers are 100%, 55%, 110%, and 200% relative to native insulin (8, 10). Second, the diiodotyrosine insulins show the same relative activities as the corresponding monoiodoinsulins, but the [^{125}I]diiodotyrosine insulins in the mixture cause problems due to their decay to $^{125}I^-$ and labelled polymers which bind non-specifically to cells or membranes. In contrast, monoiodoinsulin preparations show no non-specific binding, and any background counts are due to trapping of medium. Third, the presence of about 70% uniodinated insulin in the mixture reduces the specific activity to about 2.7×10^{16} Bq/mol compared with a value of about 8×10^{16} Bq/mol for [^{125}I]-monoiodo-insulin (see Section 3.7). These arguments hold true for most preparations of ^{125}I-labelled peptides.

3.6 Validation of tracers

The validation of tracers includes their chemical and biological properties. The chemical validation includes the yield of radioiodinated tracer, its chromatographic or electrophoretic behaviour during purification (see Section 3.5), the precipitability in 10% trichloroacetic acid, and the stability during storage. The peptide tracer should be stored at $-20°C$ in buffer containing 1% albumin at a concentration up to 3.7×10^7 Bq/ml. Small aliquots should be stored and lyophilization is not recommended.

The biological properties of the tracers include their biological activity and immunological reactivity. Various factors like oxidative damage, number of iodine atoms incorporated and iodination of specific tyrosine residues may alter the properties of the tracer. The biological activity of the labelled peptide may be tested in receptor-binding and biological assays. In a competitive receptor-binding assay, using cells or membranes, the binding of the labelled peptide at different concentrations should be measured in the absence and presence of an excess of unlabelled peptides (see Sections 4.8 and 4.9). In a cellular or cell-free bioassay the potency of the labelled peptide should be determined over a tenfold concentration range and compared with a dose–response curve of the unlabelled peptide. Based on the value of the specific activity (see Section 3.7), the relative potency of the tracer can be estimated. Finally, the immunoreactivity of the tracer can be evaluated in a radioimmunoassay (see *Peptide hormone secretion* Chapter 3) using a 'self-displacement' curve where antibody-bound tracer is measured over a tenfold concentration range and compared with the displacement curve of unlabelled peptide. It should be emphasized that the immunoreactivity and the biological activity of the tracer should be regarded as complementary parameters because the antigenicity and bioactivity of the peptide are often related to different structures of the peptide. The antigenic site may be composed of a linear peptide sequence of 5–6 residues, whereas the bioactive site is composed of residues which are brought together by the tertiary structure of the peptide. The latter may thus be more sensitive to oxidative damage or denaturation of peptide during iodination than the immunoreactivity.

3.7 Determination of specific activity of the labelled peptide

The specific activity of the tracer can be calculated from the following parameters: the specific activity of iodide, the amounts of added iodide and peptide, and the fraction of incorporated iodine according to the formula:

$$\text{sp. act. of iodide} \times \frac{\text{mol iodide}}{\text{mol peptide}} \times \text{fract. of incorp. iodine.}$$

^{125}I is sold from the suppliers at specific activities near the theoretical maximum, i.e. 'carrier-free' and with an isotopic abundance approaching 100%. The specific activity is given as 4.8–6.3×10^{11} Bq/mg iodide or $\sim 8.0 \times 10^{16}$ Bq/mol (2.1×10^6 Ci/mol). If a monoiodinated tracer is prepared with 100% isotopic abundance the theoretical maximal specific activity should thus be $\sim 8.0 \times 10^{16}$ Bq/mol. In practice, the value may be slightly lower, depending on the content of ^{127}I in the iodide and monoiodinated peptide.

The specific activity of an unfractionated iodinated peptide preparation will vary significantly with the degree of iodination. As an example, the chloramine-T iodination of insulin described above (see Section 3.2) will result in $[^{125}$I]insulin with a specific activity of about 2×10^{16} Bq/mol ($\sim 5 \times 10^5$ Ci/mol). This is based on a ratio of added iodide and insulin of about ~ 0.8. The amount of added $[^{125}$I]iodide is calculated from the amount of added radioactivity and the specific activity. The fraction of incorporated $[^{125}$I]iodine is determined by trichloro-acetic acid precipitation.

4. Receptor-binding of peptides

A receptor-binding assay involves incubation of ^{125}I-labelled peptide and cellular, membrane, or soluble receptors followed by separation of bound and free tracer by sedimentation or filtration of cells or membranes, precipitation of soluble receptors, or removal of free radioactivity by charcoal adsorption. The assay conditions may vary significantly depending on the peptide and receptor, in particular the binding affinity and capacity. These parameters determine the concentration of ^{125}I-labelled peptide, and the amount of cells, membranes or receptors to be used in the assay. Other factors are the composition of the incubation buffer, the temperature and time of incubation, additions to reduce adsorption and degradation of [^{125}I]peptide, and the separation procedure. The individual factors will be discussed below.

4.1 ^{125}I-labelled peptide

A monoiodinated tracer with full binding affinity should be used, if available, due to its stability and negligible non-specific binding. The concentration of ^{125}I-labelled peptide should be kept as low as possible to obtain an optimal ratio between receptor-bound and non-specifically bound peptide. The concentration should not exceed one-tenth of the value of K_d to avoid errors in the quantitative analysis of steady-state binding and estimation of K_d (12). Use of tracers with lower or higher binding affinity than the unlabelled native peptide in displacement experiments will result in under- or over-estimation, respectively, of the total concentration of receptors (i.e. binding capacity). The value of the dissociation constant (i.e. binding affinity) will be correctly estimated equal to that of the native peptide (see reference 13 for detailed discussion).

A major problem is to use low concentrations of tracer and still to have radioactivity enough for accurate measurements. Let us consider a 'typical' receptor-binding assay with isolated cells or membranes. The incubation volume is 250 μl, the amount of receptors is approximately 10 fmol per test tube and their binding affinity is 1 nmol/litre. Under these conditions the total binding at 'tracer' concentration is 5% of the added radioactivity. It may be desirable to measure binding at a tracer concentration corresponding to the low concentrations occurring in plasma or extracellular fluid under physiological conditions say 4×10^{-11} mol/litre. Furthermore, this concentration is less than one-tenth of the value of K_d. One test-tube of 250 μl contains 10 fmol ^{125}I-labelled peptide with a specific activity of 2×10^{16} Bq/mol. Each sample contains 200 Bq, corresponding to 12000 disintegrations per minute or approximately 10000 counts per minute (c.p.m.) in a γ-counter with 80% efficiency. The total binding, therefore, represents 500 c.p.m. and the non-specific binding 50 c.p.m. Both numbers have been corrected for the background of the counter, approximately 30 c.p.m. This example shows that receptor-binding of an ^{125}I-labelled peptide can be measured in the relevant concentration.

4.2 Receptor preparation

Receptors for peptide hormones, growth factors, or neurotransmitters are glycoproteins which are located on the cell surface. Receptor binding may thus be studied, using intact cells either in suspension or in monolayer culture, crude or purified plasma membrane preparations, or solubilized receptors which have been partially purified as glycoproteins or completely purified by ligand or antibody affinity chromatography. The receptor concentration in the assay mixture should be in the order of 50 pmol/litre to ensure sufficient amounts of receptor-bound radioactivity as discussed above (see Section 4.1). This can be obtained with cell concentrations varying from 10^5–10^8 cells/ml depending on the cell type and its receptor content, which may vary from 10^2–10^5 per cell. Plasma membrane preparations may be used in concentrations of 0.1–1 mg protein/ml depending on their degree of purification and receptor content. Partially or fully purified receptors can be added to obtain the concentration desired. In general, the cell, membrane or protein concentration should be kept as low as possible to fulfil three requirements: the magnitude of receptor-bound [^{125}I] peptide should be 3–10 times higher than non-specific binding; the receptor binding should be proportional to the concentration of cells, membranes, or receptor protein; and the proteolytic degradation of the labelled peptide should be minimal, i.e. less than 5% of total tracer. The last condition may also be achieved by decreasing the incubation temperature to 15°C or 4°C and by addition of protease inhibitors. Bacitracin 0.5–1.2 mg/ml inhibits the degradation of most peptides by intact cells or membranes, whereas phenyl-methyl-sulphonyl-fluoride (PMSF) 0.1–0.2 mg/ml can be used with solubilized receptors.

A complete description of the procedures for preparation of cells, membranes, or solubilized receptors is outside the scope of this handbook, due to the variations in techniques between different tissues and cell types. The general principles with typical examples will be briefly reviewed. A biochemical identification of receptors for regulatory peptides in a particular cell type relies on a homogenous preparation of suspended cells, membranes or solubilized receptors, whereas intact organs or tissue slices are not suitable due to high levels of non-specific binding or trapping of ^{125}I-labelled peptide. Exceptions have been reported for the demonstration of insulin receptors in incubation of mouse soleus muscle and in perfusions of liver or heart with [^{125}I]insulin (for a review, see reference 13). Furthermore, peptide receptors have been visualized at the light and electromicroscopic levels by immunochemical or autoradiographic methods using tissue sections.

4.3 Preparation of isolated cells

Apart from the circulating blood cells, the preparation of isolated cells is based on a combination of chemical and mechanical treatments of the tissue: removal of Ca^{2+}-ions, proteolytic digestion, and mechanical dispersion. The chemical treatment may be performed by perfusion of intact organs or by incubation of tissue slices with Ca^{2+}-chelators and collagenase. The isolated cells are separated from undigested tissue by gentle mechanical treatment followed by filtration and differential centrifugation. A Percoll density gradient centrifugation may be applied to separate viable cells from damaged cells. In *Protocol 4*, the procedure for the preparation and maintenance of adrenal chomaffin cells in culture is described (14). It should be emphasized that, in principle, this technique can be applied on other tissues. The techniques for animal cell culture are described in another volume in the *Practical Approach Series*.

Protocol 4. Preparation of isolated adrenal cells

Materials
(a) Intact bovine adrenal glands are obtained from the slaughterhouse and transported on ice to the laboratory.
(b) Buffer composed of NaCl 154 mM, KCl 2.55 mM, K_2HPO_4 2.13 mM, KH_2PO_4 0.88 mM, D-glucose 10 mM, Hepes 15 mM at pH 7.4.
(c) Collagenese (Sigma C 0130) 0.3 g and DNase 0.5–1 mg are dissolved in 120 ml buffer.
(d) Percoll solution is prepared by mixing 45 ml Percoll (Pharmacia Biochemicals) with 5 ml 10 × concentrated buffer (b).
(e) Culture medium consisting of a 50:50 mixture of DMEM (buffered with 15 mM Hepes and 35 mM $NaHCO_3$) and Hams F-12 (buffered with 15 mM Hepes and 11 mM $NaHCO_3$) supplemented with 10% fetal calf serum and 100 μg/ml each of penicillin and streptomycin.

Procedure
1. The glands are trimmed by removing the surrounding adipose tissue without cutting the capsule surrounding the cortext. The glands are sprayed with 70% ethanol.
2. The blood contained within the gland is removed by 'pumping up' the gland with buffer via a 10-ml plastic syringe whose tip is inserted into the exit of the adrenal vein.
3. Each gland is filled with 10 ml of collagenase solution and incubated for 15 min at 37°C in a beaker with the collagenase perfusate. This step is repeated, reusing the same collagenase solution.
4. The gland is cut in half longitudinally and the medullary tissue, which is quite soft and partially digested, is scooped out. The tissue is chopped into small pieces with a pair of scissors and incubated with fresh collagenase solution in a shaking water-bath (140 cycles/min) at 37°C for 30 min.
5. The cell suspension is filtered through a coarse nylon mesh filter (500 μm) to remove undigested material, and centrifuged 50g for 15 min. The cells are resuspended in buffer and centrifuged three times, followed by filtration through a fine nylon mesh filter (60 μm).
6. The chromaffin cells are purified on a Percoll density gradient by mixing the cell suspension with Percoll solution and centrifuging the preparation 47 000g for 20 min (20 000 r.p.m. in a centrifuge with radius of 15 cm). Three bands

Protocol 4. *Continued*

appear in the gradient: the top layer consists of adrenal cortical cells, the intermediate layer is that of the chromaffin cells and the bottom layer is a discrete layer of red cells.

7. The chromaffin cells are collected, washed once in fresh buffer by centrifugation $50g$ for 5 min and resuspended in culture medium at a density of $\sim 5-10 \times 10^5$ cells per millilitre. The cells are plated on culture dishes that have been pre-coated with rat-tail collagen or poly-D-lysine and are incubated at $37°C$ in a 10% CO_2 atmosphere for 3 days. The medium with dead cells is removed before cells are used in experiments.

4.4 Preparation of plasma membranes

Plasma membranes are used in receptor measurements for the following reasons: peptide hormone receptors are membrane glycoproteins and the membrane preparation is enriched in binding capacity; plasma membranes can be prepared from solid tissues by simple homogenization and fractionation; the binding activity of membrane receptors is stable during long-term storage under proper conditions; receptor functions other than binding, like intrinsic kinase activity, ion channel activity, non-covalent associated enzyme activities (adenylyl cyclase, phospholipase, etc.) can be studied in membrane preparations. A disadvantage of membrane preparations in receptor studies is that the receptor-coupling with cellular actions of the hormone cannot be studied.

Preparation of plasma membranes is now a fundamental technique in biochemistry and is based on a combination of mechanical disruption of the tissue and fractionation of cellular organelles by centrifugation. The mechanical treatment is performed by homogenization, using a homogenizer either with pestle or knives depending on the composition of the tissue. With softer tissues (e.g. brain, liver) a hand- or motor-driven pestle should be used, and with harder tissues (e.g. muscle, connective tissue) a motor-driven knife homogenizer is needed. In some cases, cell lysis in hypotonic buffer is also applied. The cellular organelles are separated by differential centrifugation at $1000-50\,000g$ followed by density gradient centrifugation using sucrose or Percoll. In *Protocol 5*, a procedure for preparation of placental membranes is described, but techniques for other tissues are described in the *Biological Membranes* volume in the *Practical Approach Series*.

Protocol 5. Preparation of placental plasma membranes
Materials
(a) The placenta (typical weight ~ 300 g) should be refrigerated as soon as possible after the delivery. Blood vessels and membranes are cut away from the tissue.

(b) Sucrose 250 mmol/litre, Hepes 50 mmol/litre, pH 7.6 (3 litres per placenta).

Protocol 5. *Continued*

(c) Homogenization buffer: Same as above plus protease inhibitors: bacitracin 1 mg/ml, aprotinin 100 kIE/ml, and PMSF 0.17 mg/ml.

(d) Hepes 50 mmol/litre, pH 7.6.

(e) Hepes 50 mmol/litre, pH 7.6 with protease inhibitors as in (c).

(f) NaCl.

(g) $MgSO_4$.

Procedure

1. The placenta is rinsed in ice-cold sucrose–Hepes buffer and about 50 g tissue is cut into pieces on ice.

2. The tissue pieces are suspended in 250 ml sucrose-Hepes buffer with protease inhibitors. Portions of 50–100 ml are homogenized at a time with a knife homogenizer (Polytron) at full speed for 30 sec at 4°C. Repeat three times or until all tissue is disintegrated and filter the homogenate on gauze.

3. Centrifuge 200 ml homogenate at 600g for 10 min at 4°C ($\sim$2000 r.p.m. in a centrifuge with a radius of 15 cm).

4. Collect the supernatant and set aside on ice. Resuspend the pellet in 50 ml homogenization buffer and recentrifuge as in step 3. Combine supernatants (250 ml) and centrifuge at 12 000g for 30 min at 4°C (8500 r.p.m. for a radius of 15 cm).

5. Collect the supernatant and discard the pellet. Add 0.58 g NaCl and 5 mg $MgSO_4$ per 100 ml of supernatant (final concentrations 0.1 mol/litre NaCl and 0.2 mmol/litre $MgSO_4$). Mix and centrifuge 40 000g for 40 min (17 000 r.p.m. for a radius of 15 cm).

6. Discard supernatants and resuspend the pellet in 20 ml Hepes-buffer 50 mmol/litre, pH 7.6. Wash 2–3 times by centrifugation 40 000g for 40 min (17 000 r.p.m.) until the supernatant is clear.

7. The final pellet is resuspended in 20 ml Hepes-buffer with addition of protease-inhibitors using a Potter–Elvehjem homogenizer with 10–20 strokes of a hand-held Teflon pestle. The membrane suspension is frozen and stored at -20°C in small aliquots of 0.5–1.0 ml.

4.5 Preparation of solubilized receptors

Peptide hormone receptors are integral membrane glycoproteins, and their isolation requires solubilization of the phospholipid membrane environment using detergents, i.e. bipolar amphiphilic reagents. The detergent forms a micelle with the hydrophobic membrane-spanning region of the receptor and brings it into solution in water. Cloning and sequencing of a number of different membrane receptors including those for peptide hormones, growth factors, catecholamines, and neurotransmitters has revealed different classes depending on their membrane spanning portions. Most peptide hormone receptors traverse the plasma membrane once, like the epidermal growth factor, insulin, and growth hormone receptors, whereas the β-adrenergic, muscarinic, and substance K receptors have seven transmembrane segments. Finally, the neurotransmitter receptors which are gated ion channels have several transmembrane segments which form the channel across the membrane. The transmembrane segments of integral membrane proteins are composed of 25 hydrophobic amino-acids arranged in an α-helical structure and have been designated using the hydropathy analysis (15). The non-denaturing detergent Triton X-100 1% w/v has proved efficient for the solubilization of many receptors, but in some cases other detergents like Lubrol PX, digitonin, octylglucoside, or Tween-80 may be required. For solubilization of receptors, the detergent should have access to the plasma membrane, which is achieved in suspensions of whole cells or crude plasma membranes, whereas intact or sliced tissue should not be used.

The solubilization procedure involves the following steps: suspension of cells or membranes in an appropriate buffer, pH 7.4 with detergent and protease inhibitors to prevent degradation of the soluble proteins; efficient stirring to prevent sedimentation of cells or membranes; and ultracentrifugation at 100 000 g to separate soluble and insoluble material. All steps are carried out at 4°C to minimize undesired proteolysis, denaturation, or aggregation of receptors. In many cases, a partial purification of receptors by wheat germ agglutinin affinity chromatography is applied. This is based on the fact that membrane receptors are glycoproteins with N-linked carbohydrate chains consisting of mannose and N-acetylglucosamine. Full purification of receptors requires affinity chomatography either with the specific ligand or a specific antibody. In *Protocol 6* the solubilization and partial purification of insulin receptors from plasma membranes or cell suspensions is described as an example.

Protocol 6. Solubilization of receptors

Materials

(a) Plasma membranes, cell monolayers, or cell suspensions.

(b) Solubilization buffer composed of NaCl 150 mM, Hepes 50 mM (pH 7.4), Triton X-100 1% w/v, phenylmethylsulfonylfluoride 0.17 mg/ml, aprotinin 100 kIE/ml, and bacitracin 1.8 mg/ml.

(c) Wheat germ agglutinin–Sepharose (Pharmacia) in a column (2 cm × 1 cm).

(d) Column buffer composed of NaCl 30 mmol/litre, Hepes 30 mmol/litre (pH 7.4) and Triton X-100 0.1% w/v.

(e) Elution buffer with N-acetylglucosamine 0.3 mol/litre in column buffer (d).

Procedure

1. Plasma membranes are resuspended in solubilization buffer at a protein concentration of 1–10 mg/ml, and cells at a concentration of 10^7–10^8 cells/ml.

Protocol 6. *Continued*

Cultured cells in monolayer are scraped off the flask with a rubber policeman and suspended in solubilization buffer.

2. The suspension is stirred on a magnetic stirrer or shaker, for 1 h at 4°C. The velocity should be adjusted to keep membranes or cells in suspension without formation of foam.

3. The solubilization mixture is centrifuged at 100 000*g* for 1 h and the supernatant collected.

4. The supernatant is applied on a wheat germ agglutinin–Sepharose column (2 cm × 1 cm) and washed once with 30 ml column buffer. The glycoproteins are eluted with *N*-acetylglucosamine and about 1–2 ml are collected. Protein inhibitors are added in the concentrations given above (see Section 4.4.2) and the partially purified receptor is stored at −80°C.

4.6 Conditions of binding assay

The conditions of the receptor-binding assay include the incubation buffer, temperature, and time. They are determined by the nature of the peptide hormone and its receptor and the preparation of receptor (cellular, membrane, or soluble). The incubation buffer can be varied regarding the salt composition, buffer, pH, and additions. In general, binding assays using intact cells require an isotonic salt solution which mimics the ion composition of the extracellular fluid like Krebs–Ringer, Hank's or Tyrode's. Plasma membranes can be incubated in a simple isotonic or hypotonic solution like NaCl 150 mM or $MgCl_2$ 2 mM, respectively. Solubilized receptors may show specific requirements to the ion content.

A variety of suitable buffers can be used in binding studies. Hepes has been widely used because of its physiological pH optimum and its non-interference with receptor-binding. Thus, receptor-binding in intact cells is measured in Krebs–Ringer-buffer where bicarbonate has been replaced by Hepes 25 mmol/ litre. Tris and phosphate buffers should be avoided because they are unphysiological and may have untoward effects on cellular metabolism, including receptor functions. In contrast, plasma membranes and solubilized receptors can be incubated with either Hepes, Tris, or phosphate in a concentration of 10–50 mmol/litre. The pH should be adjusted to the optimum of the interaction between peptide hormone and receptor, which is likely to be in a range around the physiological pH ~7.0–7.8. Note that the pH of Hepes-buffer is highly temperature-dependent with a coefficient of −0.01/°C. Thus, the pH should be rigorously adjusted at the temperature of the incubation or washing step. Bovine or human serum albumin (Cohn fraction V) should always be included in a concentration of 1–10 mg/ml to reduce adsorption of peptide to the surface of glass or plastic vials. The albumin preparation may be contaminated with bioactive substances like peptide hormones, citrate, or fatty acids which may interfere in assays of receptor-binding or biological activity, in particular with intact cells. Thus, treating the albumin by dialysis or charcoal to remove small molecules should be considered. Protease inhibitors like bacitracin (0.5–1.8 mg/ml) and aprotinin (100 kIE/ml) may be included to reduce degradation of free peptide hormone by proteolytic activity released to the medium or associated with the plasma membrane. For determination of tracer degradation, see Section 7.3.

The temperature of incubation can be varied between 4°C and 37°C depending on the experimental purpose. The receptor-binding properties can be studied at any temperature, but the range of 4°C to 15°C should be preferred for two reasons: in intact cells the receptor-mediated endocytosis of the hormone is reduced, and the proteolytic degradation of free and bound tracer is decreased. It may also be taken into account that the affinity of the peptide hormone-receptor interaction is temperature-dependent. In general, the binding affinity increases with lowering of the temperature because hydrophobic interactions are involved in the binding reaction. Thus, the amount of receptor-bound tracer is higher at lower temperatures. On the other hand, temperatures in the range of 30°C to 37°C should be chosen if the cellular signalling and processing of receptor and hormone are being studied. As the binding kinetics are temperature-dependent, the incubation time to obtain steady state of binding varies and increases at lower temperatures. At 37°C, 15–60 min are required to reach steady state, whereas 5–16 h are needed at 4°C. It is recommended that the time-course of binding is carefully examined at different temperatures before steady-state measurements are performed in the analysis of receptor-binding affinity, capacity, and peptide specificity.

4.7 Separation of bound and free tracer

A receptor-binding assay is terminated by separation of bound and free tracer. The choice of method depends on two factors: the receptor preparation, and the reaction rate which is determined by the incubation temperature. Cell monolayers are washed 2–3 times at 4°C. Intact cells and plasma membranes in suspension are separated from the buffer by sedimentation or filtration, whereas solubilized receptors are separated from free tracer by precipitation of receptors or adsorption of free tracer. Intact cells have a density of ~ 1.05–1.09 g/ml compared to that of the buffer of about ~ 1.01–1.02 g/ml, and can be sedimented by centrifugation at 50–200g for 2–5 min (~ 500–1000 r.p.m. in a centrifuge with a radius of 15 cm). One exception is the isolated adipocyte (density ~ 0.9) which float on the incubation buffer after centrifugation at 9000g for 30 sec in a microfuge. Plasma membranes which have a density of about ~ 1.04 g/ml can be sedimented by centrifugation at 9000g for 2–5 min (~ 7500 r.p.m. in a centrifuge with a radius of 15 cm or $\sim 10\,000$ r.p.m. in a microfuge). The centrifugations are generally performed at room temperature for practical reasons, but the subsequent washings of pellets should be performed with ice-cold buffer.

Measurements of fast binding kinetics require centrifugation in a refrigerated centrifuge.

Trapping of incubation medium with free radioactivity in the cell pellet contributes significantly to the non-specific binding and increases the background and experimental error. The trapped medium is removed by washing the cell pellet 2–3 times in ice-cold buffer by resuspending and centrifuging the cells. A rapid reduction of the free radioactivity can also be achieved by diluting the cell suspension ~ 5–20 times with ice-cold NaCl (9 g/litre), or by layering the cell suspension on ice-cold incubation buffer with 0.32 M sucrose, before centrifugation. Alternatively, the trapped radioactivity can be reduced by centrifugation in a two-phase system where the cell suspension is layered on silicone oil or dibutylphthalate (density ~ 1.05 g/ml) in a polypropylene tube. After centrifugation, the tip of the tube with the cell pellet in the oil phase is excised and the radioactivity counted in a γ-counter. This reduces the trapped buffer to about 10–15% of the cell colume and the trapped radioactivity to about 0.05% of the added tracer or about 1% of total binding. Isolated adipocytes are separated by flotation in silicone oil or dinonylphthalate (density ~ 0.99 g/ml) after dilution of the cell suspension in 10 ml NaCl (9 g/litre) at 10°C and centrifugation at 200g for 40 sec. The volume of trapped buffer is about 2% of the fat cell volume equivalent to about 0.04% of the added tracer. A major advantage of the two-phase oil centrifugation procedures is that bound and free tracer are separated in one step, which is useful in kinetic studies.

Some studies of peptide hormone receptors have applied filtration of the cell or membrane suspension for the separation of bound and free radioactivity. A major problem with the filtration method has been the non-specific adsorption of peptide tracer to the filter. The degree of absorption depends on the chemical properties of the ^{125}I-labelled peptide and the type of filter, which should be carefully evaluated. Pre-treatment of filters like pre-soaking of Millipore filter with 1 mg/ml albumin or of Whatman GF/F filter with 1 mg/ml polyethylenimine, has been used successfully to reduce non-specific binding of ^{125}I-labelled insulin or insulin-like growth factor I, respectively.

In the solubilized receptor assay, bound and free tracer are separated either by precipitation of receptors with 250 mg/ml polyethylene glycol (MW 6000) or by adsorption of peptide hormone with activated charcoal. Both methods give comparable results, but are hampered by a relatively large amount of non-specifically bound tracer (approximately 30–50% of total binding).

4.8 Cell receptor-binding assay

Protocol 7. Cell receptor-binding assay

Materials
(a) Cell suspension (2–5×10^6 cells/ml) or cultured cells in monolayer in a 24-well multidish (0.5–1×10^6 cells/well).

Protocol 7. *Continued*

(b) ^{125}I-labelled peptide (10^6 c.p.m./ml ~ 0.25–1.5 nmol/litre).

(c) Unlabelled peptide 10 μmol/litre.

(d) Assay buffer: Krebs–Ringer solution with Hepes 25 mmol/litre, pH = 7.4 at 37°C and bovine serum albumin 10 mg/ml. Bacitracin 0.5–1.0 mg/ml may be included to inhibit peptide degradation in the medium during incubation of freshly isolated cells, but is generally not necessary with cultured cells.

(e) Washing buffer: Krebs–Ringer–Hepes-buffer with albumin 2 mg/ml, pH = 7.4 at 4°C for isolated cells or phosphate-buffered saline, with albumin 2 mg/ml, pH = 7.4 at 4°C for cultured cells.

Procedure

1. Isolated cells or cell monolayers are incubated in 200 μl of assay buffer at 37°C.

2. The assay is initiated by addition of 25 μl ^{125}I-labelled peptide at a final concentration of 25 000 c.p.m. per sample (10^5 c.p.m./ml ~ 25–150 pmol/ litre) and 25 μl buffer (for determination of total binding) or 25 μl unlabelled peptide at a final concentration of 0.1–1.0 μmol/litre (for determination of non-specific binding).

3. The incubation is continued for 30–60 min at 37°C to reach steady state. Cell suspensions are stirred on a shaker (~ 100 cycles/min).

4. The assay is terminated by cooling tubes or culture plates on ice, and centrifugation of tubes at 9000g for 2 min. An aliquot of the supernatant is collected for counting of free radioactivity, and determination of ^{125}I-peptide degradation (see Section 7.3). The incubation medium is removed and the cell pellets or monolayers are washed 2–3 times with washing buffer at 4°C.

5. The tip of the tube with the cell pellet is excised and transferred to a counting vial. The cell monolayer is harvested with 0.2 mol/litre NaOH and counted in a γ-counter.

4.9 Plasma membrane receptor-binding assay

Protocol 8. Plasma membrane receptor-binding assay

Materials

(a) Plasma membranes (~ 10 mg protein/ml) in MgCl$_2$ 2 mmol/litre and Hepes 25 mmol/litre, pH = 7.4 at 4°C.

(b) ^{125}I-labelled peptide 10^6 c.p.m./ml.

(c) Unlabelled peptide 10 μmol/litre.

(d) Assay buffer: NaCl 150 mmol/litre, Hepes 50 mmol/litre, pH = 7.4 at 20°C and bovine serum albumin 10 mg/ml.

Protocol 8. *Continued*

(e) Washing buffer: NaCl 150 mmol/litre, Hepes 50 mmol/litre, pH = 7.4 at 4°C
and albumin 1 mg/ml.

Procedure

1. Plasma membranes are diluted in assay buffer to 0.1–1.0 mg protein/ml.
   Aliquots of 200 μl are incubated in 1.5 ml Eppendorf tubes at room
   temperature for 2 h with 25 μl ^{125}I-labelled peptide and 25 μl assay buffer or
   25 μl unlabelled peptide (see *Protocol 7*). Shaking is not necessary.

2. The assay is terminated by centrifugation of the tubes at 9000g for 3 min. An
   aliquot of the supernatant is removed for counting of free radioactivity. The
   pellets are washed twice by resuspension in washing buffer and centrifugation.

3. The tip of the tube with membrane pellet is cut and counted.

4.10 Solubilized receptor-binding assay

Protocol 9. Solubilized receptor-binding assay

Materials

(a) Solubilized receptor preparation, e.g. partially purified glycoproteins in
    elution buffer and protease inhibitors (see Section 4.5).

(b) ^{125}I-labelled peptide 10^6 c.p.m./ml.

(c) Unlabelled peptide 10 μmol/litre.

(d) Assay buffer: NaCl 150 mmol/litre, Hepes 50 mmol/litre, pH 7.4 at 20°C or
    4°C and bovine serum albumin 10 mg/ml.

(e) Dissolving buffer: NaCl 150 mmol/litre, Hepes 50 mmol/litre, pH = 7.4 at
    4°C and Triton X-100 0.1% w/v.

(f) Polyethylene glycol (MW 6000) in dissolving buffer at a concentration of
    250 mg/ml and 125 mg/ml. Readjust pH at 4°C.

(g) Human gammaglobulin (fraction II) 3 mg/ml.

Procedure

1. Aliquots of solubilized receptor diluted in assay buffer (total volume 200 μl)
   are incubated in 1.5 ml Eppendorf tubes at room temperature for 2 h or at 4°C
   for 16 h with 25 μl ^{125}I-labelled peptide and 25 μl assay buffer or 25 μl
   unlabelled peptide (see Section 4.8).

2. The assay is terminated by addition of 100 μl gammaglobulin solution
   followed by 300 μl chilled polyethylene glycol solution (250 mg/ml). The
   tubes are vortexed and left for 10 min at 4°C.

3. The precipitate is centrifuged at 9000g per 5 min and the supernatant

Protocol 8. *Continued*

aspirated. The pellet is rinsed once with polyethylene glycol solution (125 mg/ml).

4. Tips are excised and counted.

---

5. Binding properties of peptide hormone receptors

A description of peptide hormone receptor-binding includes measurements of the binding kinetics, ion, pH and temperature-dependence, and peptide specificity of the binding reaction. The general purpose is to obtain a quantitative basis for elucidation of the molecular mechanism of peptide hormone action. Specifically, the data may serve in analyses of structure–function relationship of peptide hormones, kinetic models of receptor interaction, and regulation of receptors in physiological and pathological conditions. A characterization of the binding kinetics includes measurements at steady state (or equilibrium) and transient state. The latter includes the time-courses of association ('forward reaction' or 'on kinetics') and dissociation ('backward reaction' or 'off kinetics') of the complex between peptide hormone and receptor. From these results the binding parameters of the hormone–receptor interaction can be calculated including: the steady-state dissociation constant (K_d), or the steady-state association constant (K_a) which is equal to $1/K_d$, the total receptor number (R_T), the association rate constant (k_1), and the dissociation rate constant (k_2). The kinetic theory is based on ordinary laws of chemical dynamics, i.e. the law of mass action. The model assumes the presence of a cellular receptor, but does not imply a specified hypothetical underlying molecular mechanism. It is therefore a functional and mechanism-free model of receptor binding.

5.1 Receptor-binding theory

The simplest model of receptor-binding of hormones is based on three assumptions: reversibility of the interaction between receptor and ligand; bimolecularity of the binding reaction; and non-cooperativity of receptors. These assumptions are expressed in the reaction scheme:

$$R + H \underset{k_2}{\overset{k_1}{\rightleftharpoons}} RH \qquad \text{[eqn 1]}$$

(R is receptor and H the hormone ligand). At steady state the binding reaction is described by two equations. First, the *steady-state equation* based on the law of mass action action:

$$[R][H]/[RH] = k_2/k_1 = K_d = 1/K_a \qquad \text{[eqn 2]}$$

(K_d, K_a k_1, and k_2 are defined in Section 5). Second, the *conservation equation* of total receptor concentration:

$$[R_T] = [R] + [RH]. \qquad \text{[eqn 3]}$$

By combining equations 2 and 3, a relationship between the concentrations of occupied receptor or bound hormone $[RH]$ and free hormone $[H]$ is obtained:

$$[RH]=[R_T][H]/([H]+K_d). \qquad \text{[eqn 4]}$$

At transient state the reaction between receptor and hormone is described by two equations. First, a second-order rate equation of the forward reaction (association):

$$d[RH]/dt=k_1[R][H]-k_2[RH]. \qquad \text{[eqn 5]}$$

If $[H]$ is constant and equal to the concentration of total ligand $[H_T]$, integration of equation 5 gives:

$$[RH]=[R_T][H_T](1-e^{-k_n})/([H_T]+K_d). \qquad \text{[eqn 6]}$$

The rate constant of the net formation of RH (k_n) equals

$$k_n=k_2+k_1[H_T]. \qquad \text{[eqn 7]}$$

For practical purposes $[H]$ may be considered constant and equal to $[H_T]$ in experiments when $[RH]$ at equilibrium is $<1/20$ of $[H_T]$. This is true when $[R_T]\ll K_d$ or about $1/20$ of K_d. In cases where this equation is not fulfilled, the integrated rate equation is a rather complicated expression (16). A limiting case of equation 5 is for $t\rightarrow0$ and the initial velocity of the forward reaction (v_i) is:

$$v_i=k_1[R_T][H_T]. \qquad \text{[eqn 8]}$$

Second, the backward reaction (dissociation) is described by a first-order rate equation:

$$d[RH]/dt=k_2[RH] \qquad \text{[eqn 9]}$$

and by integration:

$$[RH]=[RH_0]e^{-k_2t} \qquad \text{[eqn 10]}$$

where $[RH_0]$ is the initial concentration of RH at $t=0$. Equations 1–10 are fundamental in the quantitative analysis of kinetic data on receptor–hormone interactions. The binding parameters K_d and $[R_T]$ are estimated independently from results obtained at steady state (equations 2–4) and transient state (equations 5–10). In the latter, K_d is calculated as the ratio of k_2 and k_1, which are determined by measurements of k_n at different values of $[H_T]$ followed by linear regression analysis (equations 5–7). The value of $[R_T]$ is determined from measurements of v_i at different hormone concentrations (equation 8), and k_2 is also determined directly from the rate constant of dissociation (equations 9–10). Consistency between the estimates of binding parameters at steady state and transient state suggests that the simple model (equation 1) is valid.

5.2 Transient-state binding kinetics

Studies of transient-state binding kinetics include measurements of the time-courses of association and dissociation of the receptor–hormone complex and calculation of binding parameters from equations 5–10 (see Section 5.1). The time-course of association of ^{125}I-labelled hormone with the receptor should be measured at different hormone concentrations ranging from $1/10 \times K_d$ to $10 \times K_d$ in order to obtain values of k_n, i.e. the rate constant of net formation of RH for calculation of k_1 and k_2 from a plot of k_n versus $[H_T]$ (equation 7). At low hormone concentrations, i.e. for $[H] \to 0$, k_n is equal to the rate constant of dissociation, $k_2 \sim$ the intercept on the ordinate. The rate constant of association, k_1 can be calculated from the slope of the straight line. This analysis implies that the assumptions of a simple bimolecular, reversible, and non-cooperative reaction are fulfilled. Alternatively, the initial rate (v_i) can be measured at varying hormone concentrations and the value of k_1 calculated from equation 8. For this calculation, the value of $[R_T]$ determined at steady state is used.

The time-course of dissociation of the receptor–hormone complex should be measured after binding of ^{125}I-labelled hormone at a concentration equal to K_d in order to ensure a high initial value of receptor-bound hormone, $[RH_0]$. After removal of unbound hormone, the dissociation of RH is measured in order to obtain a value of the rate constant of dissociation, k_2, from equation 10. If the assumptions are fulfilled, the dissociation of receptor-bound hormone should be monoexponential.

The time-courses of association and dissociation of the receptor-bound peptide are measured using the procedures for receptor-binding assays (see Section 4 for details). It is generally advisable to perform the measurements with intact cells at 4°C in order to reduce receptor-mediated internalization and degradation of peptide which will disturb the kinetic analysis. Binding kinetics with membranes or soluble receptors are studied at 4° or 20°C. The reaction rates are highly temperature-dependent and are slower at lower temperatures, which may be convenient in the experiments. In transient-state kinetics the use of rapid separation methods is of major importance.

5.2.1 Association kinetics

(a) Samples of the receptor preparation corresponding to the number of time points and different hormone concentrations are prepared in duplicate. The incubation times should be chosen to give measurements of binding both during the initial rapid phase and at steady state. At least four hormone concentrations over a 30- to 50-fold range should be used.

(b) The reaction is initiated by addition of ^{125}I-labelled plus unlabelled hormones.

(c) After the given time period, the incubation is finished by separation of bound from free hormone and washing the sample at 4°C (see Section 4.7).

(d) In parallel, the time-course of non-specific binding is measured by incubation with tracer and an excess of unlabelled hormone (~ 0.1–1 μmol/litre). The total binding is corrected for non-specific binding to give receptor-bound tracer for calculations according to equations 6–7.

(e) The amount of receptor-bound tracer is plotted versus time for each hormone concentration and the half-time $(T_{\frac{1}{2}})$ estimated. The rate constant, k_n is calculated as $\ln 2/T_{\frac{1}{2}}$ and plotted versus the hormone concentration. This should yield a straight line (equation 7) from which the values of k_2 and k_1 are estimated as the intercept with the ordinate and the slope of the line, respectively. The association curve can also be linearized in a semi-logarithmic plot (see Section 6.1.1).

5.2.2 Dissociation kinetics

(a) Samples of the receptor preparation are pre-incubated with ^{125}I-labelled hormone in a concentration giving half saturation of receptors until steady state is achieved.

(b) The pre-incubation medium with free tracer is removed and the samples washed once with ice-cold buffer.

(c) The dissociation is initiated by addition of medium without hormone at the given temperature.

(d) The incubation is terminated after various time periods by separation of bound and free tracer and washing the samples at $4°C$.

(e) A parallel incubation with an excess of cold hormone $(0.1–1 \; \mu mol/litre)$ is performed to determine non-specific binding. The total binding is corrected for non-specific binding to give receptor-bound tracer for calculations according to equation 10.

(f) The receptor-bound tracer is plotted on a logarithmic scale as a function of time of dissociation. This should yield a straight line and the value of k_2 is estimated as the slope of the line.

5.3 Steady-state binding kinetics

Studies of steady-state binding kinetics include measurements of the dose–response relationship between free hormone and receptor-bound hormone under conditions where their concentrations remain constant. The binding should be measured at hormone concentrations ranging from $1/10 \times K_d$ to $10 \times K_d$ in order to obtain values of receptor-bound hormone which differ by a factor ten. If the assumptions of a bimolecular, reversible and non-cooperative reaction are fulfilled, the values of K_d (or K_a) and $[R_T]$ can be calculated according to equation 4 either by graphical methods (see Section 6.1) or computer programs (see Section 6.2). In many cases, however, the assumptions are not valid, due to heterogeneity or cooperativity of receptors, and the binding should be measured over at least three decades of hormone concentrations to evaluate the deviations from the simple model. In other cases, the hormone is internalized and degraded leading to non-reversibility of the reaction and declining hormone (and receptor) concentrations. This latter interference can be avoided by incubating cells at low temperatures or measure binding with membrane or soluble receptor preparations.

The incubation time required to obtain steady state of receptor-binding is determined in measurements of the time-course of association and calculation of $T_{\frac{1}{2}}$ (or k_n) (see Section 5.2.1). The time to reach steady state is then determined as 6 times the value of $T_{\frac{1}{2}}$. As the association rate is dependent on the hormone concentration (equations 6–7), steady-state binding must be established at the lowest hormone concentration used. The fact that $k_n = k_2$ for $[H] \to 0$ (equation 7) implies that the value of $T_{\frac{1}{2}}$ (or k_n) for association at low hormone concentrations can also be determined as $T_{\frac{1}{2}}$ (or k_2) for dissociation. Thus, dissociation experiments can be applied for indirect determination of the incubation time required for steady state.

5.3.1 Steady-state binding assay

(a) Samples of the receptor preparation, corresponding to the number of different hormone concentrations over a range of 2–4 decades, are prepared in triplicate.

(b) The incubation at the given temperature is initiated by addition of ^{125}I-labelled and unlabelled hormone.

(c) After the given time period for steady-state, bound hormone and free hormone are separated and the samples washed at $4°C$ (see Section 4.7).

(d) Binding is also measured with an excess of unlabelled hormone ($\sim 0.1–1.0$ μmol/litre) for determination of non-specific binding and correction of total binding to give receptor-bound tracer.

(e) The amount of receptor-bound tracer is plotted versus hormone concentration according to equation 4 or transformations of this equation, and the binding parameters K_d and $[R_T]$ are estimated as described below (see Section 6.1.3).

5.4 Peptide specificity

A major function of peptide hormone receptors is to recognize specific ligand(s) and to discriminate between the large number of extracellular peptides and proteins. In general, peptide hormone receptors show a high degree of peptide specificity in their ligand binding. The physiological ligand is bound with high affinity, whereas peptides with similarities in their structure bind with lower affinity, and unrelated substances do not interact. In biochemical studies of hormone receptors, the peptide specificity is demonstrated in the competitive radioligand-binding assay by measuring the receptor-bound ^{125}I-labelled hormone in the presence of increasing concentrations of unlabelled hormone analogues as well as other peptides. Different receptors for structurally related peptides may show significant cross-reactivity between their respective ligands, and in these cases it is very important to determine their peptide specificity (17). Examples are receptors for insulin and insulin-like growth factors I and II, receptors for the secretin–glucagon family of peptides, and receptors for growth hormone and prolactin. Apart from identification of peptide hormone receptors, the binding of hormone analogues has primarily been used in analysis of the structure–function relationship of the hormone. The purpose of these studies is to define the receptor-binding region of the peptide hormone for development of hormone antagonists and superagonists.

In experimental practice, a few peptide hormone analogues are necessary for assessment of receptor-binding specificity. These analogues may include peptide hormones of different animal species, chemically or enzymatically modified peptides, synthetic analogues, prohormones, or related peptides. The analogues should have binding affinities between 1% and 100% relative to the native hormone, and include the related hormones in order to discriminate between their respective receptors. Today, biosynthetic DNA-recombinant techniques have increased the availability of peptide hormone analogues significantly. Measurements of peptide specificity are performed with cellular membrane or solubilized receptors in a steady-state binding assay with a fixed concentration of ^{125}I-labelled peptide hormone and increasing concentrations of unlabelled peptide hormone analogues (see Section 5.3.1). The native unlabelled hormone is included in the assay as reference, i.e. its binding affinity is defined as 100%, and the binding affinities of the analogues are expressed in per cent relative to the native hormone.

6. Analysis of radioligand-binding

The binding parameters of the receptor–ligand reaction can be calculated from the experimental data in two different ways: graphical methods or computerized analysis. In both cases, the calculations are based on the best fit of the mathematical model to the data with evaluation of the statistical error. Like the binding theory, the calculation methods are general for all molecular interactions between a ligand and a receptor and not limited to peptide hormone receptors. Furthermore, similar equations, graphical representations, and calculation methods are used when studying different types of interactions like enzyme–substrate, antigen–antibody, oxygen–hemoglobin and ligand–receptor. The equation describing the receptor-binding of a ligand at steady state (see Section 5.1, equation 4) is formally similar to the Langmuir adsorption isotherm describing the adsorption of small molecules to surfaces, the Henderson–Hasselbalch equation describing the acid–base equilibrium, and the Michaelis–Menten equation characterizing the enzyme–substrate interaction, although the meaning of the parameters in these different equations is different in physicochemical terms. It reflects, however, that the graphical representations used for ligand–receptor interaction, have been developed for describing enzyme reactions, acid–base equilibrium, and adsorption of small molecules to larger molecules. In the following the most commonly used graphical representations for ligand–receptor interaction, with their advantages and pitfalls, will be described. For a detailed review of the graphical methods see reference 18. The application of computers has improved the analysis of radioligand-binding studies by avoiding the transformation of the data needed for the graphical analysis and by enabling us to describe more complex radioligand–receptor interactions. One collection of radioligand analysis programs will be introduced (19).

6.1 Graphical representations

The graphical analysis of peptide hormone receptor-binding experiments falls into two categories. In transient-state studies the time-courses of ^{125}I-labelled peptide hormone association and dissociation are measured, and the data plotted according to equations 6 and 10, respectively (see Section 5.1). The rate constants of association (k_1) and dissociation (k_2) are calculated as described below in Sections 6.1.1 and 6.1.2. In studies performed at steady state, the concentration dependence of ^{125}I-labelled hormone binding is measured and the data plotted according to equation 4 (see Section 5.1). The affinity constants (K_a or K_d) and the total receptor number (R_T) are calculated as described below in Section 6.1.3. For both categories of receptor-binding experiments, however, the graphs of the original data are curvilinear and the binding parameters can only be roughly estimated. A linear transformation of the results is needed in order to determine the binding parameters by linear regression analysis. The exponential curves of the association and dissociation experiments are transformed by a semi-logarithmic plot. The hyperbolic curve of steady-state binding is transformed by changing the mathematical expression describing the relationship between bound and free hormone.

Only the simplest model which fulfils the assumptions of a reversible, bimolecular, and non-cooperative interaction between receptor and hormone will be described. Deviations from the model in terms of irreversibility due to internalization of receptor and degradation of hormone, multiple classes of binding sites, and cooperative interactions are outside the scope of this review, and their theory and graphical representation has been described by Boyneams and Dumont (18).

It should be emphasized that the graphical representation of hormone–receptor binding kinetics described below involves the values of bound ^{125}I-labelled hormone which have been corrected for non-specific binding determined in the presence of excess unlabelled hormone.

6.1.1 Association curves

The time-course of association of ^{125}I-labelled hormone to one class of non-cooperative receptors on the cell surface, in plasma membranes or in solution is described by equation 6 in Section 5.1. A plot of the relationship between the concentration of receptor-bound hormone [RH] and the time t gives a monoexponential curve which reaches a given level of receptor occupancy at steady state [RH$_{ss}$] (*Figure 1A*). The half-time, $T_{\frac{1}{2}}$ equals the time to reach half of the value of [RH_{ss}], and from $T_{\frac{1}{2}}$ the rate constant of net formation of RH (k_n) can be calculated as $k_n = \ln 2/T_{\frac{1}{2}}$. The value of k_n can be estimated more precisely from a semi-logarithmic plot of the association curve by transformation of equation 6:

$$[RH_{ss}] - [RH] = [RH_{ss}]e^{-k_n t} \qquad \text{[eqn 11]}$$

where $[RH_{ss}]$ equals the concentration of $[RH]$ at steady state according to equation 4. A semi-logarithmic plot of $[RH_{ss}] - [RH]$ versus time yields a straight line from which k_n can be determined as the negative slope by linear regression analysis. The time-course of hormone association is measured at different hormone concentrations $[H_T]$ in the range of $1/10\ K_d$ to $10\ K_d$, and the values of k_n estimated. According to equation 7 (Section 5.1) a plot of k_n versus $[H_T]$ is a straight line (*Figure 1B*). From this plot, the rate constant of association, k_1 is determined as the slope of the line, and the rate constant of dissociation, k_2 as the intercept on the ordinate. This procedure gives values of k_1 and k_2, that are independent of the parameters determined in steady-state experiments, and the affinity constant, K_d can be calculated as k_2/k_1, and compared with the value obtained at steady state (see Section 6.1.3). In this way the model of receptor–hormone interaction can be evaluated.

6.1.2 Dissociation curves

The time-course of dissociation of the ^{125}I-labelled hormone–receptor complex is described by equation 10 in Section 5.1. A plot of the relationship between $[RH]$ and t gives a monoexponential decay curve which starts at $[RH_0]$ and reaches zero with increasing time (*Figure 2A*). The half-time, $T_{\frac{1}{2}}$ equals the time to reach half of the value of $[RH_0]$, and from $T_{\frac{1}{2}}$, the rate constant of dissociation, k_2, can be calculated as $k_2 = \ln 2/T_{\frac{1}{2}}$. A semi-logarithmic plot of equation 10 yields a straight line from which the value of k_2 can be more precisely determined as the negative slope by linear regression analysis (*Figure 2B*).

6.1.3 Steady-state binding curves

In theory, the concentration of hormone–receptor complex (RH) is determined at varying hormone concentrations (H) at steady state, and the total receptor concentration, $[R_T]$ and the affinity constant, K_a or K_d estimated graphically according to equation 4 in Section 5.1. A direct representation of $[RH]$ versus $[H]$ is a hyperbola having two asymptotes—a horizontal: $[RH] = [R_T]$ and a vertical: $[H] = -K_d$, of which the latter is not in the physically meaningful range (*Figure 3A*). The value of $[R_T]$ can be determined from the horizontal asymptote, and the value of K_d as the hormone concentration at half saturation: $0.5\ [R_T]$. As receptor-binding measurements should extend over at least three decades of hormone concentration, it is more convenient to plot $[RH]$ versus $\log[H]$, which yields a sigmoid curve.

In practice, the receptor-binding of ^{125}I-labelled hormones is measured at a fixed ^{125}I-labelled hormone concentration and increasing concentrations of unlabelled hormone. A direct plot of $[RH]$ versus $[H]$ or $\log[H]$ is not very useful, in spite of its simplicity, because the specific activity of the ^{125}I-labelled hormone decreases with increasing concentration of unlabelled hormone, and

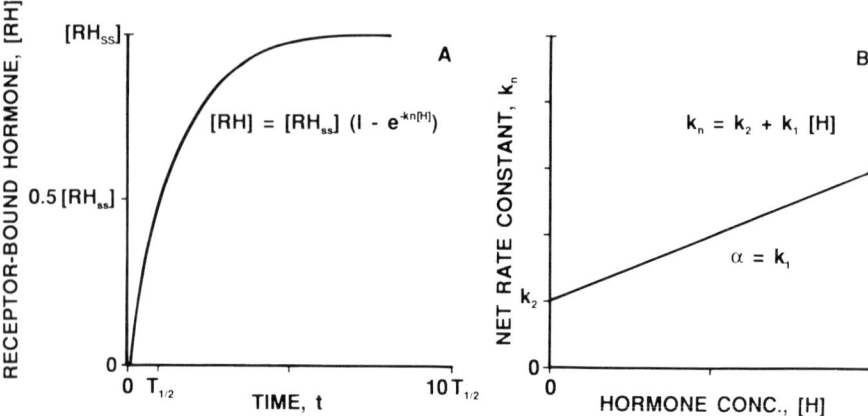

Figure 1. Time-course of association of receptor-bound hormone. A: $[RH]$ is plotted versus time according to equation 6 in section 5.1: $[RH] = [RH_{ss}] (1 - e^{k_n t})$, where $[RH_{ss}]$ is the concentration of $[RH]$ at steady state (equation 4 in section 5.1). B: k_n is plotted versus $[H]$ according to equation 7 in section 5.1: $k_n = k_2 + k_1 [H]$.

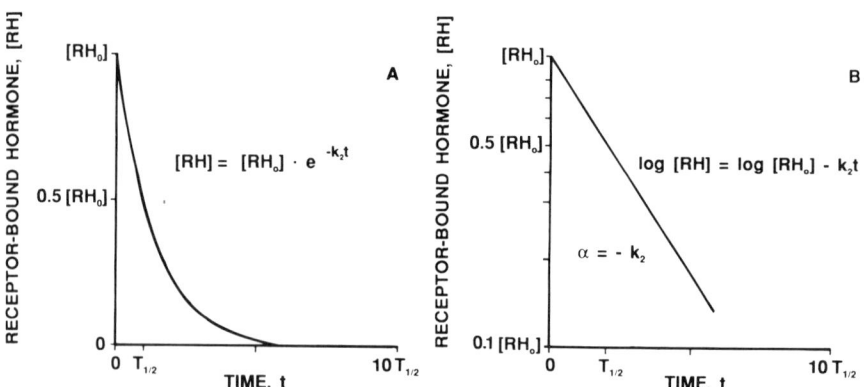

Figure 2. Time-course of dissociation of receptor-bound hormone. A: $[RH]$ is plotted versus time according to equation 10 in section 5.1: $[RH] = [RH_0] e^{-k_2 t}$. B: Semilogarithmic plot of $[RH]$ versus time according to the equation: $\log [RH] = \log [RH_0] - k_2 t$.

the value of $[RH]$ can only be plotted after correction. The difficulty can be solved by plotting the ratio of bound and free tracer versus the concentration of unlabelled hormone. The plot is based on a simple transformation of equation 4:

$$[RH]/[H] = [R_T]/([H] + K_d). \qquad \text{[eqn 12]}$$

This hyperbola has a horizontal asymptote: $[RH]/[H] = 0$ and a vertical: $[H] = -K_d$. The value of $[RH]/[H]$ tend to a limiting value of $[R_T]/K_d$ at zero hormone concentration. K_d equals $[H]$ at half-saturation, i.e. when $[RH]/[H]$ is decreased to half of its initial value, and $[R_T]$ is calculated by multiplication of the value of $[RH]/[H]$ at $[H] \to 0$ with K_d. The use of a semi-logarithmic plot allows

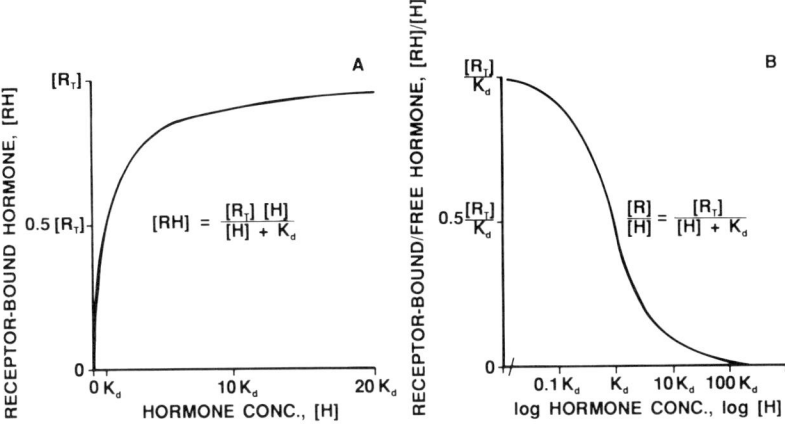

Figure 3. Steady state receptor binding of hormone. A: $[RH]$ is plotted versus $[H]$ according to equation 4 in section 5.1: $[RH] = [R_T] [H]/([H] + K_d)$. B: $[RH]/[H]$ is plotted versus log $[H]$ according to equation 12 in section 6.1.3: $[RH]/[H] = [R_T]/([H] + K_d)$.

a representation of binding over a large range of hormone concentrations (*Figure 3B*).

The presence of a non-specific binding site of low affinity ($K_d > 1$ μmol/litre) is represented as a straight line in the physiological hormone concentration range (10 pmol/litre–10 nmol/litre), around K_d for the receptor, where non-specifically bound hormone is proportional to the hormone concentration (*Figure 4A*). Determination of receptor-bound hormone requires a correction for non-specific binding. This can be easily done in a plot of the ratio between bound and free tracer versus the log$[H]$ where the non-specific binding is represented as a horizontal line (*Figure 4B*). This plot should also be used in studies of peptide specificity of receptors by plotting the ratio of bound and free hormone tracer ($\sim[RH]/[H]$) versus the concentration of unlabelled hormone analogue on a logarithmic scale. The affinity or inhibition contant of the analogue, K_i, equals the concentration at half-saturation.

The linear graph of $[RH]/[H]$ versus $[RH]$ was originally introduced by Scatchard (20) in the context of a study of small molecules binding to proteins. It is based on the following transformation of equation 4:

$$[RH]/[H] = [R_T]/K_d - [RH]/K_d. \qquad \text{[eqn 13]}$$

The Scatchard plot has been widely used for graphic estimation of binding parameters by linear regression. The abscissa-intercept equals $[R_T]$, the slope $-1/K_d$ and the ordinate-intercept $[R_T]/K_d$ (*Figure 5*). An advantage of this plot is that deviations from the simple model (equation 1) in cases of binding site multiplicity and cooperativity are more readily recognized on the Scatchard plot than on the plot of $[RH]/[H]$ versus log $[H]$ (equation 12).

Two disadvantages of the Scatchard plot are apparent: the independent

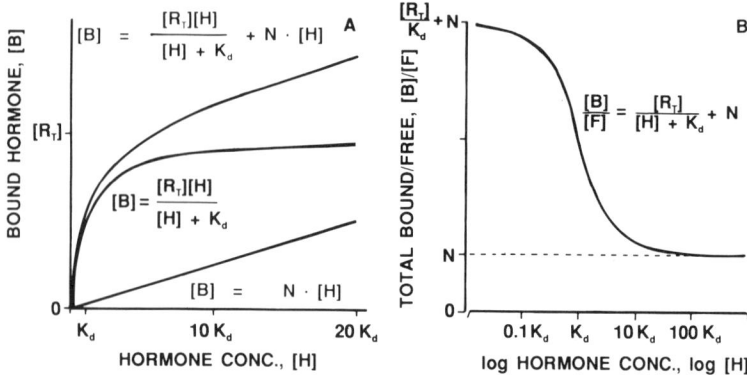

Figure 4. Receptor-bound and non-specifically bound hormone. A: Total-bound hormone is plotted versus the hormone concentration according to the equation: $[B] = [R_T] [H]/([H] + K_d) + N \cdot [H]$. The curve is composed of two curves representing receptor-bound hormone: $[B] = [R_T] [H]/([H] + K_d)$ and non-specifically bound hormone: $[B] = N \cdot [H]$. B: The ratio of total-bound and free hormone is plotted versus log $[H]$ according to the equation $[B]/[F] = [R_T]/([H] + K_d) + N$. The non-specifically bound hormone is represented by the dotted horizontal line according to the equation: $[B]/[F] = N$.

variable $[RH]$ and its experimental error are included in both coordinates; the correction of binding data for non-specific binding augments the experimental error at high receptor occupancies. If binding is determined with a constant amount of tracer and increasing amounts of unlabelled hormone, the variation coefficient of experimental error for $[RH]/[H]$ increases at low occupancy, whereas that for $[RH]$ increases at high occupancy. Thus, the evaluation of linearity is difficult at both ends, and the determinations of the two intercepts are subject to large errors, in particular the estimate of $[R_T]$.

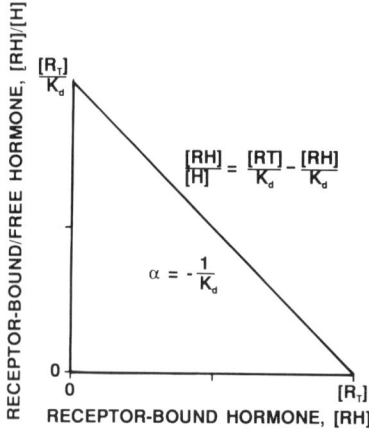

Figure 5. Scatchard plot of steady state receptor binding of hormone. $[RH]/[H]$ is plotted versus $[RH]$ according to equation 13 in section 6.1.3: $[RH]/[H] = [R_T]/K_d - [RH]/K_d$.

Other linear graphs may be used for representation of binding data. These include the double-reciprocal (Lineweaver–Burk) plot and the semi-reciprocal (Hanes) plot, which are used in enzymology. The double-reciprocal plot: $1/[RH]$ versus $1/[H]$ suffers from the disadvantage that the variation coefficient of experimental error for $1/[RH]$ increases with increasing values of $1/[RH]$ (i.e. decreasing values of $[RH]$). This results in a large error on the estimate of K_d, and the plot should accordingly not be used in binding studies. This drawback does not apply for the semi-reciprocal plot of $[H]/[RH]$ versus $[H]$, but in spite of this, it has not been used in receptor-binding studies.

For the quantitative analysis of multiple or cooperative binding sites, two graphs should be considered. The Hill plot: $\log [RH]/[R_T]-[RH])$ versus $\log [H]$, and the average affinity plot: $([RH]/[H])/([R_T]-[RH])$ versus $\log [RH]/[R_T])$. Both graphs allow estimates of the affinity constants of two classes of independent receptors or two states of cooperatively interacting receptors. For further details see reference 18.

6.2 Computer programs

The quantitative analysis of receptor-binding data has been greatly facilitated by the introduction of computers, and more recently microcomputers such as IBM PC. Although a graphical analysis as described above (see Section 6.1) is obligatory, in particular with respect to evaluation of the model for receptor–hormone interaction, the computerized analysis should be used for calculation of binding parameters and statistical test of the model. Computer programs have been developed for the analysis of steady-state and transient-state kinetic studies of radioligand-binding and one example of such programs: 'KINETIC, EBDA, LIGAND, LOWRY', developed by McPherson (19) for IMB PC will be described. The collection of radioligand-binding analysis programs was published in 1985 by Elsevier Science Publishers, Amsterdam, The Netherlands, and the software is distributed by Elsevier-Biosoft®, 68 Hills Road, Cambridge CB2 1LA, United Kingdom.

The analysis of steady-state binding data is based on equation 4 (see Section 5.1), which is generalized to conditions where the hormone is binding to two or more specific sites as well as a non-specific site, simultaneously.

$$[RH]= \sum_{i=1}^{i=n} \frac{[R_T]_i \times [H]}{K_{di}+[H]} + N[H] \qquad \text{[eqn 14]}$$

where $[R_T]_i$ and K_{di} are the constants for the ligand-binding site i, the number of which may vary between 1 and n. N is the ratio of bound and free at infinite free hormone concentration and represents the gradient of the line relating non-specific binding and free concentration of labelled hormone. The program LIGAND was originally described by Munson and Rodbard (21) for non-linear curve fitting of complex radioligand-binding site interactions. Its main features are: the exact model is used to describe the interaction; non-specific binding is treated as a fitted parameter and need not be subtracted prior to analysis; the data for more than one ligand can be co-processed simultaneously; modelling is possible; statistical tests are performed to test model hypotheses; LIGAND uses total ligand concentration rather than free. The program uses an iterative procedure for fitting the equation to the non-linear steady-state binding data and calculation of the binding parameters.

The analysis of transient-state binding data is based on equations 6 and 10 (see Section 5.1) which are generalized to multiple binding sites with different constants giving multi-exponential relationships. The programs fit the equations to the non-linear association or dissociation data by an iterative procedure and estimate the rate constants. For further details and references the reader is referred to the original articles on the specific programs (19, 21, 22).

7. Receptor-mediated endocytosis and degradation of peptide hormones

In addition to their role in intracellular signalling, peptide hormone receptors mediate endocytosis and degradation of their ligands. Following binding on the cell surface, receptor-bound hormone is rapidly translocated to endosomal vesicles ($T_{\frac{1}{2}} = 3$–10 min) where it is degraded by aminopeptidases and endopeptidases. Peptide hormone receptors share this property with polypetide growth factor receptors and receptors for plasma proteins like low density lipoproteins, and asialoglycoproteins, whereas transferrin receptors mediate translocation, but not degradation of its ligand. Receptors for plasma proteins recycle constitutively between the endosomal compartment and the plasma membrane every 15–20 min. In contrast, receptors for polypeptide hormones and growth factors are characterized by regulated, i.e. ligand-induced, endocytosis and transfer to a lysosomal compartment where both receptor and ligand are degraded. Recent studies have shown, however, that some peptide hormone receptors like the insulin receptor, are internalized by regulated and constitutive pathways followed by recycling to the plasma membrane. It should be emphasized that the molecular mechanism of peptide hormone internalization and degradation, including identification of the peptidases and description of the intracellular trafficking, has not been resolved. Degradation studies in intact organisms and in perfused organs suggest that receptor-mediated endocytosis represents the major pathway of peptide hormone degradation *in vivo*.

In vitro studies of endocytosis of peptide hormones are based on assays of surface-bound and internalized hormone and measurements of degraded hormone. In the following, the assays of endocytosis and recycling of peptide hormone receptors will be described. Various methods for the quantitation of degraded peptide are presented.

7.1 Surface-binding and internalization assay

Assay of peptide hormone endocytosis is based on a differentiation between the hormone which is associated with receptors on the cell surface and the hormone which is accumulated inside the cell. It is based on the observation that the stability of the hormone–receptor complex is sensitive to pH. Thus, treatment of cells with buffer at pH below ~5 usually results in release of receptor-bound hormone from the cell surface, whereas internalized hormone is not affected. The assay involves incubation of intact living cells with [125I]-labelled hormone at temperatures between 30 and 37°C. At lower temperatures the rate of endocytosis is significantly reduced and at 4°C the receptor-bound hormone remains at the cell surface. This may be used as a control of the efficiency of the acid treatment, which should remove all the cell-associated hormone after binding at 4°C.

The assay was originally designed by Haigler *et al.* (23) who used treatment of cells with 0.2 M acetic acid (pH = 2.5) and 0.5 M NaCl at 4°C to remove surface-bound [125I] epidermal growth factor, whereafter the internalized radioactivity was determined. The high salt concentration was applied because it reduces the affinity of the receptor–epidermal growth factor interaction. The composition of the acid buffer has been modified and others have used a more physiological buffer like Krebs–Ringer buffer with 25 mM Hepes titrated to pH = 3.5 in order to preserve the viability of the cells in continued incubation for studies of processing of internalized hormone (24). Other reagents may be applied for release of surface-bound hormone. Suramin, a drug used in therapy of trypanosomiasis and onchocerciasis, inhibits protein–protein interactions and has been used for release of receptor-bound platelet-derived growth factor (25). Treatment of cells with proteases like trypsin 0.2 mg/ml or pronase 1 mg/ml for 10 min at 37°C removes receptor-bound ligands from the cell surface (26). Addition of an excess of unlabelled ligand, like a hormone analogue, may also be used to induce a very specific dissociation of the [125I]-labelled hormone. In *Protocol 10*, the procedure for acid treatment of cells is described.

Protocol 10. Receptor internalization assay

Materials
(a) The materials used for binding are the same as for the cell receptor-binding assay (see Section 4.8.1).
(b) Acid-washing solution: C_2H_5COOH 0.2 mol/litre and NaCl 0.5 mol/litre, pH = 2.5.
(c) Alternative acid-washing buffer: Krebs–Ringer solution with Hepes 25 mol/litre, pH = 3.5 at 4°C and bovine serum albumin 10 mg/ml.

Procedure
1. The binding of [125I]-labelled peptide is carried out as described in the procedure for the cell receptor-binding assay (*Protocol 7*) except that the amount of added [125I]-labelled peptide is increased twofold to a final concentration of 50 000 c.p.m. per sample (2×10^5 c.p.m./ml).
2. After termination of the incubation by removal of the incubation medium and

Protocol 10. *Continued*

free tracer, the cells are washed twice with washing buffer (pH = 7.4) at 4°C. Acid-washing solution (or buffer) in a volume of 500 μl is added, and the cells are incubated 5–10 min at 4°C. With cell suspensions the acid-washing buffer (c) should be preferred. The acid-washing solution, containing the surface-bound ($\sim$ acid-sensitive) radioactivity is collected and counted.

3. The cell suspension or monolayer is washed twice with washing buffer (pH = 7.4) at 4°C. The tip of the tube with the cell pellet containing the internalized ($\sim$ acid-resistant) radioactivity is excised and counted. The cell monolayer is harvested with 0.2 mol/litre NaOH and the internalized radioactivity counted.

7.2 Recycling of receptors

The observation that cellular binding of some peptide hormones maintains a constant level for several hours at 37°C, indicates that cell surface receptors must be replenished after endocytotic uptake of hormone–receptor complexes. Three mechanisms could account for the receptor replacement: *de novo* receptor biosynthesis; insertion from a pre-formed receptor pool; and reinsertion of internalized receptors through recycling. *In vitro* studies of insulin receptors have shown that receptor recycling is the most important mechanism in short-term (6–12 h) incubations, whereas receptor biosynthesis has a role in long-term (24–48 h) incubations. The receptor recycles between the endosomal compartment and cell surface a large number of times before it is degraded. It appears that insulin receptors are internalized and recycled through two major pathways (27). The first has been named the retroendocytic pathway, and through this pathway $\sim$ 25% of receptors with intact insulin are returned to the cell exterior. In the second pathway $\sim$ 75% insulin–receptor complexes are uncoupled through acidification of endosomal vesicles, and insulin is delivered to lysosomes where it is degraded, while the receptors are routed through the Golgi back to the plasma membrane.

Our understanding of receptor recycling has come from both morphological and biochemical studies. In one type of experiment, the intracellular routing of photoaffinity-labelled insulin receptors has been followed by autoradiography or by subcellular fractionation. Receptors labelled on the cell surface at 4°C with photoactive [^{125}I]insulin disappeared after incubation at 37°C, and reappeared on the cell surface after about 5 h (26). In another type of experiment, the distribution of receptors between the cell surface and cell interior was measured after treatment with inhibitors of receptor-recycling like chloroquine and Tris. In these experiments, surface receptors were determined by binding to cells at 4°C, while intracellular receptors were determined in cells which have been trypsinized to remove surface receptors. Inhibition of receptor recycling caused loss of receptors on the surface and accumulation inside the cell (27). In *Protocol 11* the technique for studying receptor distribution between the cell surface and cell interior will be described. Photoaffinity labelling will not be described, since the synthetic photoactive insulin analogues are not generally available.

Protocol 11. Receptor recycling assay

Materials
(a) ^{125}I-labelled peptide hormone (10^6 c.p.m./ml ~ 0.25–1.5 nmol/litre) and unlabelled peptide hormones (1–10 μmol/litre).

(b) Cell suspension (10^6 cells/ml) or monolayer (2–5×10^5 cells/well).

(c) Incubation buffer: Krebs–Ringer–solution with 25 mmol/litre Hepes (pH 7.4 at 37°C) and 10 g/litre bovine serum albumin.

(d) Washing buffer: same as (c) but adjusted to pH 7.4 at 4°C.

(e) Recycling inhibitors: Tris (1 mol/litre) and chloroquine (100 mmol/litre) in incubation buffer (e).

(f) Trypsin (0.2 mg/ml) in incubation buffer (e).

(g) Solubilization buffer: 150 mmol/litre NaCl, 50 mmol/litre Hepes (pH 7.6 at 4°C) 10 g/litre Triton X-100, 0.17 mg/ml phenylmethyl-sulphonyl-fluoride, 100 kIE/ml aprotinin, and 1.8 mg/ml bacitracin.

Procedure
1. Receptor uptake and recycling is initiated by incubating cells with unlabelled hormone (10 nmol/litre) at 37°C for 15 min. The influence of inhibitors of recycling are studied by addition of 0.2 mmol/litre chloroquine or 35 mmol/ litre Tris.

2. Cells are rapidly cooled to 4°C and washed with cold washing buffer to remove extracellular hormone. The cells are divided into three portions for determination of total receptors, intracellular receptors, and surface receptors.

3. Total receptors are determined after solubilization of cells in Triton X-100 (*Protocol 6*).

4. Intracellular receptors are determined after treatment of cells with trypsin for 10 min at 37°C to remove surface-bound hormone, followed by solubilization.

5. The amounts of total and intracellular receptors are quantitated in a solubilized receptor-binding assay (see Section 4.10). The Triton X-100 concentration should be reduced below 0.5% w/v, because higher concentrations of Triton X-100 reduce receptor affinity (28). This could be done by diluting the solubilized receptor preparation with detergent-free buffer, by purifying receptors on a wheat germ agglutinin Sepharose column or by precipitating receptors with polyethylene glycol 250 g/litre and gamma-globulin 3 g/litre (see Sections 4.5 and 4.10).

Protocol 11. *Continued*

6. Surface receptors are determined in a cell receptor-binding assay at 4°C for 5–16 h (see Section 4.8).

7.3 Proteolytic degradation of peptide hormones

The major pathway of peptide hormone degradation *in vivo* is intracellular proteolysis following receptor-mediated endocytosis. As pointed out above (see Section 7), the cellular pathways involved in degradation of peptide hormones, neuropeptides, and polypeptide hormones are fundamentally similar. The precise molecular mechanism is not yet known, but a number of intracellular peptidases exist which may degrade peptide hormones *in vivo*. The proteases can be subdivided into exopeptidases, whose action is directed at the amino- or carboxy-terminus of the peptide, or endopeptidases, that cleave peptide bonds internally in peptides. Endopeptidases are also termed proteinases. In addition, endopeptidases are classified according to the essential catalytic residues at their active sites. Four distinct classes of proteinases have been identified: serine; cysteine; aspartic; and metalloproteinases (see review in reference 29). Few examples exist, where specific peptidases have been characterized in the degradation of a particular peptide hormone *in vivo*. In most cases, it is believed that the proteolytic breakdown of peptide hormones to amino acids is a result of the sequential action of several peptidases with varying specificity. The initial cleavage of a few peptide bonds in the hormone may be due to a specific endopeptidase, whereas the subsequent proteolysis is due to non-specific peptidases. One example is the specific insulin protease, an endopeptidase which cleaves insulin at several sites in the A and B chains resulting in a partially degraded form of insulin, which is subsequently degraded by less specific endo- and exopeptidase (see reviews in references 13 and 30).

Assays of the degradation of peptide hormone quantify changes in the physical, immunological, and biological properties of the peptide. Changes in physical properties, e.g. size and hydrophobicity, can be assessed by molecular sieve chromatography, precipitation with trichloroacetic acid, or HPLC. Changes in immunological properties can be assessed by precipitation with specific antibodies against the peptide hormone. Changes in biological properties can be assessed by receptor-binding or bioassay. In general, assays of biological activity provide more physiologically meaningful data and are more sensitive, because an initial cleavage of the peptide chain may deteriorate the activity of the molecule without changing its size. On the other hand, biological assays provide little information about the structure of the degradation products. Thus, two or more assays should be applied simultaneously in studies of peptide hormone degradation.

The techniques of three different assays are described below: trichloroacetic acid precipitation; gel filtration; and immunoprecipitation, which measure changes in physical and immunological properties of radiolabelled peptide hormones. Changes in biological properties are measured with a receptor-binding assay, as described above (see Sections 4.8–4.10), or a bioassay.

Degradation of unlabelled peptides can be quantified in a radioimmunoassay, radioreceptor assay, or bioassay. The techniques of bioassays are specific for the particular peptide under study and are outside the scope of this chapter (see *Peptide Hormone Secretion*, Chapter 5). Finally, it should be emphasized that hormone degradation assays like trichloroacetic acid precipitation or immuno-precipitation should be routinely applied in receptor-binding assays to measure a decrease in the concentration of labelled hormone during incubation.

7.4 Trichloroacetic acid precipitation

The trichloroacetic acid precipitation assay can detect degradation products of molecular weight below approximately 2000 daltons, which are soluble in trichloroacetic acid 100 g/litre, whereas larger peptides are precipitable. The assay is probably the least sensitive, but is is also the least demanding of time and resources. It is limited to peptide hormones of molecular weight larger than 2000 daltons, and the initial and partial cleavage products are not detectable. On the other hand, the proteolysis following the initial cleavage of a peptide hormone is rapid and complete, and in many cases the trichloroacetic acid precipitation is a good measure of peptide degradation. The application of the assay is limited to radiolabelled peptide hormones. It is important to note that the precipitation of a labelled peptide which is present in the concentration range of 10 pmol/litre –10 nmol/litre requires the presence of a carrier protein like serum albumin 1–10 g/litre.

Protocol 12. Trichloroacetic acid precipitability as an assay of radioligand degradation

Materials

(a) ^{125}I-labelled peptide hormone (10^6 c.p.m./ml ~ 0.25–1.5 nmol/litre) in dilution buffer: NaCl 150 mmol/litre, Hepes 50 mmol/litre (pH = 7.4) and bovine serum albumun 4 g/litre.

(b) Incubation buffer: Krebs–Ringer solution with Hepes 25 mmol/litre (pH = 7.4) and bovine serum albumin 10 g/litre.

(c) Trichloroacetic acid 100–200 g/litre.

(d) Cell or membrane preparations (see Sections 4.8 and 4.9).

Procedure

1. Aliquots (1–2 ml) of trichloroacetic acid are distributed in 5 ml plastic vials.

2. Aliquots (25–100 μl) of ^{125}I-labelled peptide hormone (approximately 10^4 c.p.m.) in dilution buffer (blank) or in incubation buffer after incubation at 20–37°C with cells or membranes (sample), are added to the vials with

Protocol 12. *Continued*

trichloroacetic acid. A white and coarse precipitate is formed by mixing the samples on a whirlymixer.

3. The samples are centrifuged at 1500g for 5 min ($\sim$3000 r.p.m. at a radius of 15 cm). The supernatants with the soluble radioactivity are decanted to counting vials. The supernatants and precipitates are counted in a γ-counter.

4. The percentage degraded ^{125}I-labelled peptide is calculated as follows:

% degraded = % soluble c.p.m. (sample) − % soluble c.p.m. (blank)

where

$$\text{\% soluble c.p.m.} = \frac{\text{c.p.m. (supernatant)} \times 100}{\text{c.p.m. (supernatant)} + \text{c.p.m. (precipitate)}}$$

The blank value should be as low as possible. In highly purified ^{125}I-labelled peptide hormone preparations which are essentially free of [^{125}I]iodine, the value should not exceed 3%. In less pure preparations a blank value of $\sim$10% is acceptable.

7.5 Immunoprecipitation assay

The immunoreactivity of a peptide hormone depends on the presence of an antigenic epitope with which the antibody reacts. The epitope consists of at least 4–6 amino acids in a proper primary sequence and tertiary structure. A decrease in the amount of immunoreactive peptide indicates that the epitope has been degraded. The initial and partial cleavage products of the peptide may still be immunoreactive and are therefore precipitated and not detected by the assay. The immunoprecipitation assay is, however, more sensitive and specific than the trichloroacetic acid precipitation assay, although it suffers from the same drawback in that it detects only extensive molecular changes. Furthermore, the assay can be used with peptides of molecular weight less than 2000 daltons which are precipitated by antibodies, but not by trichloroacetic acid. The immunoprecipitation assay is technically simple and depends only on the availability of an antibody. It is limited to ^{125}I-labelled peptide hormones.

Protocol 13. Immunoprecipitation as an assay of radioligand degradation

Materials

(a) ^{125}I-labelled peptide hormone (10^6 c.p.m./ml $\sim$0.25–1.5 nmol/litre) in dilution buffer (see Section 7.4.1).

(b) Incubation buffer (see Section 7.4.1).

(c) Antiserum or monoclonal antibody diluted appropriately $\sim$100–1000 times.

(d) Ethanol 93% v/v.

Protocol 13. *Continued*

(e) Precipitation buffer: Krebs–Ringer solution with Hepes 25 nmol/litre (pH 7.4) and bovine serum albumin (40 g/litre).

(f) Cells or membrane preparation (see Sections 4.8–4.9).

Procedure

1. An aliquot (50 μl) of ^{125}I-labelled peptide hormone diluted in incubation buffer (approx. 10^4 c.p.m.) before or after incubation with cells or membranes.

2. Antibody dilution (50 μl) is added, mixed, and incubated ∼16 h at 0°C, or ∼60 min at 20°C.

3. Precipitation buffer (100 μl) is added and the sample mixed.

4. Ethanol (93%) 1200 μl is added and the sample mixed.

5. The samples are centrifuged 1500*g* for 10 min (∼3000 r.p.m.). The supernatants containing non-immunoreactive radioactivity are decanted into counting vials. The supernatants and immunoprecipitates are counted in a γ-counter.

6. The percentage degraded ^{125}I-labelled peptide is calculated as follows:

   % degraded = % soluble c.p.m. (sample) − % soluble c.p.m. (blank)

   where

   $$\% \text{ soluble c.p.m.} = \frac{\text{c.p.m. (supernatant)} \times 100}{\text{c.p.m. (supernatant)} + \text{c.p.m. (precipitate)}}.$$

The blank value should not exceed ∼5% with most antibodies.

7.6 Gel filtration chromatographic assay

The gel filtration assay detects changes in the molecular size of the peptide hormone. The type of the gel particle depends on the initial size of the peptide. As the molecular weight of peptide hormones range between ∼1000 and 30 000 daltons, the gel type can vary between Sephadex G-10 and G-100, or equivalent products. The gel filtration assay is more informative than the precipitation assays described above, because the various degradation on products are fractionated and identified. Aggregates of ^{125}I-labelled peptides resulting from the iodination elute in the void volume of the gel and can be subtracted. ^{125}I-tyrosine is absorbed on the gel and elutes after the salt volume of the gel. Intermediate proteolytic fragments can be identified, whereas initial cleavage products elute with the native tracer. Radioimmunoassay, radioreceptor assay or bioassay can be applied to analyse the peaks of ^{125}I-labelled peptide further.

The gel filtration chromatographic assay of [^{125}I]insulin is given as an example in *Protocol 14*.

Protocol 14. Gel filtration as an assay of radioligand degradation

Materials

(a) [^{125}I]insulin (10^6 c.p.m./ml $\sim$0.25 – 1.5 nmol/litre) in dilution buffer (see Section 7.4.1).

(b) Incubation buffer (see Section 7.4.1).

(c) Elution buffer: Ammonium acetate ($NH_4C_2H_5COO$) 0.1 mol/litre (pH 7.4) with bovine serum albumin 2 g/litre.

(d) Conservation buffer: NaN_3 (1 g/litre) in elution buffer.

(e) Column of Sephadex G-50 (0.9×50 cm) which has been equilibrated with elution buffer. The column should be calibrated with markers of the void volume, V_0 (like Evans blue 1640 or serum albumin), the elution volume, V_e of the peptide [^{125}I]insulin or unlabelled insulin, and the total volume, V_t (like ^{22}Na). The elution volume of [^{125}I]tyrosine or tyrosine should also be determined, as it exceeds the total volume. Other proteins may be used as size markers. The column may be used at room temperature, or at 4°C in a cold room. After use, it is conserved with conservation buffer.

(f) Cell suspension, cell monolayer or plasma membranes (see Sections 4.8–4.9).

Procedure

1. [^{125}I]insulin, which has been incubated with cells or membranes (sample) or diluted in dilution buffer (blank) containing approximately 5×10^4 c.p.m. in a volume of 0.1–0.5 ml is applied to the column.

2. The column is eluted with elution buffer at a flow rate of 0.1–0.5 ml/min. Fractions of column effluent (0.5–1.0 ml) are collected and counted in a γ-counter.

3. The recovery of radioactivity should exceed 90%. However, during chromatography of tracer amounts of [^{125}I]insulin, recovery may be reduced due to adsorption of insulin to the column. This may be prevented by adding unlabelled insulin to samples as a carrier or by chromatographing unlabelled insulin (100 mg) on each new column before use.

4. The amount of degraded [^{125}I]insulin can be calculated by subtracting the percentage of the total radioactivity present in the insulin peak fraction of the sample from the percentage present in this fraction of the blank.

Acknowledgements

Merete Jacobsen is thanked for typing the manuscript and Lisbeth Jensen for drawing the figures.

References

1. Bolton, A. E. and Hunter, W. M. (1973). *Biochem. J.*, **133**, 529.
2. Rehfeld, J. F. (1978). *J. Biol. Chem.*, **253**, 4016.
3. Hunter, W. M. and Greenwood, F. C. (1962). *Nature, Lond.*, **194**, 495.
4. Freychet, P., Roth, J., and Neville, D. M., Jr. (1971). *Biochem. Biophys. Res. Commun.*, **43**, 400.
5. Fraker, P. J. and Speck, J. C., Jr. (1978). *Biochem. Biophys. Res. Commun.*, **80**, 849.
6. Gether, U., Nielsen, H. V., and Schwartz, T. W. (1988). *J. Chromatog.*, **447**, 341.
7. Thorell, J. I. and Johansson, B. G. (1971). *Biochim. Biophys. Acta*, **251**, 363.
8. Linde, S., Sonne, O., Hansen, B., and Gliemann, J. (1981). *Hoppe-Seyler's Z. Physiol. Chem.*, **362**, 573.
9. Jørgensen, K. H. and Larsen, U. D. (1980). *Diabetologia*, **19**, 546.
10. Linde, S., Hansen, B., Sonne, O., Holst, J. J., and Gliemann, J. (1981). *Diabetes*, **30**, 1.
11. Maceda, B. P., Linde, S., Sonne, O., and Gliemann, J. (1982). *Diabetes*, **31**, 634.
12. Gammeltoft, S., Østergaard Kristensen, L., and Sestoft, L. (1978). *J. Biol. Chem.*, **253**, 8406.
13. Gammeltoft, S. (1984). *Physiol. Rev.*, **64**, 1321.
14. Livett, B. G. (1984). *Physiol. Rev.*, **64**, 1103.
15. Kyte, J. and Doolittle, R. F. J. (1982). *J. Molec. Biol.*, **157**, 105
16. Moore, W. J. (1963). *Physical chemistry*, p. 264–265. Longmans, London.
17. Gammeltoft, S., Haselbacher, G. K., Humbel, R. E., Fehlmann, M., and Van Obberghen, E. (1985). *EMBO J.*, **4**, 3407.
18. Boyneams, J. M. and Dumont, J. E. (1980). *Outlines of Receptor Theory*, p. 1–32. Elsevier/North-Holland, Amsterdam.
19. McPherson, G. A. (1985). *J. Pharmacol. Methods*, **14**, 213.
20. Scatchard, G. (1949). *Ann. N.Y. Acad. Sci.*, **51**, 660.
21. Munsson, P. J. and Rodbard, D. (1980). *Anal. Biochem.*, **107**, 220.
22. McPherson, G. A. (1983). *Computer Prog. in Biomed.*, **17**, 107.
23. Haigler, H. T., Maxfield, F. R., Willingham, M. C., and Pastan, I. (1980). *J. Biol. Chem.*, **255**, 1239.
24. Levy, J. R. and Olefsky, J. M. (1987). *Endocrinology*, **121**, 2075.
25. Betsholtz, C., Johnsson, A., Heldin, C.-H., and Westermark, B. (1986). *Proc. Natl. Acad. Sci. USA*, **83**, 6440.
26. Heidenreich, K. A., Brandenburg, D., Berhanu, P., and Olefsky, J. M. (1984). *J. Biol. Chem.*, **259**, 1485.
27. Marshall, S. (1985). *J. Biol. Chem.*, **260**, 13517.
28. Harrison, L. C., Billington, T., East, I. J., Nichols, R. J., and Clark, S. (1978). *Endocrinology*, **102**, 1485.
29. Bond, J. S. and Butler, P. E. (1987). *Ann. Rev. Biochem.*, **56**, 333.
30. Sonne, O. (1988). *Physiol. Rev.*, **68**, 1129.

2

Receptor characterization

JUDITH L. TREADWAY and JEFFREY E. PESSIN

1. Introduction

Characterizing the structural and functional properties of peptide hormone receptors is generally an important first step in determining the molecular mechanism of hormone action and transmembrane signalling in intact cells. In this chapter we give the procedures for isolating and purifying peptide hormone receptors from various tissues and cells, as well as characterizing the physicochemical properties of these receptors. We have focused primarily on methodology for insulin and IGF-I (insulin-like growth factor-I) receptors, with the intent that this information can be applied to other peptide hormone receptor systems as well.

2. Microsomal membrane preparation

The first step in peptide hormone receptor characterization is to select a tissue source, based on abundance or biological significance, from which material can be obtained. In studies of insulin and IGF-I receptors, the human placenta and rat liver (insulin receptor only) have served as abundant and readily available sources. Certain cell lines such as FAO-hepatoma cells and IM-9 lymphocytes have relatively high endogenous levels of insulin receptors as well. More recently, multiple copies of the cDNA for particular receptors have been inserted into cells in culture, which subsequently results in a high level of receptor expression. Such is the case for the insulin receptor transfected into Chinese hamster ovary (1, 2) and NIH 3T3-fibroblast cell lines (3).

After choosing a tissue source, the next step is to isolate the receptor-enriched microsomal membrane fraction by standard differential centrifugation techniques. The procedures given in this chapter include modifications of standard techniques to minimize the degradation of membrane components. This is achieved primarily through the inclusion of a wide array of protease inhibitors (*Table 1*) in the homogenization buffer, and keeping the preparation cold (e.g. on ice) at all times. Plasma membranes generally do not need to be separated from other subcellular

Table 1. Protease inhibitor stocks (stored at $-20°C$)

Leupeptin 10 mM (Sigma No. L-2884) in water
Aprotinin, 50 TIU/ml (Sigma No. A-1153) in water
Pepstatin A, 1 mM (Sigma No. P-4265) in 100% ethanol
PMSF (Phenylmethylsulfonylfluoride, Sigma No. P-7626), 200 mM in 100% ethanol

components constituting the crude microsomal fraction, provided that receptors can be purified upon additional processing.

2.1 Human placenta (4, 5)

The instructions given in *Protocol 1* are for one full-term human placenta. The placental homogenization buffer (1 litre) and PBS (4 litres) should be prepared from stock solutions before processing the placenta (*Table 2*). Keep the placenta at 4°C until processing. For best results, the placenta should be used within 9 h of delivery.

Table 2. Buffers for membrane preparation

Homogenization Buffer
250 mM Sucrose
10 mM Tris
2 mM EDTA
10 μM Leupeptin
0.05 TIU/ml Aprotinin
1 μM Pepstatin A
25 mM Benzamidine (Sigma No. B-6506)
1 mM 1,10 Phenanthroline (Sigma No. P-9375)
1 mM PMSF (add just before use)
pH 8.0

Phosphate Buffered Saline (PBS)
150 mM NaCl
5 mM Na_2HPO_4
pH 7.4

Sodium Chloride Stock
4 M NaCl

Magnesium Sulphate Stock
500 mM $MgSO_4$

Protocol 1. Human placental microsomal membrane preparation

1. Place the placenta in a large kitchen strainer and rinse with 1 litre of ice-cold PBS to remove the residual blood. Trim away the cord and amnionic membrane, then cut the placenta into quarters, and rinse it with 1 litre PBS. Cut off the chorionic plate and other connective tissue, and then mince into

Protocol 1. *Continued*

25–50-g pieces. Rinse the minced placenta with an additional 2 litres of ice-cold PBS.

2. Place the placental tissue in a 1-litre wide-mouth Nalgene container, and add homogenization buffer which contains the protease inhibitor cocktail (*Table 2*) until the total volume reaches approximately 750 ml. Homogenize the placenta (on ice) using a Brinkman Polytron homogenizer with a large probe for 7 min at setting 5.5. It will be necessary to clear the probe of connective tissue every 2–3 min during this homogenization step.

3. Centrifuge the homogenate at 600*g* for 10 min at 4°C. Set aside the supernatant at 4°C.

4. Pool the pellets, mix with additional homogenization buffer (350 ml), and rehomogenize using the Polytron for 5 min. Centrifuge this second homogenate at 600*g* for 10 min at 4°C, and combine the resultant supernatant with that from the first spin.

5. Centrifuge the pooled supernatant at 12 000*g* for 30 min at 4°C. Take the resultant supernatant (approximate volume 800 ml) and add enough of the NaCl Stock and $MgSO_4$ Stock (*Table 2*) to give final concentrations of 0.1 M NaCl and 0.2 mM $MgSO_4$.

6. Mix by inversion several times, and then centrifuge at 30 000*g* for 60 min at 4°C. Discard the supernatant, being careful not to lose any of the pellet.

7. Resuspend the pellets (approximate volume 100 ml) with 55 ml homogenization buffer to which 2.42 g NaCl (0.75 M) has been added. Homogenize the resuspended pellets using five passes at medium speed with a Potter–Elvehjem tissue homogenizer.

8. Centrifuge the homogenate at 48 000*g* for 45 min at 4°C to obtain the microsomal membrane pellet. Discard the supernatant, resuspend the pellets in approximately 50 ml homogenization buffer (without NaCl), and repeat the homogenization using the Potter–Elvehjem apparatus.

9. Recentrifuge the homogenate at 48 000*g* for 45 min at 4°C. Discard the supernatant, and resuspend the final pellets in ≤20 ml of homogenization buffer (protein concentration 10–20 mg/ml).

10. Pass the resuspended membranes serially through 18, 22, and 26 gauge needles, being careful not to cause foaming. Store the placental membranes at −70°C until use.

2.2 Liver (6)

Protocol 2. Rat liver microsomal membrane preparation

1. Rinse fresh rat livers with PBS in order to remove residual blood. If necessary,

Protocol 2. *Continued*

   If necessary, the liver can be perfused with isotonic saline prior to removal
   from the animal (7).

2. Following the rinse step, dilute the tissue with 2–5 ml/g of ice-cold
   homogenization buffer (*Table 2*) and mince finely. Homogenize, using 10
   passes at maximal speed with a Potter–Elvehjem apparatus.

3. Centrifuge the homogenate at 600g for 10 min at 4°C.

4. Transfer the supernatant to a second centrifugation tube, and recentrifuge at
   12 000g for 30 min at 4°C.

5. Take the resultant supernatant and add enough of the NaCl Stock and
   MgSO$_4$ Stock to achieve final concentrations of 0.1 M and 0.2 mM,
   respectively. Mix well, then centrifuge at 40 000g for 40 min at 4°C. This yields
   a loosely packed pellet.

6. Resuspend the pellets using 10 vol. of homogenization buffer, and homogenize using the Potter–Elvehjem apparatus as described above. Centrifuge the
   homogenate at 40 000g for 40 min at 4°C. Resuspend the pellet, homogenize,
   and centrifuge at 40 000g once more.

7. Resuspend the final pellet with homogenization buffer as described above to
   give a protein concentration of 10 mg/ml.

2.3 Adipose tissue (6, 8)

Fresh adipose tissue, or adipocytes isolated from fat pads by the collagenase
digestion procedure (9), should be homogenized directly in homogenization buffer
using a Potter–Elvehjem apparatus. Centrifuge the homogenate at 3000g for
10 min at 4°C. Collect the infranatant by aspiration, and centrifuge it at 48 000g for
45 min at 4°C. Discard the supernatant, and resuspend the pellet in homogenization buffer as described in *Protocol 1* to give a protein concentration of 10 mg/ml.

2.4 Cultured cells

Protocol 3. Microsomal membrane preparation from cultured cells

1. Cells grown to confluent monolayers in standard tissue culture dishes should
   be rinsed once with ice-cold PBS, scraped, and then pooled with additional
   PBS.

2. Centrifuge the cells at 300g for 5 min, discard the supernatant, and resuspend
   the cell pellet with 2–5 ml of PBS.

3. Add the cell suspension dropwise to a stirring solution of 1 mM Tris plus
   protease inhibitors (pH 7.4) at 4°C (final dilution of cell suspension
   approximately 1 : 50). Stir for 5 min.

Protocol 3. *Continued*

4. Centrifuge the solution at 3000*g* for 15 min at 4°C. If the supernatant appears cloudy, set it aside at 4°C to recentrifuge later. Otherwise, discard the supernatant.

5. Resuspend the pellet in homogenization buffer, and homogenize using 5–10 passes at 75% maximal speed with a Potter–Elvehjem apparatus. Centrifuge the homogenate at 3000*g* for 15 min at 4°C.

6. Combine the supernatant with that set aside previously, and recentrifuge at 48 000*g* for 45 min at 4°C.

7. Resuspend the pellet in homogenization buffer as described above to give a final protein concentration of 2–5 mg/ml.

3. Solubilization

Most peptide hormone receptors are soluble in non-denaturing, non-ionic detergents, such as Triton X-100. By adding such detergents to a crude membrane preparation, receptors can be extracted from the lipid bilayer of the plasma membrane and separated from detergent-insoluble components. Solubilized receptors can be concentrated upon subsequent purification. The study of solubilized receptors is important since properties intrinsic to the receptor can be partitioned from the regulatory effects by the plasma membrane or other membrane-associated components.

A number of tissues can be solubilized directly, omitting the need for preparing microsomal membranes. This approach is particularly useful when examining tissues in which standard subcellular fractionation techniques are difficult to execute (e.g. muscle). A second application for direct solubilization is the isolation of receptors as they exist *in vivo*. By including the appropriate agents in the solubilization media, receptors can be preserved as structurally or covalently modified in response to various stimuli. Recently, much interest has focused on the *in vivo* alterations in phosphorylation state of protein–tyrosine kinase hormone receptors in response to ligand and/or agents which modulate the binding and kinase properties of these receptors. This has led to the inclusion of protease, phosphatase, and kinase inhibitors in the solubilization media. These methods have been applied to the isolation of receptors from whole tissues (10) and cells (11).

3.1 Membranes

Before solubilization, it is necessary to measure the protein concentration and volume of the membrane preparation. The concentration of Triton X-100 required for solubilization depends on the protein concentration of the membrane preparation: for example, if the starting material is < 10 mg/ml, use 1.0% Triton X-100; if 10–19 mg/ml, use 1.5%; if > 20 mg/ml, use 2.0%. Add an aliquot of a 20% Triton X-100 stock solution to achieve the appropriate final concentration. Note that the use of excess detergent does not improve the efficiency of receptor solubilization and can result in the inactivation of receptor function. We also routinely add an aliquot of the 10 mM leupeptin stock solution to give a final concentration of 10 μM. Mix the membrane-detergent solution for 60 min at 4°C, then centrifuge at 100 000g for 60 min at 4°C. The supernatant contains the solubilized membrane proteins, and generally is processed immediately for further purification. Typically, the solubilization procedure results in $\geq 80\%$ recovery of receptor binding activity (12).

If the protein concentration of the solubilized membrane preparation is to be determined by standard colorimetric assays, 1–2% sodium dodecylsulphate (SDS) should be added to the assay reagent to remove the turbidity caused by Triton X-100 at concentrations exceeding 1% (13). Also note that Triton X-100 gives a high absorbance at 280 nm.

3.2 Tissue (14, 15)

The example given will be for preparing solubilized receptors from skeletal muscle tissue. Skeletal muscle should be rapidly dissected from anesthetized animals, blotted to remove any residual blood, then transferred to a vessel containing liquid nitrogen. Care must be taken to assure that the muscle tissue remains frozen until the step of Polytron homogenization. Use a large ceramic mortar and pestle to grind the muscle under liquid nitrogen until a fine powder results. Next, let the liquid nitrogen evaporate, then quickly weigh the powdered muscle in a pre-tared beaker containing homogenization buffer (2.5 ml/g powdered tissue) (*Table 3*). Once the frozen powdered muscle has been added to the homogenization buffer, immediately homogenize it at 4°C using a Polytron homogenizer (large probe) at maximal speed. It is necessary to work rapidly to ensure that the tissue mixes well with the protease inhibitors and metal ion chelators as it thaws. Continue to homogenize until the solution achieves a creamy texture. Next, add the appropriate amount of the 20% Triton X-100 to give a final concentration of 1%. Allow the homogenate to stir for 60 min, then

Table 3. Muscle homogenization buffer

25 mM Hepes
4 mM EDTA
4 mM EGTA
1000 TIU/ml Aprotinin
1 mM Bacitracin (Sigma No. B-0125)
2 μM Leupeptin
2 μM Pepstatin A
25 mM Benzamidine
2 mM PMSF (add just before use)
pH 7.4

Table 4. *In vivo* phosphorylation state solubilization buffer

20 mM Hepes
1% Triton X-100
2.5 mM PMSF
800 TIU/ml Aprotinin
8 mg/ml Bacitracin
8 mM EDTA
160 mM NaF
10 mM Na-pyrophosphate
200 μM Na-vanadate
2 mM Dichloroacetate
10 μM Leupeptin
2 μM Pepstatin A
pH 7.4

centrifuge at 150 000g for 90 min at 4°C. Save the supernatant for subsequent purification.

3.3 Preservation of *in vivo* phosphorylation state (11)

To isolate receptors from cells while simultaneously preserving the *in vivo* phosphorylation state, a solubilization buffer which includes protease, phosphatase and kinase inhibitors should be used (*Table 4*). Terminate the *in vivo* pretreatment by rapidly aspirating the cell media, placing the cells on ice, and adding Solubilization Buffer. Cells are then pooled and stirred for 60 min at 4°C. The solubilized cells are next centrifuged at 100 000g for 60 min at 4°C, and the supernatant is processed immediately for receptor purification. Until the receptors reach a purified/partial purified state, buffers should include 100 μM Na-vanadate, 100 mM NaF, 10 mM Na-pyrophosphate, and 4 mM EDTA. Once purified, 100 μM Na-vanadate is the only recommended addition to standard buffers used.

4. Partial purification of receptors

Before examining the physicochemical properties of peptide hormone receptors, it is necessary to separate the receptors from other components of the detergent-solubilized membrane fraction for the following reasons: (a) receptors generally constitute $\leq 0.05\%$ of the protein present in the detergent-solubilized extract, making this source too dilute for effective study; and (b) the concentration of detergent required for solubilization (e.g. 1%) may have an inhibitory effect upon receptor function. In this section we present some common chromatographic techniques used to purify and concentrate receptors from detergent-solubilized membrane extracts. These methods are also effective in lowering the detergent concentration of the sample from that required for initial membrane solubilization.

Partial purification can be accomplished by a relatively rapid separation of solubilized membrane proteins on the basis of size or general structural/functional characteristic. In some cases, the enrichment of receptor following partial purification is sufficient for subsequent analyses (e.g. lectin affinity chromatography of insulin receptors). Partial purification is also useful in removing proteases and other degradatory elements from the membrane/tissue solubilate and isolating receptor fractions for subsequent homogenous purification.

4.1 Immobilized lectins

Lectins are a group of proteins which react reversibly with specific sugar residues of glycosylated proteins. Immobilized lectins are extremely useful for separating many glycosylated proteins, including insulin (16), IGF-I (17), IGF-II (18), EGF (19), and PDGF (20) receptors, from the majority of detergent-solubilized membrane components. The glycoprotein fraction can be specifically eluted from the resin by washing with simple sugars, typically resulting in a 20-fold purification over the detergent-soluble membrane extract. The most commonly used lectin for receptor purification is wheat germ agglutinin (WGA), which specifically binds with *N*-acetylglucosaminyl residues. Other lectins used for the partial purification of hormone receptors are concanavalin A (Con A) and the related lectin *Len culinaris* hemagglutinin which bind specifically with α,D-mannopyranosyl and α,D-glucopyranosyl residues. Triton X-100 (1%) and sodium deoxycholate (1%) do not affect the binding capacity of these lectins; WGA binding capacity is also unaffected by 0.07% sodium dodecyl sulphate (SDS) (21). Lectins coupled to Sepharose beads (immobilized lectins) are available commercially or can be prepared by standard methods (22).

References 21 and 23, available from Pharmacia Fine Chemicals (Uppsala, Sweden), provide extensive practical information on designing and performing affinity chromatography and gel filtration chromatography experiments. These references should be consulted so that equipment and procedures can be selected to meet individual needs.

The amount of lectin resin to use depends on the binding capacity of the resin. This information is usually specified in the technical information from the supplier. We routinely use 20 ml of Wheat Germ Lectin–Sepharose 6MB (Sigma No. L-6257) to adsorb insulin/IGF-I receptors from the solubilized membrane proteins of one human placenta.

Wash the lectin resin with the appropriate Equilibration Buffer prior to use (*Tables 5* and *6*). Dilute the solubilized membrane sample with 2–9 volumes of Equilibration Buffer, then add the sample to the resin by either: (a) recycling the sample over the resin column using a peristaltic pump; or (b) batch-mixing the resin with the sample end-over-end in a Nalgene container. The sample and resin should mix for a minimum of 4 h at 4°C or until specific hormone receptor binding activity is no longer detected in the column effluent. It is sometimes convenient to allow the resin to mix overnight (approximately 16 h) at 4°C.

Table 5.

WGA Equilibration Buffer
50 mM Hepes
10 mM $MgCl_2$
0.1% Triton X-100
0.02% NaN_3
pH 7.4

WGA Elution Buffer
50 mM Tris
0.1% Triton X-100
0.02% NaN_3
0.3 M *N*-acetylglucosamine
  (Sigma No. A-8625)
pH 8.0

Table 6.

Con A Equilibration Buffer
50 mM Hepes
10 mM $MgCl_2$
0.1% Triton X-100
0.02% NaN_3
2 mM $CaCl_2$
100 mM NaCl
pH 7.4

Con A Elution Buffer
50 mM Tris
0.1% Triton X-100
0.02% NaN_3
0.3 M methyl α,D-mannoside
  (Sigma No. M-6882)
pH 8.0

At the end of the mixing period, pour the resin into a column of sufficient capacity. Partial purification by lectin affinity chromatography can be performed using a short, large-diameter (e.g. 2.5 cm) column for rapid separation and elution. Collect the flow-through, and then wash the resin with 10 column volumes of Equilibration Buffer. Run the Equilibration Buffer from the column so that the top of the resin bed is exposed (but not dried or cracked), then add the Elution Buffer (approximately 1 column volume), being careful not to disturb the resin bed. Open the column briefly so that the Elution Buffer permeates the resin, then stop the column flow and allow it to sit for approximately 30 min. Restart the column and collect the eluate fraction(s) by continuously adding more Elution Buffer until all hormone receptor binding activity has been recovered.

Following the elution procedure, the column should be washed with 10 column volumes of Regeneration Buffer A, followed by 10 column volumes of Regeneration Buffer B (*Table 7*), to remove residual proteins and to desorb the eluting sugar. The resin should then be washed with 10 column volumes of Equilibration Buffer, and stored at 4°C for future use.

Immobilized lectins can also be used to separate subclasses of glycoprotein receptors on the basis of lectin binding affinity using a gradient (0.1–0.5 M) of the eluting sugar, or pulse elution using several different sugars.

Table 7. Lectin column regeneration buffers

Regeneration Buffer A	Regeneration Buffer B
50 mM Tris	50 mM Na-acetate
0.1% Triton X-100	0.1% Triton X-100
0.02% NaN$_3$	0.02% NaN$_3$
1 M NaCl	1 M NaCl
pH 8.5	pH 4.5

4.2 Preparative gel filtration chromatography

Sephacryl S-400 superfine (Sigma No. S-9759) is a general purpose gel filtration medium which allows for a wide range of separation (Mr 2×10^4–8×10^6). The structural rigidity of Sephacryl S-400 makes it suitable for use in preparative gel filtration chromatography. In this section we describe the general procedure for Sephacryl S-400 gel filtration chromatography of solubilized placental membrane proteins. For specific information on how to perform a gel filtration experiment (e.g. choosing equipment, preparing columns, etc.), consult reference 23.

To obtain good resolution of receptor-binding activity by gel filtration chromatography, long columns should be used, and sample volume should be limited to 1–5% of the column volume. Toward this end, we use a 1500-ml Sephacryl S-400 resin column (Pharmacia K 50/100, 5×100 cm) for partially purifying the detergent-soluble membrane extract from two placentae (≤ 40–50 ml); for one placenta (≤ 20 ml), a 600 ml column (Pharmacia K 16/70)

Table 8. Sephacryl S-400 Buffer

50 mM Tris
0.1% Triton X-100
0.02% NaN_3
1 μM Leupeptin
pH 7.8

---

is used. The resin should be washed and pre-equilibrated with Sephacryl S-400 Buffer (*Table 8*) for at least 24 h before use.

The sample can be loaded manually, or through use of a sample applicator. Manual loading is suited primarily for small columns where flow is determined by gravity. If the sample is more dense than the eluent buffer (e.g. through the addition of glucose, salt, glycerol, etc.), it can be loaded under the eluent on to the resin bed. Typically, the Triton X-100 solubilized placental membrane extract is sufficiently dense that addition of these agents is not necessary. Draw the sample into a syringe fitted with fine capillary tubing. Close the column outlet, and place the tip of the tubing a few millimetres above the bed surface. Slowly dispense the sample under the eluent buffer. When removing the tubing, draw a small amount of the buffer into the syringe to prevent sample mixing with the overlaying buffer.

A second method for manual loading is to apply the sample directly on to a drained resin bed. If this method is selected, extreme care must be taken not to disturb the resin bed and/or to let it run dry. Disconnect the top of the column from the buffer reservoir and open the column outlet to drain the buffer to the top of the resin bed. Close the column outlet and then apply the sample by running it gently down the side of the column wall on to the resin bed. Resume flow and let the sample enter the resin bed. When the sample has completely permeated the resin bed, stop the column flow and layer buffer gently on top of the resin bed, rinsing any residual sample from the column wall. When sufficient buffer has been overlayed, reconnect the top of the column to the buffer reservoir and resume flow.

A sample applicator system provides a more convenient method when using large columns and sample volumes. For the 600-ml Sephacryl S-400 column, a three-way valve is placed in series with the tubing connecting the buffer reservoir and the top of the column. A syringe (minus plunger) is attached to the three-way valve, and is set on bypass until sample loading is required. To load a sample, the syringe is filled with sample, and the valve is turned to bypass the buffer reservoir. When the sample has drained gravimetrically on to the column, the valve is turned to resume flow from the buffer reservoir. Care must be taken to assure that air is not introduced into the lines.

For the 1500 ml Sephacryl S-400 column set-up (*Figure 1*), a peristaltic pump (Pharmacia P-3) is used to maintain flow rate at 75 ml/h. The column is run upward to counteract the force of gravity when resolving samples over a large resin bed. A four-way valve (Pharmacia SRV-4) is connected in series with the

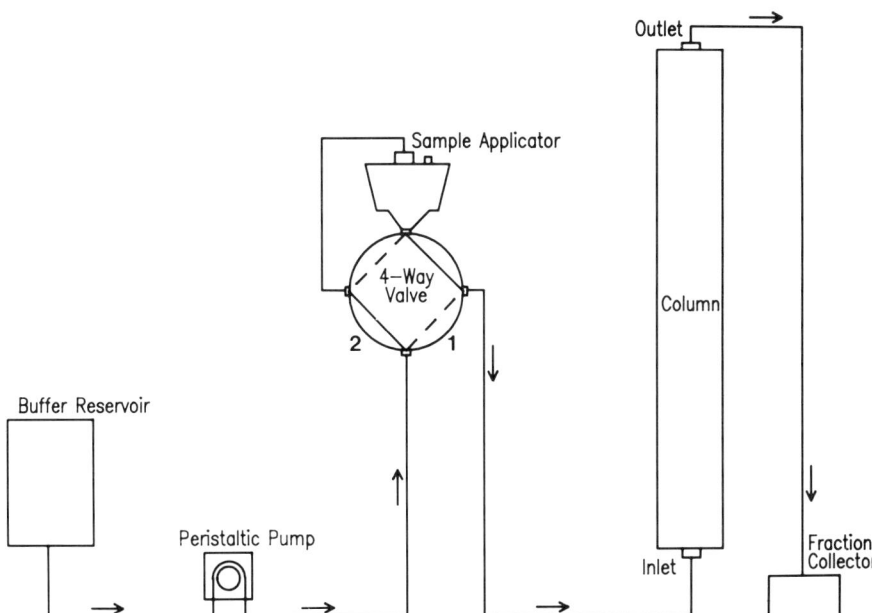

Figure 1. Schematic representation of the Sephacryl S-400 preparative gel filtration chromatography set-up. Position 1 of the four-way valve indicates the direction of flow under sample applicator bypass conditions. Position 2 indicates the direction of flow during sample loading.

buffer reservoir and the bottom of the column (now serving as the column inlet). The other two positions on the valve are occupied by a sample reservoir (Pharmacia SA-50), and tubing which connects the buffer reservoir with the top of the sample reservoir. The connecting tubing and sample reservoir, filled with S-400 buffer (*Table 8*), are set on bypass until loading is required. To load the sample, release the vacuum pressure in the sample reservoir, and use a syringe and needle to remove a volume of buffer equal to the amount of sample to be loaded. Next, gently inject the sample as a layer below the eluent buffer, then reseal the sample reservoir. Turn the valve to direct flow from the buffer reservoir to the sample reservoir, and from the sample reservoir to the column inlet. The buffer entering the sample reservoir chamber will force the sample on to the column. When sample loading is completed, turn the valve to resume flow directly from the buffer reservoir.

For Sephacryl S-400 gel filtration chromatography, column flow rate and sample loading rate should be between 40–80 ml/h. Once the sample has been loaded on to the column, collect a volume of eluent buffer equivalent to the column void, then connect the column outlet tubing to a fraction collector. Collect fractions (7.5 ml/tube for the 600-ml and 1500-ml Sephacryl S-400 columns) until the reddish-pink haemoglobin-containing fractions are eluted

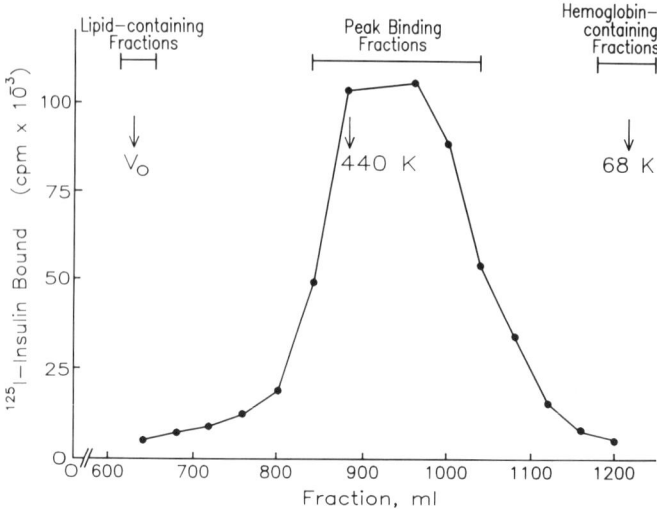

Figure 2. Sephacryl S-400 gel filtration profile of ^{125}I-insulin binding to partially purified, detergent-soluble placental membrane extract. The Triton X-100 soluble extract of membranes from one human placenta was subjected to preparative Sephacryl S-400 gel filtration chromatography (*Figure 1*). After voiding the first 600 ml of the column effluent, fractions (8 ml) were collected and assayed (50 μl) for ^{125}I-insulin binding. At the top of the figure, the fractions containing lipid and haemoglobin originating from the placental membrane extract are indicated. The elution positions of protein molecular weight markers (V_0 = Dextran blue; 440 K = ferritin; 68K = haemoglobin) are also indicated.

from the column. Assay for ^{125}I-insulin binding (as described in Chapter 1) and pool the peak binding fractions ($V_t = 160$–200 ml). The binding profile for a typical Sephacryl S-400 gel filtration purification procedure is given in *Figure 2*. Following the elution of the haemoglobin-containing fractions, let the column run for an additional 24 h to wash the column.

Sephacryl S-400 gel filtration results in approximately six- to tenfold purification of the insulin receptor. The pooled peak binding fractions are usually processed immediately for homogeneous receptor purification.

5. Homogeneous purification of receptors

Analyses of structure and function generally require concentrated receptor preparations that are free of impurities which can alter the properties being examined (e.g. receptor kinase activity), or confound interpretation of the results. Homogeneous receptor samples can be prepared directly from the detergent-soluble extract of membranes or tissues, or from material which has already undergone partial purification. Homogeneous purification is typically based upon the specificity by which immobilized ligands and monoclonal antibodies adsorb receptors from solution. In this section we present the methods for preparing immunoaffinity and ligand-affinity resins, and the procedures for their use in the homogeneous purification of peptide hormone receptors.

5.1 Preparation of affinity resin

Ligands and antibodies are immobilized by covalent interaction with the reactive groups of a coupling gel. Two common coupling gels are CNBr-activated Sepharose 4B (Sigma No. C-9142), and Affi-Gel 10 (Bio-Rad No. 153-6046). These activated gels react specifically with the primary amine groups of peptide ligands and monoclonal antibodies. CNBr-activated Sepharose 4B is synthesized by reacting the hydroxyl groups of Sepharose with CNBr. When ligand or antibody is added to CNBr-activated Sepharose, covalent isourea bonds form between the primary amine groups and Sepharose. Affi-Gel 10 consists of Sepharose with covalently attached 10-carbon spacer arms. The terminal carboxyl groups of the spacer arms are joined by an ester linkage to *N*-hydroxysuccinimide. Upon addition of ligand or antibody, the primary amine groups displace the *N*-hydroxysuccinimide and form stable amide bonds with the spacer arm. Coupling gels with spacer arms may be advantageous if steric hindrance of receptor adsorption occurs as a result of immobilizing the ligand/antibody directly to Sepharose. We have routinely found Affi-Gel 10 to be superior over CNBr–Sepharose for the coupling of peptide ligands.

The methods for preparing Insulin–Sepharose from CNBr-activated Sepharose 4B and Affi-Gel 10 are presented in this section. Similar procedures have been described for immobilized EGF (24), IGF-I (25), IGF-II (18), bombesin (26), and monoclonal anti-receptor antibodies (5). The manufacturer's instructions should be consulted for choosing the appropriate activated resin and coupling conditions based on individual need.

The procedure for immobilizing monoclonal antibodies to coupling gels is essentially as described for ligands, except that (a) coupling is more effective at a lower protein concentration (e.g. 1–5 mg antibody/ml gel), and (b) 0.5% Tween-80 should be included in the buffers to eliminate non-specific protein binding (21, 30).

Protocol 4. Coupling of insulin to Affi-Gel 10 (27, 28)

1. Pour 25 ml of the Affi-Gel 10 slurry into a Buchner funnel, and wash with three volumes of cold (4°C) water. It is necessary to perform the wash step rapidly because the reactive groups of Affi-Gel 10 undergo hydrolysis in aqueous solution at neutral pH.

2. Immediately mix the Affi-Gel 10 with 50 ml of coupling buffer (*Table 9*) which includes bovine insulin (4 mg/ml gel) and 1 μCi ^{125}I-insulin. Mix the gel suspension end-over-end for 16 h at 4°C.

3. Sediment the Affi-Gel 10 by low-speed centrifugation (300g), and discard the supernatant.

Protocol 4. *Continued*

4. Add 1 vol. of 1 M ethanolamine, pH 9.0 to the resin, and mix overnight at 4°C to block any remaining active groups.

5. Settle the Affi-Gel 10 in a column, and wash with Affi-Gel Urea Buffer (*Table 9*) until ^{125}I-insulin is no longer detected in the column eluate.

6. The column should then be washed at 4°C, alternating between 2 litres of Insulin-Sepharose Urea Buffer and 2 litres of Insulin–Sepharose Equilibration Buffer (*Table 11*) to remove non-covalently bound proteins.

7. Due to the large amount of insulin bound to the resin, the column should be washed continuously (7–10 days) until the level of insulin leakage from the column is <100 pM. This can be readily determined using a standard radioimmunoassay kit for insulin (e.g. Micromedic Systems Inc., Horsham, PA, USA, Kit No. D1804).

8. This procedure typically results in the coupling of 0.8–1.5 mg insulin per millilitre Affi-Gel 10 resin. The column should be equilibrated with Insulin–Sepharose Equilibration Buffer and stored at 4°C until use.

Table 9	Table 10
Affi-Gel Coupling Buffer 100 mM Hepes 4.5 M Urea 80 mM $CaCl_2$ pH 7.6	*CNBr–Sepharose Coupling Buffer* 100 mM $NaHCO_3$ 1 M NaCl pH 8.3
Affi-Gel Urea Buffer 50 mM Tris 1 M NaCl 4.5 M Urea pH 8.0	*CNBr–Sepharose Borate Buffer* 100 mM Na-borate 1 M NaCl pH 8.5 *CNBr–Sepharose Acetate Buffer* 100 mM Na-acetate 1 M NaCl pH 4.1

Protocol 5. Coupling of insulin to CNBr-activated Sepharose 4B (21, 29)

1. Mix CNBr-activated Sepharose 4B with 1 mM HCl (200 ml/g) and allow to swell for 15 min (1 g dry gel = 3.5 ml swollen gel).

2. Apply the resin to a scintered glass filter, and wash with 10 volumes of 1 mM HCl, followed by 1 vol. of coupling buffer (*Table 10*).

3. Immediately add the activated Sepharose to 2 vol. of coupling buffer which includes bovine insulin (4 mg/ml gel) and 1 μCi ^{125}I-insulin. Mix the gel suspension end-over-end for 16 h at 4°C.

4. Sediment the activated Sepharose by low-speed centrifugation (300*g*), and discard the supernatant.

Protocol 5. *Continued*

5. Add 1 vol. of 1 M ethanolamine, pH 9.0 to the resin, and mix overnight at 4°C to block any remaining active groups.

6. Wash the resin with CNBr–Sepharose Coupling Buffer until ^{125}I-insulin is no longer detected in the column eluate.

7. Wash the resin with CNBr–Sepharose Borate Buffer for 48 h, followed by CNBr–Sepharose Acetate Buffer for 24 h (*Table 10*) to remove any non-covalently adsorbed proteins. Continue washing by alternating between 2 litres Insulin–Sepharose Equilibration Buffer and 2 litres Insulin–Sepharose Urea Buffer (*Table 11*) until insulin leakage from the column is negligible.

8. Store the column in Insulin–Sepharose Equilibration Buffer at 4°C until use.

Table 11. Insulin–Sepharose column buffers

Equilibration buffer 50 mM Tris 1.0 M NaCl 0.1% Triton X-100 pH 7.8	*Octylglucoside (β-OG) Wash Buffer* 50 mM Hepes 1.0 M NaCl 0.6% n-octyl-β,D glucopyranoside (β-OG, Sigma No. 0-8001) 0.02% NaN$_3$ pH 7.8
Elution Buffer 50 mM Na-acetate 1.0 M NaCl 0.6% β-OG 10% Glycerol 0.02% NaN$_3$ pH 5.0	*Urea Buffer* 50 mM Na-acetate 1.0 M NaCl 0.1% Triton X-100 4.5 M Urea 0.02% NaN$_3$ pH 6.0

5.2 Purification procedure

5.2.1. Insulin receptor (31, 32)

The most common procedure for purifying insulin receptors to homogeneity is through the use of Insulin–Sepharose affinity chromatography.

The procedure given is for purifying insulin receptors using a 15-ml Insulin–Sepharose column (1.6 × 10 cm). Starting material for this procedure is the partially purified sample (80–100 ml pooled peak binding fractions) obtained from Sephacryl S-400 filtration or alternatively the WGA–Sepharose eluant of detergent-soluble membrane extract from one human placenta. All steps are performed at 4°C.

Other methods reported for the homogeneous purification of insulin receptors include immunoaffinity chromatography (5, 33, 34), and biotinylinsulin–avidin affinity chromatography (35, 36).

Protocol 6. Affinity purification of insulin receptors

1. Equilibrate the column by washing for 16 h at 4°C in Insulin–Sepharose Equilibration Buffer (*Table 11*).

2. Load the column by recycling the sample 4–6 times using a peristaltic pump set at a flow rate of 20–25 ml/h.

3. Collect the flow-through, then wash the column with *n*-octyl-β,D-glucopyranoside (β-OG) Wash Buffer (*Table 11*). The use of the β-OG Wash Buffer provides for the effective exchange of the Triton X-100 detergent for the β-OG detergent. This allows for the rapid concentration of the purified insulin receptor without concentration of the detergent at a subsequent step. Wash the column with β-OG buffer until detection of Triton X-100 by absorbance at 280 nm is negligible. Typically, the exchange for a 15-ml column is complete following washing with 60–80 ml of the β-OG Wash Buffer.

4. To elute the receptors from the column, pass 25–30 ml of the Insulin–Sepharose Elution Buffer (*Table 11*) over the column, and collect the eluate using a fraction collector. The fractions must be collected into tubes containing 1.5 M Hepes, pH 8.0 (1 vol. Hepes/10 vol. eluate), in order to neutralize the sample and prevent the loss of receptor kinase activity.

5. Assay the fractions for insulin binding activity, and pool the peak binding fractions for concentrating by the procedure given in Section 2.4.3.

The elution procedure can be modified to batch-collect the column eluate once the receptor-containing fractions have been identified through routine use. For the 15-ml Insulin–Sepharose column, we generally pass 20 ml of Elution Buffer over the column, discard the first 5 ml of eluate, and collect the remaining 15 ml of eluate into a tube containing 1.5 ml of the 1.5 M Hepes, pH 8.0.

Regenerate the Insulin–Sepharose column by washing for 3–4 h at 4°C using Insulin–Sepharose Urea Buffer (*Table 11*). The column should then be washed overnight with Insulin–Sepharose Equilibration Buffer, and stored at 4°C.

5.2.2 IGF-I receptor

There are two common approaches for purifying IGF-I receptors to homogeneity: (a) IGF-I Sepharose affinity chromatography (25); and (b) immunoaffinity chromatography using the IGF-I receptor specific monoclonal antibody, α-IR3 (37–39).

Material which is partially purified by either WGA-chromatography or gel filtration chromatography is generally used as the starting material for homogeneous purification of IGF-I receptors. Another source of starting material for IGF-I receptor purification is the flow-through from an Insulin–

Sepharose affinity resin column, since this material is enriched in IGF-I receptors and essentially devoid of the homologous insulin receptor protein. To perform IGF-I receptor purification using IGF-I–Sepharose, use the same buffers and procedures given for Insulin–Sepharose affinity chromatography.

The procedure and buffer system for α-IR3 immunoaffinity chromatography is identical to that for Insulin–Sepharose affinity chromatography, except that receptors can be eluted using either alkaline (50 mM sodium bicarbonate, 0.6% β-OG, 0.02% NaN_3, 1 M NaCl, 10% glycerol, pH 11.0) or acidic (e.g. Insulin–Sepharose Elution Buffer) conditions. Alkaline elution conditions do not affect the binding or kinase properties of the IGF-I receptor (39). Acidic elution results in slightly lower yield, but binding and kinase activities are fully retained (39). α-IR3 immunoaffinity chromatography results in a 3200-fold purification of the IGF-I receptor over WGA-chromatography, with a typical yield of 5–10 µg of purified receptor protein per human placenta.

5.2.3 EGF receptor

There are two common approaches for purifying EGF receptors to homogeneity: (a) EGF–Sepharose affinity chromatography (24, 40, 41); and (b) immunoaffinity chromatography using anti-EGF receptor antibodies (42–45).

EGF is extracted and purified from mouse submaxillary glands by standard procedures (46), and coupled to Affi-Gel 10 as described in Section 2.4.1. This procedure results in a ligand–resin coupling ratio of approximately 1 mg/ml when the coupling is performed at 3–4 mg EGF/g gel (24).

Detergent-soluble membrane protein (20 mg) is used as the starting material. Typically, membranes are obtained from A-431 cells, a human epidermoid carcinoma cell line which expresses $2-3 \times 10^6$ EGF receptors per cell (24). The resin should be pre-equilibrated by washing overnight at 4°C using EGF–Sepharose Equilibration Buffer (*Table 12*). Add the detergent-soluble membrane extract directly to a 1.0 ml EGF–Sepharose column and recycle the sample 2–3 times (3 h) at 4°C. Collect the column flow-through, then wash the resin with 10 column volumes of EGF–Sepharose Equilibration Buffer. The EGF receptors are batch-eluted using 1.5 ml of EGF–Sepharose Elution Buffer (*Table 12*), and collected into a tube containing 1 M Hepes, pH 7.4 for neutralization. The yield of EGF receptor by this method is approximately 1 µg purified EGF receptor/mg

---

Table 12

EGF–Sepharose Equilibration Buffer
20 mM Hepes
0.1% Triton X-100
pH 7.4

EGF–Sepharose Elution Buffer
20 mM NH_4OH
0.1% Triton X-100
pH 9.8

---

detergent-soluble membrane protein. EGF–Sepharose affinity resin is regenerated by the procedure and buffers given for Insulin–Sepharose chromatography (Section 5.2).

Monoclonal anti-EGF receptor antibody should be coupled to CNBr-activated Sepharose 4B to achieve a final concentration of approximately 10 μM. Prepare the resin for use by washing overnight at 4°C using Equilibration Buffer. Dilute the detergent-soluble membrane extract 1:10 with Equilibration Buffer (*Table 13*), then apply the sample to the anti-EGF receptor antibody column, and recycle for 1–3 h at 4°C. Collect the column flow-through, then wash the resin with 10 vol. of Salt Buffer (*Table 13*). Re-equilibrate the column by washing with 2 volumes of Equilibration Buffer. Elute the EGF receptors by adding Elution Buffer (*Table 13*) to the column, and immediately neutralize the eluate with 1.5 M Hepes pH 8.0.

In addition to procedures for ligand-affinity and immunoaffinity purification, homogeneous purification of the EGF receptor has been reported using Tyrosine–Sepharose affinity chromatography (47, 48).

Table 13. Anti-EGF receptor immunoaffinity chromatography buffers

Equilibration Buffer
20 mM Hepes
0.1% Triton X-100
10% Glycerol
pH 7.5

Elution Buffer
50 mM Glycine
10% Glycerol
0.1% Triton X-100
pH 2.5

Salt Buffer
50 mM Borate
0.7 M NaCl
0.3 M $MgCl_2$
10% Glycerol
0.1% Triton X-100
pH 8.3

5.3 Concentration of purified receptors

The Centricon-30 miniconcentrator (Amicon No. 4208) is used to concentrate the purified receptor preparation (represented as an affinity resin eluate) approximately 25-fold, while simultaneously reducing the concentration of salt (1 M) present. The procedure involves ultrafiltration of the sample through a concentrator membrane (30 kd exclusion limit), which results in the removal of low-molecular weight solutes and solvent from the receptor sample (49). It is important to note that this concentration method is effective only for samples in β-OG buffers, as β-OG is freely permeable to the concentrator membrane.

For detailed information on the use of the Centricon-30 miniconcentrator, consult the technical bulletin provided by the manufacturer (49). In brief, load a 2 ml aliquot of the dilute receptor preparation into the sample reservoir of the miniconcentrator, and centrifuge the apparatus at 3000g for 20 min at 4°C. Discard the filtrate which collects in the filtrate cup, and add another 2 ml aliquot of the dilute receptor preparation to the sample reservoir. Repeat the loading/centrifugation process until the original sample (e.g. 15 ml Insulin–Sepharose eluate) has been concentrated to a volume of 1–2 ml.

The concentrated receptor sample is next reconstituted with Centricon Wash Buffer (*Table 14*) to exchange the solvent and further desalt the sample. Add 2.5 ml of Centricon Wash Buffer to the sample reservoir, and centrifuge for 25 min at 3000g at 4°C. Repeat the wash step, then recover the sample by capping the sample reservoir with a retentate cup, inverting the centricon apparatus, discarding the lower reservoir, and applying 50 μl of Centricon Wash Buffer directly to the membrane. Centrifuge the inverted apparatus at 1000g for 10 min at 4°C. The concentrated receptor sample, recovered in the retentate cup, is now ready for analysis.

Table 14. Centricon Wash Buffer

50 mM Hepes
10% Glycerol
0.6% β-OG
pH 7.4

---

6. Preparing insulin and IGF-I receptor heterodimers

Functional insulin and IGF-I receptor $\alpha\beta$ heterodimeric subunits are prepared from native heterotetrameric complexes by a selective reduction of Class I disulphide bonds under non-denaturing conditions. The procedure employed in our laboratory consists of alkaline pre-treatment followed by reduction using a low concentration of DTT (50, 51). This method is 90% effective in converting $\alpha_2\beta_2$ holoreceptors into $\alpha\beta$ heterodimeric subunits. The insulin and IGF-I receptor $\alpha\beta$ heterodimeric subunits prepared by this technique retain full binding activity at tracer ligand concentrations, and under appropriate conditions display ligand-dependent activation of the receptor kinase. The $\alpha\beta$ heterodimeric subunits provide a useful model for studying the interactions responsible for transmembrane signalling and regulation of the functional properties of the native receptor complex.

The insulin and IGF-I receptor subunits are covalently linked by two general classes of disulphide bonds. The class I disulphides are operationally defined as those responsible for the covalent linkage between the two $\alpha\beta$ heterodimeric receptor halves which form the $\alpha_2\beta_2$ heterotetrameric receptor structure. The class II disulphides have been defined as those which covalently link the individual α and β subunits together. We (50, 51) and others (52, 53) have recently taken advantage of the difference in DTT sensitivity to reduce these two classes of disulphide bonds within the $\alpha_2\beta_2$ heterotetrameric holoreceptor structure in order to generate functional $\alpha\beta$ heterodimeric insulin and IGF-I receptor molecules.

6.1 Solubilized receptors

The $\alpha\beta$ heterodimeric subunits of insulin and IGF-I receptors can be prepared from solubilized $\alpha_2\beta_2$ heterotetrameric receptor complexes following partial or homogeneous purification. The first step involves the alkaline pretreatment of native receptor complexes, achieved by adding 1 vol. of 1.0 M Tris (pH 10.5) to 10 vol. of sample, and incubating the mixture for 25 min at 23°C (final pH 8.5). At the end of the incubation period, add a volume of a concentrated (100 mM) DTT stock solution to the sample to yield a final concentration of 2 mM DTT. Mix the sample, then incubate for an additional 5 min at 23°C. Control receptors (heterotetramers) are prepared similarly, except that an equal volume of H_2O is substituted for DTT.

Following the alkaline-DTT treatment, a rapid gel filtration step is performed to remove the reductant and to return the pH to 7.6. Load the sample on to a Sephadex G-50 column (1.6 × 20 cm) at 23°C which has been pre-washed with Equilibration Buffer (*Table 15*). Once the sample has entered the resin bed, connect the column to the reservoir containing additional Sephadex G-50 Equilibration Buffer, and resume flow. Collect and discard the column void volume, then set to collect 0.4 ml effluent fractions (approximately 30 fractions). Assay the fractions for ligand binding activity, then pool the peak binding fractions (approximate volume 1–1.5 ml). Next, samples are subjected to Bio-Gel A-1.5-m gel filtration at 4°C (Section 7.1.2) to separate and isolate the $\alpha\beta$ and $\alpha_2\beta_2$ receptor complexes. *Figure 3* illustrates a typical Bio-Gel A-1.5-m column

---

Table 15. Buffers for heterodimer preparation

Sephadex G-50 Equilibration Buffer
50 mM Tris
150 mM NaCl
0.1% Triton X-100
0.1% BSA
0.02% NaN_3
pH 7.6

TEN Buffer
50 mM Tris
2 mM EDTA
150 mM NaCl
pH 8.5

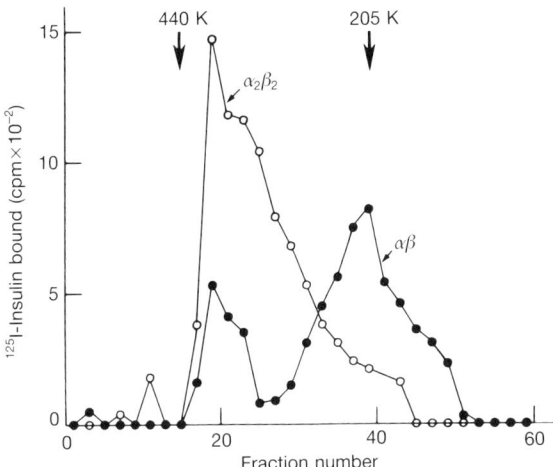

Figure 3. Bio-Gel A-1.5-m gel filtration profile of ^{125}I-insulin binding to $\alpha_2\beta_2$ heterotetrameric and $\alpha\beta$ heterodimeric insulin receptor complexes. Purified human placental insulin receptors (14 μg) were incubated at pH 8.5 for 25 min followed by an additional 5 min incubation with (●) or without (○) 2.0 mM DTT at 23°C. The samples were neutralized with the simultaneous removal of the DTT by rapid Sephadex G-50 gel filtration. The voided insulin receptors were pooled and applied to a Bio-Gel A-1.5-m column at 4°C and the first 20 ml was discarded before collecting 0.4 ml fractions at a flow rate of 15 ml/h. Every other column fraction (50 μl) was then assayed for specific insulin binding activity. The elution positions of protein molecular weight markers (440 K = ferritin; 205 K = myosin) are also indicated. (Reproduced from reference 50.)

profile for ^{125}I-insulin binding to receptor $\alpha\beta$ heterodimers and control $\alpha_2\beta_2$ heterotetramers prepared by this method (50). Peak binding fractions of the $\alpha\beta$ and $\alpha_2\beta_2$ preparations are now ready for use in structural/functional analyses.

Others (52, 53) have reported slightly different methods for preparing $\alpha\beta$ heterodimeric subunits from native insulin receptor complexes. In brief, the detergent soluble $\alpha_2\beta_2$ heterotetrameric insulin receptors are co-incubated with 1.25 mM DTT and 75 mM Tris (pH 8.5) for 30 min at 23°C. The reaction is terminated by centrifuging samples for 5 min at 1000*g* through 3-ml syringes packed with desalting gel (Bio-Rad Laboratories) equilibrated with 30 mM Hepes, 0.1% Triton X-100, 0.02% NaN$_3$, pH 7.6. The $\alpha\beta$ and $\alpha_2\beta_2$ receptor complexes are then separated using sucrose density gradient centrifugation (Section 7.2).

6.2 Membrane receptors (54, 55)

By slight modification, the method described above is employed to prepare $\alpha\beta$ heterodimeric subunits from $\alpha_2\beta_2$ heterotetrameric insulin and IGF-I receptor complexes in membranes. Mix placental membranes (5 mg/ml) with 0.1 volume of 1.0 M Tris (pH 10.5) and incubate for 25 min at 23°C (final pH 8.5). Next, add

DTT (final concentration 2 mM), and incubate for an additional 5 min at 23°C as described above. Following the alkaline pH–DTT treatment, dilute the sample 1:9 with TEN buffer (*Table 15*), and centrifuge at 48000*g* for 30 min at 4°C. Discard the supernatant, then wash the pellet and resuspend to 5 mg/ml with TEN Buffer. This procedure is effective in reducing the DTT concentration below detectable levels.

Next, the membrane-associated $\alpha\beta$ heterodimeric subunits are solubilized by adding 1.0% Triton X-100 and mixing end-over-end for 30 min at 4°C. Centrifuge the sample at 12000*g* for 15 min at 4°C to remove the insoluble material. The supernatant is then subjected to Bio-Gel A-1.5-m gel filtration to separate and isolate the $\alpha\beta$ and $\alpha_2\beta_2$ receptor complexes.

7. Hydrodynamic properties

The determination of hydrodynamic properties of growth factor receptors has been useful in estimating the minimum molecular size of covalent-associated holoreceptor complexes as well as examining the oligomeric state of non-covalent-associated receptor complexes under a variety of experimental manipulations. In addition, since these properties are determined under nondenaturing conditions, the potential interaction with endogenous effector molecules and their regulation of receptor function can be studied.

7.1 Analytic gel filtration chromatography

Analytical gel filtration chromatography allows for the sizing and separation of different species of receptor proteins without the denaturation which occurs during SDS-polyacrylamide gel electrophoresis (SDS-PAGE). Resins used for analytic gel filtration are typically characterized as rigid, highly cross-linked polymers of fine (or superfine) particle-size grade, and having low absorptivity for the solvent or proteins being examined. A wide variety of analytic resins are available commercially. Some basic considerations for selecting a resin are: (a) choosing a particle-size grade that will separate proteins in the fractionation range (Mr) being examined; (b) assuring that the physical stability of the resin is sufficient to meet the flow-rate demand; and (c) assuring that the resin is chemically stable in the solvent system to be used. In general, the best resolution and separation of receptor proteins by analytic gel filtration chromatography is achieved when experiments are performed with very long, small diameter columns at a low flow rate (10–25 ml/h), and sample volumes which do not exceed 1–2% of the column volume.

Gel filtration columns should be calibrated using molecular weight standards (e.g. Pharmacia Kit No. 17-0441-01) before being used for analytical purposes. It is important to note that gel filtration columns separate proteins on the basis of diameter (size), which is a function of mass for uniform globular proteins. Since the crystalline structure for many peptide hormone receptors is not known, this assumption may not be valid. In addition, factors such as the level of hydration and the amount of detergent bound to the receptor may dramatically affect gel filtration properties. Receptor size/mass determinations should therefore be made by several independent means in order to fully characterize a particular receptor protein. The advantage offered by gel filtration chromatography is that intact receptor species can be isolated and collected relatively rapidly for subsequent functional analyses.

7.1.1 Sephacryl S-400 resin

Preparative gel filtration chromatography using Sephacryl S-400 resin has been described previously in Section 2.3.2. We sometimes employ Bio-Gel A-1.5-m resin in place of Sephacryl S-400 resin in the initial step of insulin receptor purification.

7.1.2 Bio-Gel A-1.5-m resin

Analytical separation of the purified insulin and IGF-I receptor complexes in our laboratory has been typically performed using 1.6 × 50 cm columns of Bio-Gel A-1.5-m resin (Bio-Rad No. 151-0450). The Bio-Gel A-1.5-m resin is supplied preswollen, and is prepared for use by pouring into a 1.6 × 50 cm column and washing extensively (16 h) with distilled water at 4°C. The column is next washed for 4–6 h with Bio-Gel Equilibration Buffer at 4°C (*Table 16*). After draining the buffer level to the top of the resin bed, the sample (1.0 ml) is loaded and allowed to permeate the resin. Following this, additional Bio-Gel Equilibration Buffer is added and the top of the column is reconnected to the buffer reservoir to resume flow at 15 ml/h. Since the excluded volume of the column is approximately 25 ml, after the sample is loaded on to the column, 20 ml of the column effluent is voided and discarded before collecting 0.4 ml fractions ($n = 80$).

Bio-Gel A-1.5-m resin columns can be reused indefinitely if drying, cracking, and/or contamination are prevented. After performing an analytic chromatography experiment, the resin should be washed extensively (16 h) with distilled water and stored at 4°C to prevent microbial growth.

Table 16. Bio-Gel Equilibration Buffer

50 mM Tris 150 mM NaCl 0.1% BSA 0.1% Triton X-100 0.02% NaN$_3$ pH 7.6	*Note:* Tris buffers have a large temperature coefficient (approximately 0.05 pH unit/°C). Therefore, buffers must be adjusted for pH at the temperature of intended use (e.g. 4°C for Bio-Gel A-1.5 m analytic gel filtration chromatography).

7.2 Sucrose gradient centrifugation (52)

The hydrodynamic properties of insulin receptors solubilized from rat liver, human placenta, IM-9 lymphocytes, and turkey erythrocyte membranes have been previously examined by sucrose velocity sedimentation (56–58). These studies have indicated that the $\alpha_2\beta_2$ heterotetrameric insulin receptor solubilized in Triton X-100 displays a Stokes radius of 70–85 Å and with a sedimentation coefficient of approximately 11S. Calculation of the frictional coefficient ratio (f/f_0) has ranged from 1.2 to 1.6 (relatively globular to highly asymmetric), due to differences in estimates of the amount of hydrated bound detergent associated with the insulin receptor complex. Nevertheless, sucrose gradient velocity sedimentation is an effective method for separating the $\alpha_2\beta_2$ heterotetrameric insulin receptor complex from the $\alpha\beta$ heterodimeric species (52, 53, 56). This method has also been used to examine the monomer–dimer equilibrium properties of the EGF receptor (40, 59, 60).

For detailed information on performing sucrose gradient centrifugation experiments, consult references 61 and 62. These references provide excellent technical information on selecting gradient conditions and equipment to suit individual needs. The procedure (in brief) given here has been adapted for characterizing the insulin receptor species.

Using a gradient maker, pour a linear 5–20% sucrose gradient (*Table 17*) into centrifugation tubes. Thinwall, non-wettable polyallomer centrifuge tubes should be used so that fractionation can be performed by puncturing the bottom of the tube (e.g. Beckman No. 331372 for the SW41 Ti rotor, or No. 326819 for the SW65 Ti rotor). Generally the gradient is poured on to a 20% Sucrose Solution cushion (0.5 ml for 11-ml gradient; 0.1 m for 4-ml gradient) to improve uniformity. Solubilized receptor samples (0.25 ml for a 4-ml gradient in the SW65 Ti Beckman rotor or 0.5 ml for a 11 ml gradient in the SW41 Ti Beckman rotor) are gently layered on to the sucrose and centrifuged at 100 000g for 19 h at 4°C. The centrifuge tubes are pierced through the bottom with an 18-gauge needle and 0.1-ml aliquots are collected at a flow rate of 0.3 ml/h using a peristaltic pump. Fractions containing insulin receptors are identified based on specific ^{125}I-insulin binding activity. Control experiments have demonstrated that the presence of sucrose does not interfere with insulin binding or the tyrosine-specific protein kinase activity of the insulin receptor.

Table 17. Sucrose solutions

50 mM Hepes
0.1% Triton X-100
0.02% NaN$_3$
5% or 25% (w/v) Sucrose
pH 7.6

---

8. Affinity labelling

Affinity labelling of macromolecules is a generally applicable technique which provides detailed structural information under appropriate conditions. Ideally, the reactive groups of the affinity labelling agent should covalently attach to the receptor in a highly predictable and reproducible manner. This is necessary in order to assure specificity of labelling in preparations where the receptor protein is present in relatively low amounts. Highly specific receptor labelling requires ligands which have either: (a) high affinity for the receptor with a relatively slow dissociation time; or (b) reactive groups with a short half-life, assuring that reactivity is confined to residues near the binding site.

Two general approaches have been highly successful in the identification of the insulin receptor: photoaffinity labelling and chemical cross-linking. In photoaffinity labelling, reactive (e.g. azidobenzoyl) derivatives of radiolabelled insulin are activated by UV light to produce highly unstable nitrene intermediates, which result in covalent linkage of the ligand and receptor. Since photoaffinity derivatives of insulin or other peptide hormone receptor ligands are not commercially available, we will limit our discussion to chemical cross-linking, which is more generally used. A detailed description of photoaffinity labelling can be found in reference 63.

In the procedures for chemical cross-linking, described below, several features of the homobifunctional cross-linking agent disuccinimidyl suberate (DSS) should be noted. The reactive group of DSS is a succinimidyl ester which is water labile and highly reactive with primary amino groups. Therefore, the use of this reagent is precluded in buffers containing primary amines (e.g. Tris buffers), and stock solutions of DSS should be prepared fresh before use. The characteristic reactivity of DSS with primary amines is taken advantage of by using Tris-buffers to quench the ligand–receptor cross–linking reaction. Several peptide hormones such as EGF, IGF-I and IGF-II are single polypeptide chains, making cross-linking analyses possible under both non-reducing and reducing conditions. Insulin, however, is composed of two disulphide-linked polypeptide chains which dissociate upon reduction in SDS-polyacrylamide gels. DSS cross-linking with the insulin receptor occurs predominantly via the B-chain of insulin; therefore, reducing gels of receptors cross-linked with insulin iodinated on the A-chain result in very poor labelling (64). To avoid this problem, most investigators routinely perform insulin receptor affinity cross-linking experiments using ^{125}I-monoiodo$[B_{26}]$insulin (Amersham Product No. IM.167).

8.1 Intact cells (65)

Protocol 7. Affinity labelling cells with ^{125}I-insulin

1. Incubate intact cells with KRH–BSA Buffer (*Table 18*) and 0.25–5 nM ^{125}I-monoiodo$[B_{26}]$insulin, in the presence or absence of 1–5 μM unlabelled insulin. Typically, the incubation is allowed to proceed for 30 min at 23°C.

2. Cool the cells to 4°C by placing on ice for 5 min.

3. Add freshly prepared 50 mM DSS Stock Solution (*Table 18*) at a 1:100

Protocol 7. *Continued*

dilution to yield a final concentration of 0.5 mM DSS. Allow the cross-linking reaction to proceed for 5–15 min at 4°C.

4. Terminate the reaction by adding 3 vol. of Stop Solution (*Table 18*). Wash the cells with PBS (*Table 2*) and homogenize to prepare membranes as described previously.

5. The membranes are subjected to SDS-polyacrylamide gel electrophoresis (SDS-PAGE) using 3–10% gradient resolving gels and autoradiography.

Table 18. Buffers for affinity cross-linking

KRH–BSA Buffer
130 mM NaCl
5.1 mM KCl
1.3 mM $CaCl_2$
1.3 mM $MgSO_4$
50 mM Hepes
0.1% BSA
pH 7.6

DSS Stock Solution: prepared immediately before use
  (Pierce Chemical Company No. 21555)
Cell assay: 50 mM DSS in 100% DMSO
Membrane assay: 5 mM DSS in 100% DMSO (1 : 10 dil of 50 mM stock in 50% ethanol)
Solubilized receptor assay: 5 mM DSS

Stop Solution
50 mM Tris
1 mM EDTA
250 mM Sucrose
pH 8.0

8.2 Membranes (64, 66)

Protocol 8. Affinity labelling membranes with ^{125}I-insulin

1. Incubate 0.2–0.4 mg of membrane protein with KRH–BSA Buffer (*Table 18*) and 0.25–5 nM of ^{125}I-monoiodo[B_{26}]insulin, in the presence or absence of 1–5 μM unlabelled insulin, for 60 min at 23°C (final volume = 0.2 ml).

2. Cool the samples to 4°C by placing on ice for 5 min, then centrifuge for 20 min at 40 000*g*.

3. Resuspend the membranes in 0.2 ml ice-cold KRH–BSA Buffer.

4. Add freshly prepared 5 mM DSS Stock Solution at a 1 : 50 dilution to yield a final concentration of 0.1 mM DSS. Allow the reaction to proceed for 5 min at 4°C.

5. Terminate the reaction by adding 10 vol. of Stop Solution, and centrifuge at 40 000*g* for 20 min at 4°C.

Protocol 8. *Continued*

6. Subject the membranes to SDS-PAGE using 3–10% resolving gels and autoradiography.

8.3 Solubilized receptors (32)

Protocol 9. Affinity labelling of solubilized insulin receptors

1. Incubate detergent-soluble partially purified or purified insulin receptors with KRH–BSA Buffer (*Table 18*) and 0.25 nM ^{125}I-monoiodo[B_{26}]insulin, in the absence or presence of 1–5 μM unlabelled insulin, for 60 min at 23°C (final volume = 0.2 ml).

2. Add freshly prepared 5 mM DSS Stock Solution at a 1:100 dilution to achieve a final concentration of 50 μM. Allow the reaction to proceed for 5 min at 4°C.

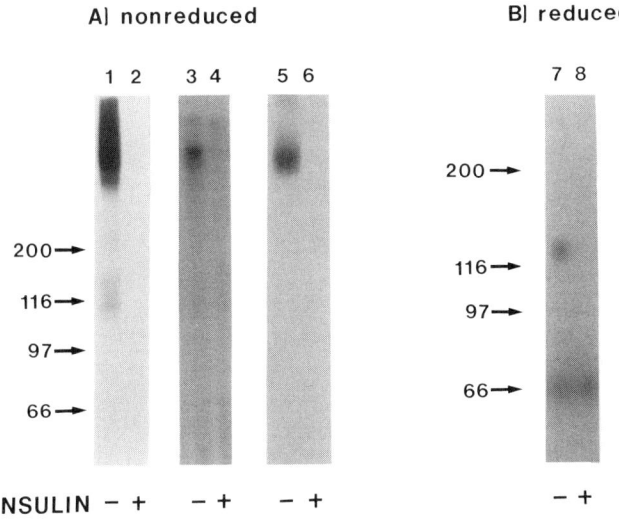

Figure 4. SDS-polyacrylamide gel electrophoresis of ^{125}I-insulin affinity cross-linked insulin receptors. (A) Human placental membranes (lanes 1 and 2), WGA-purified rat skeletal muscle insulin receptors (lanes 3 and 4) and the Insulin–Sepharose purified placental insulin receptors (lanes 5 and 6) were incubated with 0.25 nM ^{125}I-insulin for 60 min in the presence (lanes 2, 4, and 6) and absence (lanes 1, 3, and 5) of 5 μM unlabelled insulin. The samples were then affinity cross-linked with DSS, resolved by SDS-PAGE and subjected to autoradiography. (B) The WGA-purified insulin receptors from rat skeletal muscle were affinity cross-linked with 0.25 nM ^{125}I-monoiodo[B_{26}]-insulin in the absence (lane 7) or presence (lane 8) of 1 μM unlabelled insulin. The samples were cross-linked and subjected to reducing (100 mM DTT) SDS-PAGE. The ^{125}I-labelled band at Mr = 68,000 is bovine serum albumin which was present in the cross-linking buffer and was labelled in a non-specific manner. The protein Mr markers used were: 200 K = myosin, 116 K = β-galactosidase, 97 K = phosphorylase b, 68 K = bovine serum albumin.

Protocol 9. *Continued*

3. Terminate the reaction by adding 50 μl of 1 M Tris, pH 10.5, and incubate at 4°C for an additional 15 min.

4. Finally, subject samples to SDS-PAGE and autoradiography.

Figure 4 illustrates the results of typical affinity cross-linking experiments using membrane (55), partially purified (67), and homogeneous purified (50) insulin receptor preparations. Note that in some experiments (lanes 1–6), SDS-PAGE gels were run under non-reducing conditions. Identification of the insulin binding α subunit is readily detected by the inclusion of 50–100 mM DTT in the SDS-polyacrylamide sample loading buffer (e.g. *Figure 4*, lanes 7 and 8). Also note from *Figure 4* that by the procedure given here, the affinity-labelling is specific to the insulin receptor as 5 μM unlabelled insulin completely blocked cross-linking with ^{125}I-insulin.

References

1. Ellis, L., Clauser, E., Morgan, D. O., Edery, M., Roth, R. A., and Rutter, W. J. (1986). *Cell*, **45**, 721.
2. Chou, C. K., Dull, T. J., Russell, D. S., Gherzi, R., Lebwohl, D., Ullrich, A., and Rosen, O. M. (1987). *J. Biol. Chem.*, **262**, 1842.
3. Whittaker, J., Okamoto, A. K., Thys, R., Bell, G. I., Steiner, D. F., and Hofmann, C. A. (1987). *Proc. Natl. Acad. Sci., USA*, **84**, 5237.
4. Harrison, J. C., Billington, T., East, I. J., Nichols, R. J., and Clark, S. (1978). *Endocrinology*, **102**, 1485.
5. Harrison, L. C. and Itin, A. (1980). *J. Biol. Chem.*, **255**, 12066.
6. Cuatrecasas, P. (1972). *Proc. Natl. Acad. Sci. USA*, **69**, 318.
7. Schweitzer, J. B., Smith, R. M., and Jarett, L. (1980). *Proc. Natl. Acad. Sci. USA*, **77**, 4692.
8. Swanson, M. L., Dudley, D. T., Walker, P. S., Boyle, T. R., and Pessin, J. E. (1988). *Endocrinology*, **122**, 967.
9. Rodbell, M. (1964). *J. Biol. Chem.*, **239**, 375.
10. Burant, C. F., Treutelaar, M. K., and Buse, M. G. (1986). *J. Biol. Chem.*, **261**, 8985.
11. Klein, H. H., Freidenberg, G. R., Kladde, M., and Olefsky, J. M. (1986). *J. Biol. Chem.*, **261**, 4691.
12. Hollenberg, M. D. and Cuatrecasas, P. (1976). *Methods in Receptor Research*, pt. II (ed. M. Blecher), p. 429. Marcel Dekker, New York.
13. Wang, C.-S. and Smith, R. L. (1975). *Anal. Biochem.*, **63**, 414.
14. Burant, C. F., Treutelaar, M. K., Landreth, G. E., and Buse, M. G. (1984). *Diabetes*, **33**, 704.
15. James, D. E., Zorzano, A., Boni-Schnetzler, M., Nemenoff, R. A., Powers, A., Pilch, P. F., and Ruderman, N. B. (1986). *J. Biol. Chem.*, **261**, 14939.

16. Cuatrecasas, P. and Tell, G. P. E. (1973). *Proc. Natl. Acad. Sci. USA*, **70**, 485.
17. Bhaumick, B., Bala, R. M., and Hollenberg, M. D. (1981). *Proc. Natl. Acad. Sci. USA*, **78**, 4279.
18. Morgan, D. O., Edman, J. C., Standring, D. N., Fried, V. A., Smith, M. C., Roth, R. A., and Rutter, W. J. (1987). *Nature, Lond.*, **329**, 301.
19. Yarden, Y. and Schlessinger, J. (1987). *Biochemistry*, **26**, 1434.
20. Claesson-Welsh, L., Ronnstrand, L., and Heldin, C.-H. (1987). *Proc. Natl. Acad. Sci. USA*, **84**, 8796.
21. Pharmacia Fine Chemicals (1984). *Affinity Chromatography*. Pharmacia Fine Chemicals AB, Uppsala, Sweden.
22. Allenmark, S. and Bomgren, B. (1985). *Affinity Chromatography* (ed. P. D. G. Dean, W. S. Johnson, and F. A. Middle), p. 114. IRL Press, Oxford and Washington DC.
23. Pharmacia Fine Chemicals (1984). *Gel Filtration*. Pharmacia Fine Chemicals AB, Uppsala, Sweden.
24. Cohen, S. (1983). *Methods in Enzymology* (ed. J. D. Corbin and J. G. Hardman), Vol. 99, p. 379. Academic Press, London and New York.
25. Tollefsen, S. E., Thompson, K., and Petersen, D. J. (1987). *J. Biol. Chem.*, **262**, 16461.
26. Gaudino, G., Cirillo, D., Naldini, L., Rossino, P., and Comoglio, P. M. (1988). *Proc. Natl. Acad. Sci. USA*, **85**, 2166.
27. Bio-Rad Chemical Division (1984). Publication No. 84-0478 385, *Activated Affinity Supports*. Bio-Rad Chemical Division, Richmond, CA.
28. Cuatrecasas, P. (1972). *Proc. Natl. Acad. Sci. USA*, **69**, 1277.
29. Porath, J., Axen, R., and Ernback, S. (1967). *Nature, Lond.*, **215**, 1491.
30. Eveleigh, J. W. and Levy, D. E. (1977). *J. Solid-Phase Biochem.*, **2**, 45.
31. Fujita-Yamaguchi, Y., Choi, S., Sakamoto, Y., and Itakura, K. (1983). *J. Biol. Chem.*, **258**, 5045.
32. Boyle, T. R., Campana, J., Sweet, L. J., and Pessin, J. E. (1985). *J. Biol. Chem.*, **260**, 8593.
33. Roth, R. A. and Cassell, D. J. (1983). *Science*, **219**, 299.
34. O'Brien, R. M., Soos, M. A., and Siddle, K. (1986). *Biochem. Soc. Trans.*, **14**, 316.
35. Finn, F. M., Titus, G., Horstman, D., and Hofmann, K. (1984). *Proc. Natl. Acad. Sci. USA*, **81**, 7328.
36. Kohanski, R. A. and Lane, M. D. (1985). *J. Biol. Chem.*, **260**, 5014.
37. Kull, F. C., Jr., Jacobs, S., Su, Y-F., Svoboda, M. E., Van Wyk, J. J., and Cuatrecasas, P. (1983). *J. Biol. Chem.*, **258**, 6561.
38. Fujita-Yamaguchi, Y., LeBon, T. R., Tsubokawa, M., Henzel, W., Kathuria, S., Koyal, D., and Ramachandran, J. (1986). *J. Biol. Chem.*, **261**, 16727.
39. LeBon, T. R., Jacobs, S., Cuatrecasas, P., Kathuria, S., and Fujita-Yamaguchi, Y. (1986). *J. Biol. Chem.*, **261**, 7685.
40. Biswas, R., Basu, M., Sen-Majumdar, A., and Das, M. (1985). *Biochemistry*, **24**, 3795.
41. Slieker, L. J. and Lane, M. D. (1985). *J. Biol. Chem.*, **260**, 687.
42. Cochet, C., Gill, G. N., Meisenhelder, J., Cooper, J. A., and Hunter, T. (1984). *J. Biol. Chem.*, **259**, 2553.
43. Cochet, C., Kashles, O., Chambaz, E. M., Borrello, I., King, C. R., and Schlessinger, J. (1984). *J. Biol. Chem.*, **263**, 3290.
44. Yarden, Y., Harari, I., and Schlessinger, J. (1985). *J. Biol. Chem.*, **260**, 315.
45. Yarden, Y. and Schlessinger, J. (1987). *Biochemistry*, **26**, 1443.
46. Savage, C. R., Jr. and Cohen, S. (1972). *J. Biol. Chem.*, **247**, 7609.

47. Akiyama, T., Kadooka, T., and Ogawara, H. (1985). *Biochem. Biophys. Res. Commun.*, **131**, 442.
48. Koland, J. G. and Cerione, R. A. (1988). *J. Biol. Chem.*, **263**, 2230.
49. Amicon Corporation (1984). Publication No. 1-259A, *Centricon Microconcentrators*, Amicon Corporation Scientific Systems Division, Danvers, MA.
50. Sweet, L. J., Morrison, B. D., and Pessin, J. E. (1987). *J. Biol. Chem.*, **262**, 6939.
51. Sweet, L. J., Morrison, B. D., Wilden, P. A., and Pessin, J. E. (1987). *J. Biol. Chem.*, **262**, 16730.
52. Boni-Schnetzler, M., Rubin, J. B., and Pilch, P. F. (1986). *J. Biol. Chem.*, **261**, 15281.
53. Boni-Schnetzler, M., Scott, W., Waugh, S. M., DiBella, E., and Pilch, P. F. (1987). *J. Biol. Chem.*, **262**, 8395.
54. Feltz, S. M., Swanson, M. L., Wemmie, J. A., and Pessin, J. E. (1988). *Biochemistry*, **27**, 3234.
55. Swanson, M. L. and Pessin, J. E. (1989). *J. Membr. Biol.*, **108**, 217.
56. Aiyer, R. A. (1983). *J. Biol. Chem.*, **258**, 14992.
57. Baron, M. D., Wisher, M. H., Thamm, P. M., Saunders, D. J., Brandenburg, D., and Sonken, P. H. (1981). *Biochemistry*, **20**, 4156.
58. Pollet, R. J., Haase, B. A., and Standaert, M. L. (1981). *J. Biol. Chem.*, **256**, 12118.
59. Basu, M., Sen-Majumdar, A., Basu, A., Murthy, U., and Das, M. (1986). *J. Biol. Chem.*, **261**, 12879.
60. Boni-Schnetzler, M. and Pilch, P. F. (1987). *Proc. Natl. Acad. Sci. USA*, **84**, 7832.
61. Rickwood, D. (ed.) (1984). *Centrifugation*, 2nd edn. IRL Press, Oxford and Washington, DC.
62. Griffith, O. M. (1976). Publication No. DS-468B, *Techniques of Preparative Zonal. and Continuous Flow Ultracentrifugation*, 2nd edn. Beckman Instruments Inc., Palo Alto, CA.
63. Oppenheimer, C. L. and Czech, M. P. (1984). *Growth and Maturation Factors* (ed. G. Guroff), Vol. 2, p. 193. Wiley, New York.
64. Pilch, P. F. and Czech, M. P. (1980). *J. Biol. Chem.*, **255**, 1722.
65. Massague, J., Pilch, P. F., and Czech, M. P. (1981). *J. Biol. Chem.*, **256**, 3182.
66. Massague, J., Pilch, P. F., and Czech, M. P. (1980). *Proc. Natl. Acad. Sci. USA*, **77**, 7137.
67. Treadway, J. L., James, D. E., Burcel, E., and Ruderman, N. B. (1989). *Am. J. Physiol.*, **256**, E138.

3

Cyclic nucleotides

TREVOR LAKEY, NIGEL PYNE, and GREG MURPHY

1. General introduction

Peptide hormones bind to specific receptors on the exterior of the cell, and activate particular signal transduction systems, thus eliciting a physiological response. One type of signal which has been shown to mediate the effects of many hormones on many cells types involves the production or degradation of cyclic nucleotides [primarily cyclic AMP (cAMP) and cyclic GMP (cGMP)] inside the cell.

To establish the involvement of a cyclic nucleotide as a putative second messenger in a particular cellular system, it is necessary to demonstrate that an agonist can alter intracellular concentration of the cyclic nucleotide in a dose-dependent manner.

Such alterations in cyclic nucleotide levels may be brought about by changes in either (or both) the rate of synthesis or the rate of degradation of the cyclic nucleotide. These changes should:

- Show comparable kinetics upon hormonal stimulation;
- Attain physiologically effective levels (e.g. allow activation of the appropriate kinase) in response to physiological concentrations of the agonist;
- The effects of the agonist should be mimicked by adding exogenous permeable cyclic nucleotides directly to cells.

This chapter details standard methodology used in establishing such criteria. Details for assay of cyclic nucleotides, adenylate and guanylate cyclase, cyclic nucleotide phosphodiesterases, and AMP-dependent protein kinase will be presented.

2. Measurement of cyclic nucleotides

2.1 Introduction

Cyclic nucleotides can be measured in a variety of tissue preparations including perfused organs, tissue slices, isolated and cultured cells, and membrane preparations. This section will concentrate on the methodology for measurement of the accumulation of cyclic nucleotides in cells and on their production by membrane preparations.

There are many diverse methods for assessing production of cyclic nucleotides in cell preparations. One technique, which has been widely used, is to pre-label pools of nucleotide triphosphate (ATP or GTP) by preincubating the cells with a labelled precursor, for example tritiated adenine or adenosine. Following the test incubation and addition of unlabelled carrier cyclic nucleotide, the radiolabelled reaction product is re-isolated and quantified by liquid scintillation counting. While there may be circumstances where this assay technique is the most suitable (possibly where there are very low yields of cells) it should be used with caution: results obtained from a variety of preparations indicate that labelling procedures using adenine and adenosine do not result in a uniform distribution of specific activity among the adenine nucleotides within the cell. The radiometric method of cyclic nucleotide determination will not be described here in full.

The methods that will be described here in detail are two techniques utilizing a non-radiometric incubation: a binding protein assay for measurement of cAMP and a radioimmunoassay, suitable for determination of either cAMP or cGMP.

2.2 Incubation conditions

With cells grown in monolayer culture the use of 6-well plates is convenient, especially when performing triplicate determinations of hormonal effects. For cells in suspension the use of small conical flasks is recommended. The flasks should either be plastic or siliconized glass.

When investigating the effects of certain agents it may be necessary to perform a pre-incubation. For example, a pre-incubation may be necessary with certain phosphodiesterase inhibitors to allow them to enter the cell, especially when using short incubation times. Other agents which require a pre-incubation period include cholera toxin, pertussis toxin, and TPA (phorbol 12-myristate 13-acetate). In hepatocytes, the pre-incubation times required for maximal expression of the effects of these agents are 45 min, 1 h and 15 min respectively (refer to Section 3.4.2).

The length of incubation chosen will depend on the type of study being performed, and each system should, of course, be thoroughly characterized before routine use. For some cell preparations the duration of incubations is restricted by the length of time that cell preparations remain viable. Other considerations when planning incubation length include properties of the test agents being investigated. If the hormone or drug is metabolized rapidly, either the incubation time must be restricted accordingly or some means of maintaining its concentration must be found.

The cell system being studied should be thoroughly characterized so that optimum pH, temperature, and medium composition are known. With mammalian cells, the desired temperature will almost certainly be 37°C. A suitable buffering system should be chosen so that a stable pH can be maintained.

One of the most commonly used buffers is Krebs–Henseleit, which has an identical ionic composition to that of plasma. It has a low buffering capacity, however, since the buffering ability depends on the bicarbonate ion. However, HEPES (25 mM) can be added to the Krebs–Henseleit buffer to stabilize the pH.

With this buffer and some other buffer systems, carbon dioxide produced by metabolizing cells will decrease the pH. Cells must therefore be frequently gassed with 95% O_2/5% CO_2. The best method of gassing the cells is to use a gassing incubator, which will give a continuous supply of O_2/CO_2 and will also maintain the cells at the desired temperature. With cells being incubated in conical flasks it is also possible simply to place the in a shaking water-bath and to gas them every 5–10 min.

2.3 Termination of incubation

Various methods can be employed to terminate the cell incubation for subsequent determination of cyclic nucleotide content. The basic technique used will depend primarily on whether the cells are in suspension or in monolayer culture.

For cells in monolayer culture, one method of stopping the incubation is to rapidly remove the medium, rinse the cells (for example, with phosphate-buffered saline) and add ice-cold ethanol (80% w/v). With 6-well plates the volume required is 1 ml per well. With cells in suspension the most efficient means of stopping the incubation is to employ a technique which simultaneously ends the incubation and separates the cells from their incubation medium. One such technique, which uses perchloric acid (PCA) as the stopping reagent, is described in the next section.

2.4 Preparation of samples for cyclic nucleotide determination

2.4.1 Extraction of intracellular cyclic nucleotide

The basis of extracting cyclic nucleotides from a cell preparation is a separation of the cell from the medium in which the incubation was performed followed by disruption of the cell to allow the cyclic nucleotide to be isolated for subsequent assay.

One extraction procedure suitable for use with cells in monolayer is an ethanol extraction procedure (*Protocol 1*).

Protocol 1. Extraction of intracellular cyclic nucleotide using ethanol

1. After rinsing and addition of 1 ml of ice-cold ethanol (80% v/v) to each well, scrape off the cells using a rubber policeman or 1-ml syringe barrel, and transfer to plastic tubes using a plastic pasteur pipette.

Protocol 1. *Continued*

2. Rinse each well with a further 1 ml of ice-cold ethanol and combine the two 1-ml of ethanol for each sample.

3. Centrifuge the cells for 10 min at 800g at 4°C and decant off the resulting supernatant.

4. Resuspend the pellet in 1 ml of ethanol (80% v/v) and recentrifuge for 10 min at 800g at 4°C.

5. Combine the supernatants from the two centrifugations and lyophylize.

6. When assaying these samples for cyclic nucleotide content, reconstitute in a small volume (e.g. 250 μl) of distilled water.

An extraction procedure suitable for use with isolated animal cells in suspension, which also acts as a rapid means of terminating the incubation, is based on a centrifugation technique using a dense perchloric acid phase beneath a non-aqueous phase. The density of the non-aqueous phase must be high enough to exclude the medium but low enough to permit sedimentation of the cells (1). This procedure is suitable for use with several cells types, including hepatocytes and adipocytes. With hepatocytes the most suitable non-aqueous phase is bromododecane; a detailed protocol for this method of incubation termination and cyclic nucleotide extraction is presented in *Protocol 2*.

Protocol 2. Extraction of intracellular cyclic nucleotides using the bromodecane/perchloric acid method

1. At the end of the incubation period carefully pour the cell suspension on to bromododecane (2 ml) layered on to 0.5 ml of a PCA (0.62 M) and sucrose (0.25 M) mixture, in a 15 ml polypropylene centrifuge tube (polycarbonate tubes are dissolved by the reagent);

2. Sediment the cells at 4000g for 30 sec. Aspirate off the aqueous and bromododecane layers, resuspend the PCA precipitate and recentrifuge at 4000g for 5 min;

3. Decant the supernatant and either neutralize the PCA with an ethanolamine (0.5 M)/KOH (2 M) mixture; or quantitatively remove by the method of Khym (2), as reported by Sharps and McCarl (3). The latter method avoids dilution of the sample;

4. Store the supernatant at -20°C prior to assay for cyclic nucleotides.

A diagrammatic representation of this technique is presented in *Figure 1*. With either of the assays for cyclic nucleotides described in this chapter it is necessary to run reagent blanks with this isolation technique, otherwise interference with

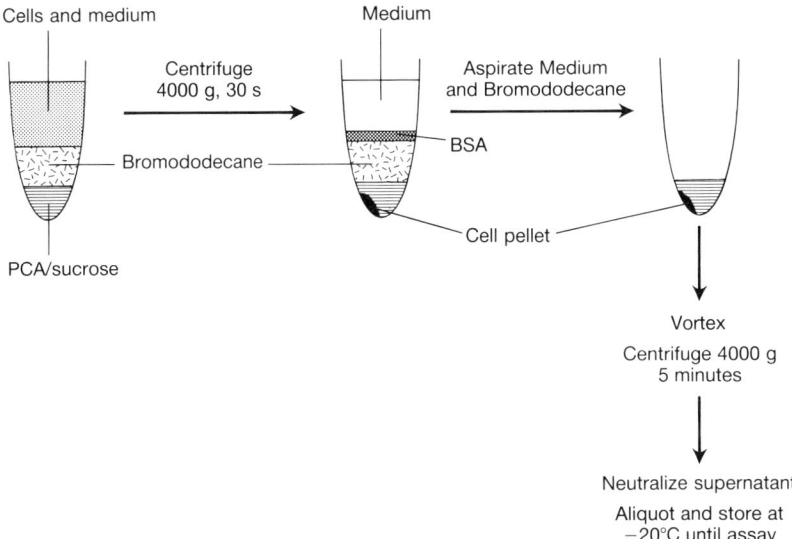

Figure 1. Extraction and isolation of intracellular cyclic AMP.

ligand binding will result in over-estimation of the amount of cyclic nucleotide present.

2.4.2 Determination of extracellular cyclic nucleotides

As with the preparation procedures for intracellular cyclic nucleotides, the choice of protocol for the preparation of samples for extracellular cyclic nucleotide determination will depend on whether the incubation uses cells in suspension or in monolayer culture.

With cells incubated in monolayer the procedure is very straightforward. Firstly, at the end of the incubation period remove the medium using a plastic pasteur pipette and boil for 2 min to precipitate protein. Centrifuge at 800*g* for 10 min. Then decant supernatant and assay undiluted.

It is advisable to perform an antibody or binding protein dilution curve in the appropriate medium (boiled as for samples), as the binding will vary from the assay for intracellular cyclic nucleotide.

For incubations using cell suspensions the determination of extracellular cyclic nucleotide concentration is achieved by assaying a sample of medium. Where a stopping and isolation procedure similar to the bromododecane/PCA procedure described above has been used, simply take a sample from the medium remaining above the bromododecane after the centrifugation step. Preparation of this medium for assay is then identical to the procedure described above: boiling followed by centrifugation. An alternative is to take a sample of cells prior to their extraction on bromododecane/PCA, add to PCA, and assay for total cyclic nucleotide content. This can then be compared to the value obtained from the intracellular determination.

2.5 Assay of cyclic nucleotide levels by radioimmunoassay

2.5.1 Theory of assay

The radioimmunoassay of cAMP is an established method (4) using antibody raised to acetylated cAMP, which can equally be applied to the assay of cGMP. Basic theory involves competition between the cAMP of the sample and between ^{125}I-labelled cAMP, as depicted in *Figure 2*. After the incubation period unbound cAMP is removed using charcoal. Quantitation of cAMP levels is achieved by comparison with a standard curve of acetylated cAMP.

Adaptation of the radioimmunoassay protocol to allow measurement of cGMP is straightforward. The only requirements are substitution of ^{125}I-labelled cGMP for cAMP label, use of antibody raised to acetylated cGMP, and use of cGMP for the standard curve. The assay is highly sensitive—it is capable of detecting less than 0.125 pmol of cGMP. In most tissues this level of sensitivity should allow measurement of cGMP in triplicate on as little as 20–40 mg of starting material.

2.5.2 Preparation of antibody

The antigen is prepared by coupling acetylated cAMP to either protein (for example human serum albumin) or poly-L-serine polymers and is injected into a rabbit or other suitable animal (see reference 4 for details).

Most cAMP antibodies exhibit relatively minor cross-reactivity with structurally related nucleoside and nucleotides, and other tissue constituents. To produce

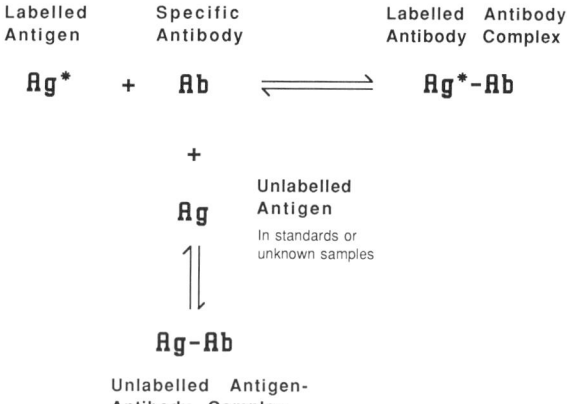

Figure 2. Reaction scheme for competitive binding assay. Unlabelled antigen (Ag) in the unknown samples or standards competes against labelled antigen (Ag*) for binding to the antibody (Ab), and thus reduces the binding of labelled antigen.

displacement equal to that of cAMP at least 50 000-fold greater concentration of ATP or other mono-, di-, or triphosphate nucleotides is required. The degree of cross-reactivity of anti-sera with ATP must be determined before routine use; this value must be less than 0.002% since ATP levels are generally around 10 000-fold greater than cAMP. Cross-reactivity with cGMP is less of a problem: tissue levels of cGMP are usually only 2–10% of cAMP levels; therefore cross-reactivity up to 1% is still acceptable.

With antibody raised to acetylated cGMP cross-reactivity with all purine and pyrimidine nucleotides is minimal, except for cIMP. This latter nucleotide cross-reacts at the 1% level.

Protocol 3. Acetylation of samples

1. Reconstitute freeze-dried samples in 250 μl of distilled water, then dilute as appropriate in distilled water. It is standard procedure to include two dilutions of each sample in each assay.

2. Add 200 μl of each dilution to 6 μl of acetylating mixture and mix rapidly. The composition of the acetylating mixture is: 4 μl triethylamine and 2 μl of acetic anhydride per 6 μl of acetylation mixture. In addition, acetylate 2 × 200 μl of distilled water and also acetylate the cAMP standard curve.

The assay details described below are specific for cAMP, but may be readily adapted for measurement of cGMP, as described in Section 2.5.1.

2.5.3 Cyclic AMP standard curve

Cyclic AMP stock should be stored as 40 pmol/ml at $-20°C$, in 400-μl aliquots. The standard curve consists of twofold serial dilutions, starting with the 40 pmol/ml standard. The cAMP concentrations of the standard curve are as follows: 40, 20, 10, 5, 2.5, 1.25, 0.65, 0.312, and 0.156 pmol/ml.

The standards should be made up in a volume of 200 μl using distilled water, and acetylated using 6 μl of acetylation mixture per tube—in the same manner as the samples.

2.5.4 Assay design

All determinations should be carried out in duplicate in a final assay volume of 300 μl. The assay buffer is 50 mM acetate, pH 4.8, which has the following composition:

- 54.3 ml 0.1 M acetic acid
- 70.8 ml 0.1 M sodium acetate
- made up to 250 ml with distilled water
- plus 8 mM theophylline.

Keep the antibody stock at $-20°C$ and dilute the amount required for use in assay buffer containing 3% (w/v) bovine serum albumin. Keep the antibody solution at 4°C at all times. Label (adenosine 3',5'-cyclic phosphoric acid 2'-O-succinyl 3-[^{125}I] iodotyrosine methyl ester) is required at a concentration of approximately 5 nCi per tube. With a stock concentration of 1 μCi per 100 ml, the dilution used is 5 μl per millilitre of buffer.

Samples and standards should be prepared as shown in *Table 1*.

After making all the additions, mix the tubes and Incubate the assay overnight at 4°C to allow binding of antibody to the acetylated cAMP of standards or samples.

Table 1. Preparation of standards and samples

	Acetylated Standards	Sample	H_2O	*cAMP	Buffer	Antibody
1. NSB			50	50	200	
2. Total counts				50	250	
3. Max. binding			50	50	100	100
4. 7.8 f	50			50	100	100
5. 15.6 f	50			50	100	100
6. 31.2 f	50			50	100	100
7. 62.5 f	50			50	100	100
8. 125.0 f	50			50	100	100
9. 250.0 f	50			50	100	100
10. 500.0 f	50			50	100	100
11. 1000.0 f	50			50	100	100
12. 2000.0 f	50			50	100	100
13. Sample 1, Dil 1		50		50	100	100
14. Sample 1, Dil 2		50		50	100	100
etc.						

* cAMP = ^{125}I-labelled cAMP.   NSB = non-specific binding.   f = fmol cAMP per tube.   All volumes are in microlitres.

2.5.5 Separation of unbound cyclic AMP

Separation of unbound cAMP of standards or samples is achieved using charcoal. Add the charcoal, to all tubes except total counts, as 1 ml per tube of 2 mg/ml Norit A charcoal and 2.5 mg/ml bovine serum albumin in 100 mM potassium phosphate buffer, pH 6.3. The composition of the charcoal mixture is as follows:

- 0.2 g Norit A charcoal
- 0.25 g bovine serum albumin
- 77.5 ml of buffer A (1.36% (w/v) KH_2PO_4)
- 22.5 ml of buffer B (1.75% (w/v) K_2HPO_4)

Mix all tubes immediately following addition of charcoal and incubate for 20 min at 4°C. Pellet the charcoal by centrifugation at 1700g for 10 min at 4°C.

2.5.6 Quantification of cAMP accumulation

Decant the supernatants from the centrifugation and count for ^{125}I for 4 min, preferably using a radioimmunoassay programme on a suitable γ-counter. Cyclic AMP content of samples is generally expressed relative to the protein content of each well, which should be measured using the method of Lowry *et al.* (29).

2.5.7 Summary of radioimmunoassay

Radioimmunoassay of cAMP or cGMP is sensitive, straightforward and cheap. The assays are sensitive to the femtomolar range and cross-reactivity is generally very low. No chromatographic steps are required, which makes for a very straightforward operation, and expensive radioisotopes are not required. Furthermore, the wide availability of specially designed RIA programmes on many γ-counters greatly simplifies the quantification of cyclic nucleotide levels by this method.

2.6 Assay of cAMP by binding protein technique

2.6.1 Introduction

The receptor protein binding displacement assay for cAMP is based on competition for protein binding sites between radiolabelled cAMP and the unlabelled cAMP to be quantified; the theoretical basis of the assay is analogous to the radioimmunoassay, except that a cAMP binding protein preparation is substituted for the antibody (see *Figure 2*). Use of a naturally-occurring binding protein (usually a crude preparation of the regulatory subunit of cAMP-dependent protein kinase) which interacts with cAMP with high affinity results in a comparatively simple, but highly-sensitive and specific assay. The assay method used is essentially a modification of those developed by Gilman (5) and Brown *et al.* (6).

A binding protein method also exists for the measurement of cGMP concentrations. It is identical in principle and virtually identical in operation to that for cAMP, described below, the specific binding protein being partially purified from lobster muscle (7). However, the assay is seriously compromised in several respects: assay sensitivity of the cGMP binding protein techniques is lower than that for the cAMP technique, while tissue levels of cGMP are generally lower. Furthermore, the assay is susceptible to interference from cAMP and 5'AMP. As a result, some separation of cGMP from contaminating materials is required before use of the assay. Because of these difficulties, a radio-immunoassay for cyclic GMP is clearly preferable.

2.6.2 Preparation of cAMP binding protein

The procedure for preparation of the cAMP binding protein (*Protocol 4*) is based on the method reported by Rubin *et al.* in 1974 (8), as far as Step 3, DEAE–Sephadex batch chromatography.

Protocol 4. Preparation of cAMP binding protein from heart

1. Obtain a fresh beef heart (approximately 4 kg) from a local abattoir, place on ice, and mince after removal of the pericardium and fat tissue. All subsequent steps should be carried out at 4°C, with all buffers containing 4 mM β-mercaptoethanol.

2. Mix the heart muscle with 40 mM potassium phosphate buffer, pH 6.1, containing 2 mM EDTA (2 litres/kg wet weight), and homogenize in small batches in a blender.

3. Centrifuge the homogenate at 10 000*g* for 10 min and filter the supernatant through Whatman No. 54 paper. Extract the pellets twice more with 1 litre of buffer, and combine the extracts.

4. Bring the pooled extracts to 55% saturation by the addition of solid $(NH_4)_2SO_4$(320 g/l), with the pH maintained between pH 7 and 8 by addition of NH_4OH (approx. 5 ml of concentrated NH_4OH/litre).

5. Allow the protein to precipitate for 2.5 h, then collect by centrifugation at 10 000*g* for 10 min. Discard the supernatant, dissolve the pellet in 500 ml of 50 mM Tris–Cl, pH 7.6, containing 10 mM NaCl (buffer A). Dialyse this solution overnight against two 4-litre volumes of buffer A.

6. Stir the dialysed enzyme preparation for 1 h with DEAE–Sephadex (alternatively DEAE–cellulose may be substituted as the resin) which has been pre-equilibrated with buffer A (2.25 g protein/100 ml resin). Under these conditions the kinase is adsorbed by the resin. Collect the resin by filtration on a Buchner funnel with Whatman No. 54 paper, and wash with about 3 litres of buffer A until the filtrate becomes colourless.

7. Suspend in 800 ml of 50 mM Tris–Cl, pH 7.6, containing 0.30 M NaCl (buffer B), and stir for 45 min. Collect the DEAE–Sephadex by filtration and wash twice with 400 ml of buffer B.

8. Bring the pooled filtrates to 35% saturation by the addition of solid $(NH_4)_2SO_4$ (199 g/litre), maintaining the pH of the solution as described above. After 1 h, collect the precipitate by centrifugation at 10 000*g* for 10 min and discard.

9. Bring the supernatant to 75% saturation by adding solid $(NH_4)_2SO_4$ (258 g/litre). Collect the precipitate, which forms after 1 h, by centrifugation at 10 000*g* for 10 min, suspend in a minimal volume of potassium phosphate buffer, pH 7.0, and dialyse against 2 litres of the same buffer overnight.

10. Dispense the dialysed fraction (about 240 ml; 10.6 g protein) into 250 μl aliquots and store at -70°C. This should be diluted 90-fold for use in the assay, and is stable for two to three days at 4°C under these conditions. A

Protocol 4. *Continued*

typical preparation lasts about nine to twelve months, with no significant change in binding activity during that time.

2.6.3 Other reagents

Tritiated cAMP is obtained in an ethanol:water (1:1 v/v) solution from Amersham. Use 37 MBq (1 mCi) batches of [5',8'-^{3}H]-adenosine 3',5'-cyclic phosphate, ammonium salt, of specific activity 1.92 TBq/mmol (52 Ci/mmol). This is stable for more than 6 months stored at $-20°$C. For use in the assay, this is diluted 1:1500 as a stock solution, which is stable for up to 4 weeks at 4°C. This stock solution is diluted 1:3 in the assay, resulting in a final concentration of 1.28 pmol cAMP/assay tube, containing 66.7 nCi (32 900 d.p.m.). Bovine serum albumin should be essentially fatty-acid free, while all other reagents are of analytical grade from standard sources.

2.6.4 Assay procedure

The incubation buffer for the binding assay is 50 mM Tris-Cl, pH 7.4, containing 4 mM EDTA. Make up standard solutions of cAMP comprising 0.06, 0.12, 0.25, 0.5, 1.0, 2.0, 4.0, 8.0, and 16.0 pmol/50 μl in boiled incubation medium. The binding protein and the tracer (tritiated cAMP) should be diluted appropriately in assay buffer. Set up the incubation in 1.5-ml plastic tubes at room temperature as shown in *Table 2*, to a total volume of 300 μl.

The amount of sample assayed should reflect their expected cAMP content. For example, when using this assay to measure levels of cAMP stimulated by various ligands, take into consideration the expected level of stimulation when diluting the samples. A routine method of validating the assay is to include blanks of boiled incubation medium.

Allow the samples to reach equilibrium by incubation at 4°C for 2 h. After this period add a 250 μl aliquot of a 1% BSA/2% charcoal (w/v) mixture (made up in assay buffer), to each tube. The charcoal should be placed on ice and stirred constantly during addition. Vortex the tubes and sediment the charcoal in a microfuge (4 min, maximum speed). It is crucial to perform the charcoal

---

Table 2. Sample preparation for cAMP assay with binding protein

	Sample	Buffer	[^{3}H]-cAMP	Binding protein
Background		200 μl	100 μl	—
Total bound		100 μl	100 μl	100 μl
Standards: 0.06–16 pmol	50 μl	50 μl	100 μl	100 μl
Samples	5–50 μl	95–50 μl	100 μl	100 μl

extraction step rapidly, and in the cold, otherwise the competition equilibrium will be displaced and erroneous results obtained.

Add a 400 μl aliquot of the supernatant to 4 ml of scintillation fluid, (e.g. Ecoscint or Liquiscint) and count for 3 min, preferably using a purpose-designed RIA-style programme.

3. Measurement of adenylate cyclase activity

3.1 Introduction

Adenylate cyclase [ATP pyrophosphate lyase (cyclizing) E.C.4.6.1.1.] is the enzyme which catalyses the formation of cyclic adensine 3′,5′-monophosphate (cAMP) from ATP, in the presence of Mg^{2+}. It is a multi-component complex, found in the plasma membrane of almost all mammalian cells.

The three main components of the complex are: the receptors, on the outer face of the plasma membrane, which bind hormones and autocoids which in turn may trigger either stimulatory or inhibitory signals; the guanine–nucleotide binding regulatory proteins (G-proteins), G_s and G_i, which mediate, respectively, stimulatory and inhibitory inputs from the appropriate receptors to the final component—the catalytic unit. Detailed descriptions of the structure, function and regulation of these components may be found in reviews by Birnbaumer *et al.* (9) and Gilman (10).

The two most widely used methods for the assay of adenylate cyclase activity are a radioisotope assay using ^{32}P-labelled substrate or a non-labelled incubation followed by assay of the reaction product (cAMP) using either the binding protein assay or radioimmunoassay, as described in the previous section. The incubation conditions required for this latter type of assay are presented in Section 3.8.

3.2 Assay of adenylate cyclase using ^{32}P-ATP

One method for assaying adenylate cyclase activity *in vitro* measures the production of radioactively labelled cAMP from the substrate $[\alpha\text{-}^{32}\text{P}]$ ATP. This method depends on the ability to separate product (^{32}P-labelled cAMP) from unreacted substrate and radioactive contaminants.

A major problem of assaying adenylate cyclase activity is that only a very small proportion of the substrate is converted to cAMP. This can be as little as 0.05% under basal conditions (14). When using a radioisotope-based assay a highly efficient means of separating substrate from product is thus called for. Early assays of adenylate cyclase used chromatography on either Dowex cation exchange resin (11) or aluminium oxide (12, 13). However, neither technique resulted in a complete elimination of ^{32}P from the assay blanks.

To overcome this problem Salomon and co-workers (14) developed an assay system which combined both of these chromatography steps. The improved removal of radioactive substrate and contaminants permits highly sensitive assay of adenylate cyclase.

3.2.1 Characterization of adenylate cyclase assay

The rate of production of cyclic AMP under a defined set of conditions is used as a measure of adenylate cyclase activity. These conditions include pH, temperature, and Mg^{2+} ion concentration. Characterization of each of these parameters must be carried out for the particular adenylate cyclase preparation under study, so that optimal assay conditions can be determined.

While such characterization studies cannot be avoided, the concentration and pH of the assay buffer should not critically affect the outcome of the Salomon *et al.* (14) assay method. These authors investigated enzyme activity over the pH range 6.3 to 7.9 and using 20 to 200 mM imidazole buffer and found essentially equivalent results under all these conditions. Furthermore, this assay protocol allows flexibility in the choice of buffering system: Tris–HCl has commonly been used for this kind of adenylate cyclase assay (see for example (15, 16).

Other characterization studies which should be performed for each enzyme preparation include checks on reaction linearity and a determination of the effect of enzyme concentration on assay performance. Salomon and co-workers (14) used membrane preparations in the range 10 μg to 250 μg per 100 μl assay, while Hunt *et al.* (15) found 150 μg to 400 μg per assay to be optimal in their system. In essence the lower limit is determined by the sensitivity of the assay. If the enzyme concentration is too high the maximal response becomes diminished.

Sensitivity of the assay is dependent in part upon the volume in which the assay is performed: decreasing the assay volume results in an increase in sensitivity. The volume chosen should be large enough to allow accurate and easy pipetting but small enough to give the sensitivity required with a given enzyme preparation. 100 μl is a suitable volume for a highly sensitive assay. Increasing the amount of label is not the most appropriate means of increasing sensitivity, since there will be a corresponding increase in the activities of contaminating enzymes, and thus an elevated blank value.

3.2.2 Enzyme preparation

A range of enzyme preparations can be used in this assay, from crude membrane preparations to solubilized enzyme. Crude or partially purified adenylate cyclase can be stored for long period of time in $-70°C$ freezers if suspended in a suitable buffer. One such storage buffer is 5 mM MES (2[*N*-morpholino] ethanesulphonic acid), pH 6.5, containing 30% (v/v) ethylene glycol, 1 mM dithiothreitol, 5% (w/v) glucose, 50 mM NaCl and 50 mg/litre PMSF (phenylmethylsulphonyl fluoride). Before use in an adenylate cyclase assay samples should be washed once in preparation buffer.

3.2.3 Contaminating enzyme activities

Preparations of adenylate cyclase other than purified or solubilized enzyme will contain a variety of contaminating enzyme activities capable of interfering with the assay of activity. In fact, adenylate cyclase activity is generally small compared to the activities of enzymes such as nucleoside triphosphatases, and to a lesser extent nucleotidases and deaminases. Strategies which minimize these problems include use of high substrate concentration and presence of an ATP-regenerating system. Cyclic nucleotide phosphodiesterases, the enzymes responsible for degrading cAMP, are also likely to be present in most membrane preparations. A falsely low measure of adenylate cyclase activity would be made if the action of phosphodiesterase was not prevented. Degradation of the cAMP can be inhibited by greater than 90% by including a phosphodiesterase inhibitor (such as 3-isobutyl-1-methylxanthine) in the adenylate cyclase incubation. An alternative method is to include a high concentration of unlabelled cAMP in the incubation, which effectively prevents the hydrolysis of the ^{32}P-labelled cAMP produced. An added advantage of this latter method is that it will also tend to reduce degradation of product by other pathways.

3.2.4 ^{32}P-ATP adenylate cyclase incubation

i. Assay composition
The details of the composition of the incubation can be varied to suit the particular characteristics of the enzyme preparation being used. As a guide, the incubation conditions found to be optimal for a particulate preparation of adenylate cyclase from human thyroid tissue are presented below (see reference 17).

100–250 μg of membrane protein/tube incubated in assay volume of 100 μl with the following constituents:

- 50 mM Tris–HCl, pH 7.8 at 37°C
- 4.5 mM MgSO$_4$
- 1 mM cAMP
- 2.5 mM phosphoenolpyruvate
- 1.5 units of pyruvate kinase
- 130 μg of bovine serum albumin
- 30 mM KCl
- 0.125 M sucrose.

Phosphoenolpyruvate and pyruvate kinase act as an ATP-regenerating system; an alternative is to use creatine phosphate and creatine kinase. The 1 mM cAMP prevents hydrolysis of product (^{32}P-labelled cAMP), as discussed earlier. Magnesium ions must be present in any assay of adenylate cyclase activity since it acts as a substrate cofactor. Bovine serum albumin prevents the adsorption of hormones and other agents to assay tubes and helps to stabilize the enzyme preparation [α-^{32}P]. ATP is chosen in preference to other radiolabelled species of

ATP since it generates few labelled reaction products and can be prepared with a high specific activity.

ii Incubation set-up

Adenylate cyclase activity should be measured in triplicate for each sample. The total volume of the incubation is 100 μl and the most convenient way of setting up the incubation is:

- enzyme preparation (e.g. membrane suspension) 50 μl
- ATP 'reaction mixture' 25 μl
- hormones or other agents, or buffer 25 μl.

ATP 'reaction mixture' consists of 4 mM ATP, 4 mM cAMP and ^{32}P-labelled ATP plus the two components of the ATP regenerating system: pyruvate kinase and phosphoenolpyruvate. Initially, a 4 mM ATP/4 mM cAMP stock should be made, with the compounds being dissolved in appropriate assay buffer. The correct dilution of ^{32}P-labelled ATP should be added to the required amount of the ATP/cAMP cocktail which should then be added to pyruvate kinase and phosphoenolpyruvate. The short half-life of ^{32}P (14 days) means that the amount of ^{32}P-ATP required will change from 4% to 10% over a four-week period.

Table 3 shows the amounts of the constituents required for a range of final volumes of 'ATP reaction mixture'. This ATP reaction mixture should be made up fresh before the start of the experiment.

Table 3. Reaction mixture preparation for adenyl cyclase assay

Total volume (μl)	Pyruvate kinase* (μl)	Phosphoenol-pyruvate (mg)	4 mM ATP +4 mM cAMP +AT^{32}P (μl)
170	10	1.5	160
340	20	3.0	320
510	30	4.5	480
680	40	6.0	640
850	50	7.5	800
1020	60	9.0	960
1190	70	10.5	1120
1360	80	12.0	1280
1530	90	13.5	1440
1700	100	15.0	1600
1870	110	16.5	1760
2040	120	18.0	1920
2210	130	19.5	2080
2380	140	21.0	2240
2550	150	22.5	2400
2720	160	24.0	2560

* volumes given are for Boehringer protein kinase suspension in ammonium sulphate solution, specific activity approx. 200 U/mg at 25°C.

iii Incubation procedure

Prior to starting the assay the enzyme preparation and other incubation components should be kept on ice. The best way to start the incubation is by addition of enzyme, although ATP reaction mixture can be used instead. If the enzyme preparation is particulate it should be swirled regularly during its addition to the incubation, to prevent settling out of membranes and therefore inaccuracies in the measurement of activity.

Addition to one set of triplicates should be made every 10 or 15 sec and, following a 'flick mix' the assay tubes immediately placed in a water bath (preferably shaking) set at the desired temperature. A sample of enzyme preparation should be retained for protein determination.

Stopping the incubation is achieved by the addition of 100 μl of 40 mM ATP followed by boiling of the sample for 1 to 2 min; this is the protocol adopted by Hunt *et al.* (15) which omits sodium dodecylsulphate, present in the 'stopping solution' described in the Salomon *et al.* (14) method. Addition of unlabelled ATP to all samples prior to boiling prevents an accelerated formation of product during heating. Triplicates should be stopped using the same time interval chosen for additions.

iv Assay blanks and total counts

Assay blanks are constructed by addition of 50 μl of enzyme preparation, immediately followed by addition of 100 μl of 40 mM ATP and boiling for 1 to 2 min. Two to three aliquots of ATP reaction mixture, equivalent to the amount added to each incubation tube, should be retained as total counts. Both ^{32}P-ATP total counts and the total counts for the 3H-cAMP recovery label must be appropriately quenched (see Section 3.2.6i).

3.2.5 Isolation of reaction product

i Preparation and calibration of Dowex columns

The Dowex used for the cAMP isolation procedure is Dowex AG 50 WX 4 resin (200–400 mesh, H$^+$ form). To prepare the resin for use it should be washed by stirring in 1 N HCl and allowed to settle. After decanting the supernatant and the fines this wash process should be repeated several times using distilled water. Finally, resuspend the resin in distilled water and pour the slurry into pre-wetted 0.7 × 20 cm plastic columns, plugged at the base with glass wool. The volume of gel slurry added should be adjusted to give a packed bed volume of 2 ml per column.

Variations in the elution profiles between different batches of Dowex resin mean that the cAMP elution profile must be determined after each new preparation of the columns. Elution profiles are determined by loading on samples containing ^{32}P-ATP and ^{3}H-cAMP and collecting the eluate in 0.5 ml or 1 ml fractions. A typical elution profile is shown in *Figure 3*.

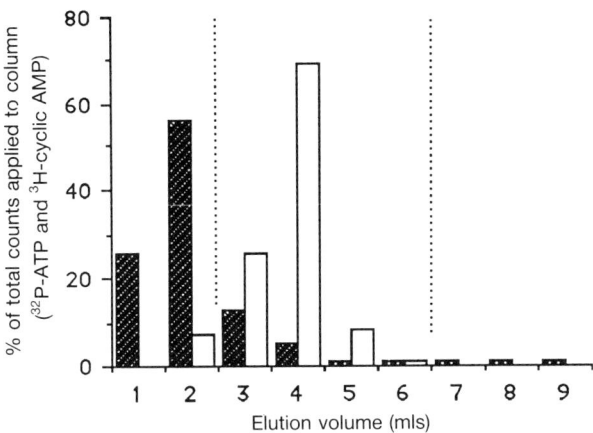

Figure 3. Elution profiles for Dowex columns. 1 ml of sample (containing ^{32}P-ATP and ^{3}H-cAMP) applied to Dowex columns and eluted using standard procedure. ^{32}P-counts are shown as hatched bars while ^{3}H-counts are open bars. The vertical lines indicate the portion of the eluate that was routinely collected.

ii Preparation and calibration of alumina columns

Alumina columns are used for the second stage of the separation and should be set up in racks in the same layout as the Dowex columns. Each 0.7×20 cm plastic column should be plugged with glass wool and filled with 1 g of neutral alumina; before use the columns should be washed with 10 ml of 0.1 M imidazole/HCl buffer, pH 7.5.

As with Dowex columns, it is necessary to characterize the elution profiles for each batch of alumina. This characterization only needs to be performed for cAMP as substrate, as contaminating ATP should remain permanently bound to the column.

iii Cyclic AMP isolation procedure

After termination of the adenylate cyclase incubation 100 μl of ^{3}H-cAMP (4000–6000 c.p.m./sample) should be added to each tube, including assay blanks. A further 700 μl of distilled water can be added to each sample. The purpose of the tritiated cAMP is to act as a recovery label so that differences in performance of individual columns can be corrected for, to allow a highly accurate assay of adenylate cyclase activity. Total counts for the ^{3}H-cAMP should also be kept, and must be suitably quenched (see Section 3.2.6i).

A modification of the Salomon et al. method for cAMP isolation (14) described by Hunt et al. (15) is to separate insoluble material from the reaction mixture by centrifugation (2000g, 10 min) before application of the samples to the Dowex columns, which must be pre-wetted with distilled water before sample application.

Protocol 5. Isolation of adenylate cyclase reaction products

1. Each sample should be carefully layered on to the surface of the resin using Pasteur pipettes; the first eluate should be discarded.

2. Add an appropriate volume of distilled water to the Dowex, so that the volume of sample applied and this wash volume equal the volume determined from calibration to be necessary to elute the bulk of the ^{32}P-ATP (2 ml in the example shown; *Figure 3*). Again discard this eluate. Extreme caution is necessary in handling the eluates at this stage, since they will contain the majority of the ^{32}P-label present in the incubations.

3. The rack containing the Dowex columns should be placed over the rack of alumina columns and the ^{3}H-cAMP eluted from the Dowex using distilled water. The volume required for this elution is determined from the Dowex column calibration profiles. In the example shown (*Figure 3*) the appropriate volume is 4 ml.

4. Following removal of Dowex racks the ^{3}H-cAMP can be eluted from the alumina columns using an appropriate volume of 0.1 M imidazole/HCl buffer (pH 7.5). Again, the quantity of imidazole required is determined from calibration profiles. The eluate should be collected directly into scintillation vials containing scintillant (see Section 3.2.6*i*).

5. Regenerate the Dowex columns with 0.1 N HCl and the alumina columns with 0.1 M imidazole/HCl (pH 7.5) columns. Immediately prior to re-use wash the Dowex columns with distilled water. Recovery of cAMP using newly packed Dowex columns is approximately 80%. The columns can be used repeatedly until recovery falls to around 40%, when they should be discarded. The major factor in determining the life of the alumina columns is a safety consideration. Since virtually all the ^{32}P-ATP binds to alumina, repeated use of these columns will result in a progressive accumulation of radioactivity. Caution should be taken in handling columns and eluates; Perspex screens will reduce the risk of exposure to radioactivity.

3.2.6. Quantification of adenylate cyclase activity

i Scintillation counting

An appropriate scintillant to use is Triton-toluene (4 ml per vial), which has the following composition:

- 1,4-bis (2-[5-phenyloxazolyl]) benzene (POPOP)   50 mg
- 2,5-diphenyloxazole (PPO)   15 mg
- toluene   1 litre
- Triton X-100   1 litre

Count samples for 4 min using a scintillation counter arranged to count both ^{32}P and ^{3}H simultaneously. Total counts must be quenched with an appropriate volume of imidazole.

ii Calculations

There are two basic calculations involved in the determination of the amount of cAMP produced during the adenylate cyclase assay. The first is a correction of each sample for variations in recovery from the sequential chromatography step. The ability to take into account such fluctuations is due to the presence of the ^{3}H-labelled cAMP, added after termination of the incubation. This correction is performed in the following way:

If X is the mean of the ^{3}H-cAMP total counts in c.p.m.,

Y is the number of c.p.m. of a given sample for ^{3}H-cAMP,

and Z is the number of c.p.m. for $[\alpha\text{-}^{32}P]$-cAMP in the same sample

then the number of c.p.m. for $[\alpha\text{-}^{32}P]$-cAMP for a column recovery

$$\text{of } 100\% = \frac{Z \times Y}{X}$$

The second stage is to calculate the amount of cAMP formed during the assay. The amount of $[\alpha\text{-}^{32}P]$-ATP used per assay tube is usually arranged to be 10^6 c.p.m. and this represents a substrate concentration of 1 mM ATP. Therefore, the total counts for $[\alpha\text{-}^{32}P]$-ATP (1 000 000 c.p.m.) = 100 nmol ATP/100 μl.

Thus, 10 c.p.m. = 1 pmol ATP = 1 pmol cAMP (since every moles of ATP hydrolysed produces 1 mole of cAMP). If a sample gave W c.p.m. of $[\alpha\text{-}^{32}P]$-cAMP and assay blanks gave a mean of V c.p.m.,

$$\text{then cAMP present in the sample} = \frac{W - V}{10} \text{ pmol.}$$

Results will generally be expressed relative to the amount of protein present in the incubation.

3.2.7 Summary of ^{32}P-ATP adenylate cyclase assay method

The assay of adenylate cyclase activity using $[\alpha\text{-}^{32}P]$-labelled ATP and separation of product from substrate using sequential chromatography steps has been described. The major advantages of this method are high sensitivity and the straightforward nature of the assay (once Dowex and alumina columns have been prepared and calibrated). A complete assay, from incubation to scintillation counting, can be carried out readily in one day.

Drawbacks are the expense of $[\alpha\text{-}^{32}P]$-ATP and the risks associated with the use of this radiolabel. Extreme care should be taken at all stages of the assay: a hot-room should be used if possible. Particular attention should be paid to disposal of the eluates and washes from the column steps. Some of these will be highly radioactive. It is also advisable to construct Perspex screens to surround the columns, and to wear finger monitors while performing the assay.

3.3 Non-labelled incubation for assay of adenylate cyclase

3.3.1 Basic incubation conditions

Adenylate cyclase incubation, where activity is to be determined by subsequent binding protein or radioimmunoassay of reaction product, will possess the following basic components: ATP (assay substrate), Mg^{2+} (an essential co-factor), buffer, a nucleotide triphosphate regeneration system, and some means of inhibiting cyclic nucleotide phosphodiesterase activity. Additionally, a thiol-reducing agent may be included in the incubation medium to prevent inactivation of components of the adenylate cyclase complex (G_s and C) by oxidation of –SH groups. 1 mM dithiothreitol is the usual choice through 1 mM β-mercaptoethanol has also been used.

ATP is usually present at 1.0–1.5 mM concentrations. Free Mg^{2+} ion is usually present at about 2.5 mM, in excess of added nucleotides and chelators. Suitable buffers for adenylate cyclase include Tris, Triethanolamine, or Hepes in the pH range 7.0–7.8. The most commonly used inhibitors of phosphodiesterase are xanthine derivatives such as theophylline (10 mM) or 3-isobutyl-1-methylxanthine (IBMX), 1 mM. Unlike the radioisotope assay of adenylate cyclase, an excess of unlabelled cAMP cannot be used to inhibit product degradation.

For a more complete discussion of the requirements of the adenylate cyclase incubation refer to the sections on assay characterization, enzyme preparation, and contaminating enzyme activities (Sections 3.2.1, 3.2.2, and 3.2.3 respectively).

3.3.2 Incubation procedure

To illustrate the requirements of the adenylate cyclase incubation, the protocol used in the assay of adenylate cyclase of membranes prepared from hepatocytes (18, 19) or purified rat liver plasma membranes (20) is presented. The final incubation volume is 100 μl containing constituents at the following final concentrations:

- 1.5 mM ATP
- 5 mM $MgSO_4$
- 10 mM theophylline
- 1 mM EDTA

- 7.4 mg phosphocreatine/ml, 1 mg creatine kinase/ml (ATP regeneration system)
- 25 mM triethanolamine/KOH buffer, pH 7.4.

Keep the samples on ice until initiation of the reaction, which is achieved by the addition of membranes (40–60 µg/tube). Incubate the samples at 30°C for 10 min. Terminate the assay by boiling at 90–100°C for 3 min, then cool the samples on ice, and sediment the precipitated protein in a bench microfuge. An aliquot of the deproteinated supernatant can then removed for assay of cAMP, as described earlier.

3.4 Use of cholera and pertussis toxins

3.4.1 Background

Bacterial toxins isolated from *Vibrio cholera* (cholera toxin) and *Bordetella pertussis* (islet activating protein) have provided popular tools as a result of their ability to catalyse ADP-ribosylation and therefore modify the function of certain G-proteins involved in the transduction of hormonal signals. Such effects were first characterized using the G-proteins involved in regulation of adenylate cyclase, G_s and G_i, as well as the retinal G-protein, transducin. It was thought that the toxins allowed convenient, specific covalent labelling of these particular G-proteins, if $[^{32}P]NAD$ was used as the substrate. It is now clear that such experiments should be viewed with caution, as a number of substrates for each toxin have been shown to exist, often in the same cell. This is particularly a problem when using pertussis toxin, as a closely-related family of G-proteins of similar molecular weight are substrates (10). However, the toxins are still of use, provided that it is borne in mind that more than one signal transduction system, or G-protein, may be affected.

Both toxins consist of two subunits, a B (or binding) subunit complex of 6 polypeptides (cholera) or 5 polypeptides (pertussis); and the catalytic A subunit. The B-subunit binds to receptors on the cell surface, and so allows transfer of the A-subunit across the plasma membrane into the cell. The A subunit then catalyses the transfer of an ADP-ribose moiety from NAD to the particular G-protein substrate. The α-subunit of G_s is substrate for ADP-ribosylation by the A-subunit of cholera toxin, which results in its activation, while the α-subunit of G_i is ADP-ribosylated by the A-subunit of pertussis toxin, which results in its inactivation; the net result of both toxin treatments being activation of adenylate cyclase (9). Cholera toxin is obtained from Sigma; Pertussis toxin from Porton Products Ltd, Wiltshire, UK.

3.4.2 Cell treatment

Incubate the cells (for example, hepatocytes, adipocytes, or platelets) with the holotoxin (i.e. a preparation containing both A and B subunits), at a final concentration of 1 µg/ml for 45 min (cholera) or 100 ng/ml for 60 min (pertussis), which is maximally effective for the cell types listed above. The exact conditions used for other systems would need to be established.

3.4.3 Membranes

Isolated membranes, rather than intact cells, may be treated with either cholera or pertussis toxin, provided that the toxins are thiol-preactivated (21, 22). Briefly, add an aliquot of the stock toxin (50 μl, 1 mg/ml) to an equal volume of 50 mM dithiothreitol. Pre-incubate this at 30°C for 20 min before adding to the assay. A suitable assay mixture (in final volume of 500 μl) is: membrane protein (0.5 mg), 10 μg/ml of thiol-activated cholera or pertussis toxin, 10 mM NAD$^+$, 1.0 mM ATP, 100 μM GTP, 7.4 mg/ml phosphocreatine, 1 mg/ml creatine kinase, and 5.0 mM MgCl$_2$ in 25 mM triethanolamine/KOH, pH 7.4.

Incubate for 30 min at 30°C, then stop the reaction by diluting with 15 ml of ice-cold triethanolamine/KOH. Collect the membranes by centrifugation and resuspend in 400 μl of triethanolamine/KOH buffer and assay adenylate cyclase activity as described earlier.

3.4.4 Protein labelling using toxins with [^{32}P]-NAD$^+$

Incubation of membranes with cholera or pertussis toxin in the presence of [^{32}P]-NAD$^+$ results in the incorporation of a [^{32}P]ADP-ribose moiety into the particular G-protein substrate(s) of the toxin. This can be used to detect and, under appropriate conditions, quantify the protein which has been modified. The assay conditions are based on those described by Hildebrandt *et al.* (22), and are similar to those described for treating membranes above, with the inclusion of thymidine which decreases non-specific labelling. The procedure is listed in *Protocol 6*.

Protocol 6. ADP ribosylation assay

The assay mixture (final volume 100 μl) contains: membrane protein (200 μg), 20 μg/ml of thiol-activated cholera toxin, or 10 μg/ml thiol-activated pertussis toxin, 20 μM [^{32}P]-NAD$^+$ (specific radioactivity 500 Ci/mol), 1.0 mM ATP, 0.5 mM GTP, 15 mM thymidine, 5.0 mM dithiothreitol, and 5.0 mM MgCl$_2$ in 100 mM potassium phosphate buffer, pH 7.5. When using cholera toxin, there is no ATP present, and add 10 μM Ca^{2+} to the 'ribosylation mixture' (23).

Procedure

1. Initiate the reaction by adding the toxin, and incubate for 30 min at 30°C.

2. Stop the reaction by diluting fivefold with ice-cold 10 mM Tris–Cl, pH 7.5, containing 1 mM EDTA.

3. Collect the membranes by centrifugation (14 000g for 10 min at 4°C), discard the supernatant, and resuspend the membranes in Laemmli's sample buffer (24) with 5% β-mercaptoethanol and allow to stand at room temperature overnight.

Protocol 6. *Continued*

4. Electrophorese the samples in 10% acrylamide slab gels after the method of Laemmli (24), dry the gels and autoradiograph using Kodak X17R-5 X-ray film and Kodak intensifying screens.

3.5 Use of forskolin to assay G_i function

The diterpene, forskolin, is a potent activator of all eukaryote adenylate cyclases (25). It is freely soluble in ethanol, in which it is usually added to membranes or cells, and is readily available (e.g. from Calbiochem.). The inhibition of forskolin-stimulated adenylate cyclase activity by guanine nucleotides has been used to detect functional G_i in isolated plasma membranes (26, 27, 28). This takes advantage of the fact (see reference 9) that G_s and G_i have very different affinities for guanine nucleotides, which allows them to be activated selectively. Isolated membranes are simply incubated with 100 μM forskolin and concentrations of guanylyl-imidodiphosphate (Gpp[NH]p) from 10^{-12} to 10^{-6} M, in addition to the standard assay components. The results of one such experiment performed on hepatocyte plasma membranes are shown in *Figure 4*, where activation of G_i (guanine nucleotide-dependent inhibition) can be seen to precede activation of G_s, as the concentration of Gpp[NH]p increases.

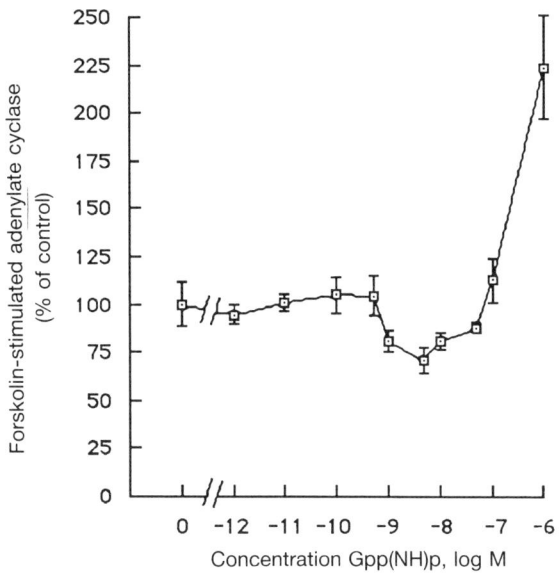

Figure 4. G_i function in hepatocyte membranes. Hepatocyte membranes (0.6 mg/ml) were incubated with the indicated concentrations of Gpp(NH)p in the presence of 100 μM forskolin, and adenylate cyclase activity measured as described in the text.

4. Measurement of guanylate cyclase activity

4.1. Introduction

Guanylate cyclase [E.C. 4.6.1.2] is the enzyme which catalyses the hydrolysis of guanosine triphosphate (GTP) to cyclic guanosine 3',5'-monophosphate (cGMP), in a reaction analogous to that catalysed by adenylate cyclase. As with adenylate cyclase, a divalent cation is required for enzyme activity, but manganese (Mn^{2+}) is the preferred metal ion for most guanylate cyclase preparations. The substrate for the enzyme appears to be a complex of GTP and divalent cation, i.e. $Mn^{2+} \cdot GTP$. Additional free Mn^{2+} appears necessary for maximal enzyme activity.

Guanylate cyclase activity can be determined using either labelled or non-labelled substrate. Use of unlabelled substrate is generally more labour-intensive since the product formed must then be assayed to determine its cGMP content. However, in certain circumstances measurement of unlabelled product may be the most appropriate means of determining guanylate cyclase activity. This may particularly be the case in tissue slice preparations or when working with cultured cells. In addition, assays performed without use of radiolabelled substances will generally be cheaper. Refer to Section 2.5 for details of such methods of cGMP determination. Where guanylate cyclase activity is assayed using a radioactively labelled substrate the most commonly used label is $[\alpha\text{-}^{32}P]GTP$; the reaction product in this case will be $[\alpha\text{-}^{32}P]cGMP$.

4.2 Assay overview

4.2.1 Principle of guanylate cyclase assay

Assay of guanylate cyclase activity using radioactively labelled substrate is identical in principle to that for adenylate cyclase (described in detail previously, see Section 3). Thus, the basic steps involved in assay of guanylate cyclase activity are incubation, purification of cGMP and liquid scintillation counting.

As with adenylate cyclase, the assay is dependent upon effective separation of the cGMP from its substrate and all possible degradation products. This is because cyclase activities are usually small compared to the activities of enzymes that are capable of interfering with the assay. Isolation of cGMP is a two-step procedure using Dowex cation exchange followed by chromatography on alumina. Although the principle is the same as that described earlier for adenylate cyclase, there are significant differences in column profiles, with the guanylate cyclase assay generally requiring larger elution volumes.

4.2.2 Assay methodology

The methodology adopted in the assay of guanylate cyclase is essentially the same throughout as that used for adenylate cyclase. Thus, the strategies employed to overcome the effects of contaminating enzymes are identical to those described in

Section 3.2.3. A substrate regeneration system and a high substrate concentration are used to minimize the effects of nucleoside triphosphatases, nucleotidases and deaminases. The effects of cyclic nucleotide phosphodiesterases are overcome either through presence of high concentration of unlabelled cGMP, or by adding phosphodiesterase inhibitors such as 3-isobutyl-1-methylxanthine.

4.2.3 Characterization of guanylate cyclase assay

As with adenylate cyclase, all aspects of the guanylate cyclase assay must be characterized thoroughly before routine use. Parameters which need to be investigated include temperature, linearity, pH and buffering conditions, metal ion requirement, and enzyme concentration. Optimal pH, for example, will vary between enzyme preparations. Beef lung guanylate cyclase has a pH optimum of 7.6, while that of liver is 7.4.

Manganese is the metal ion necessary for full expression of activity. It is possible to replace Mn^{2+} with other divalent cations, but this will in general cause a considerable reduction in activity. Guanylate cyclase activity in the presence of Mg^{2+} will be around 5 to 20% of that measured using Mn^{2+} as co-factor.

4.3 Enzyme preparation

4.3.1 Tissue distribution

Guanylate cyclase is found in many tissues. Of mammalian tissues, platelet as one of the highest levels of activity. Small intestine and lung also express high amounts of guanylate cyclase activity. The enzyme appears to be low in skeletal muscle and in the fat cell. A number of invertebrate tissues, especially sperm, have extremely high levels of guanylate cyclase activity; guanylate cyclase has been studied in a range of invertebrates, including sea-urchins, tube worms, and abalone.

4.3.2 Subcellular distribution

Enzyme activity is associated with both particulate and soluble fractions of tissue homogenates. Distribution between these subcellular fractions varies from one tissue to another: in some cases guanylate cyclase activity occurs mainly in the soluble fraction (for example, liver and platelet). In others, such as intestinal mucosa and sea-urchin sperm, most of the activity is in a particulate form. The soluble and particulate forms seem to represent separate isoenymes; the enzyme preparation used in assays can range from crude particulate preparations to apparently pure enzyme. A procedure for preparation of particulate and soluble guanylate cyclase from human platelet (3) is presented in *Protocol 7*.

Protocol 7. Preparation of platelet guanylate cyclases

1. Mix fresh human blood with 0.1 vol. of 3.8% (w/v) sodium citrate/100 mM Na_4 EDTA.

2. Centrifuge blood at 200g for 10 min at 4°C, remove the platelet-rich supernatant and centrifuge at 600g for 10 min at 4°C.

3. Resuspend the pellet in 50 mM Tris–HCl (pH 7.4)/1 mM Na_4EDTA/ 140 mM NaCl and wash twice by centrifugation at 600g for 10 min at 4°C.

4. Resuspend the pellet in 50 mM Tris–HCl (pH 7.4 at 4°C)/250 mM sucrose, and then lyse the platelets by rapid freezing in liquid N_2 followed by thawing at 37°C (twice).

5. Soluble preparations prepared by centrifugation of this lysate at 12 000g for 15 min at 4°C. Wash particulate preparations twice with Tris–sucrose by centrifugation at 12 000g (as described above) and finally resuspend in the appropriate assay buffer. Use both preparations immediately in the assay of guanylate cyclase.

4.4 Activators and inhibitors

A range of substances have been found to act either as activators or inhibitors of guanylate cyclase activity. One major class of enzyme activators are substances capable of producing nitric oxide under certain conditions. These include azide, nitrite, hydroxylamine, nitroglycerin, nitroprusside, nitrosamines, and nitrosureas. Some of these agents may also be capable of activating guanylate cyclase through mechanisms other than production of nitric oxide. For example, azide can inhibit GTPase activity and thereby stimulate activity several-fold.

Recently, much attention has been focused on atrial natriuretic factor (ANF) as an activator of guanylate cyclase. ANF, released from the atria, is able to activate particulate guanylate cyclase from a variety of tissues, including kidney, liver, and lung. However, no activation of soluble guanylate cyclase by ANF has yet been demonstrated. Indeed, it has not been possible to demonstrate hormonal activation of any soluble guanylate cyclase in cell-free preparations.

Other naturally occurring agents which can stimulate guanylate cyclase activity include calcium and calmodulin, fatty acids, and endothelium-derived relaxant factor. For a comprehensive review of stimulatory and inhibitory agents' effects on guanylate cyclase refer to Waldman and Murad (31).

4.5 Incubation conditions

4.5.1 Assay composition

As with assay of adenylate cyclase activity, incubation conditions can be varied to suit individual requirements. Incubation temperature will generally be in the range 30–37°C and either physiological or optimal pH should be chosen. [α-^{32}P]

GTP sometimes contains radioactively-labelled contaminants which co-chromatograph with cGMP, resulting in high assay blanks. Where this is the case, it is necessary to purify the labelled GTP, which can be achieved by chromatography on Dowex 50W-X4.

Presented below is the composition of a guanylate cyclase incubation medium developed for the assay of particulate and soluble enzyme preparations from human platelet (33). The figures are all final assay concentrations.

- 65 mM Tris–HCl, pH 7.5 at 30°C
- 1 mM [α-^{32}P] GTP (approx. 9–11 c.p.m. per pmol)
- 4 mM $MnCl_2$
- 4 mM cGMP
- 1 mM 3-isobutyl-1-methylxanthine
- 5 mM creatine phosphate
- 20 U/ml creatine phosphokinase
- 1 mg/ml bovine serum albumin.

Final assay volume is 100 μl and incubations would normally contain 20–30 μg of protein per tube. The Mn^{2+} is the co-factor necessary for full expression of activity. The unlabelled cGMP and the 3-isobutyl-1-methylxanthine are to prevent the degradation of radioactively labelled product by cyclic nucleotide phosphodiesterases, while creatine phosphate and creatine phosphokinase constitute the GTP regenerating system.

4.5.2 Incubation procedure

Incubations are carried out in triplicate and should be set up using the same basic layout described for adenylate cyclase (Section 3.4). With a final volume of 100 μl a suitable incubation composition is:

- enzyme preparation   50 μl
- GTP 'reaction mixture'   25 μl
- test agents, or buffer   25 μl.

GTP 'reaction mixture' should be made up in the same way as ATP 'reaction mixture'. Start the incubation by addition of enzyme preparation, 'flick mix', and add the triplicates to water-bath set at desired temperature at 10- or 15-sec intervals. Terminate the reaction by adding 100 μl of 1 M HCl followed by boiling for 2 to 3 min. The HCl prevents non-enzymatic formation of cyclic GMP during boiling.

Assay blanks should be included in each experiment by adding 100 μl of 1 M HCl to tubes containing the incubation mixture. Next add the enzyme preparation and boil the tubes for 2 to 3 min. Total counts should be prepared as described for adenylate cyclase assay.

4.6 Isolation of reaction product

4.6.1 Preparation and calibration of Dowex and alumina columns

The method used to isolate $[\alpha-^{32}P]$ cGMP from other labelled species is based on the sequential chromatography over Dowex 50 (H^+ form) and neutral alumina; it was originally described by Karczewski and Krause in 1978 (33). Preparation of and calibration procedures for Dowex and alumina are as described for the separation of cAMP in the adenylate cyclase assay (*Protocol 5*), with the exception that a Dowex column of 6 to 7 cm should be used. Regeneration procedures are identical to those described for adenylate cyclase assay. Elution profiles of $[\alpha-^{32}P]$ GTP and $[8-^{3}H]$ cGMP should be determined for each new set of columns.

Protocol 8. Isolation procedure and quantification of cGMP

1. After termination of the guanylate cyclase incubation add 100 μl of $[8-^{3}H]$ cGMP (2000 c.p.m.) solution in distilled water. As with assay of adenylate cyclase, the purpose of the tritiated label is to monitor the recovery of ^{32}P-cGMP from the Dowex and alumina columns. Aliquots of the $[8-^{3}H]$ cGMP should be retained as Total Counts.

2. Make up the volume of each sample to 1 ml by adding 700 μl of distilled water, mix the tubes and centrifuge for 10 min at 2500g to precipitate denatured proteins.

3. Apply supernatants from this centrifugation to pre-wetted Dowex columns. Allow samples to pass through the columns, wash with 0.1 N HCl and discard the eluate. The volume required for this wash should be determined from elution profiles, performed in the same way as the adenylate cyclase column profiles. It should be in the region of 3 ml.

4. Place the Dowex columns over the alumina columns and elute the cGMP from the Dowex to the alumina using 0.1 N HCl. Again, the precise volume required must be determined form elution profiles.

5. Add 10 ml of distilled water to each alumina column, discard the eluate.

6. Elute the cGMP with 4 ml of 0.2 M ammonium formate. Collect directly into glass scintillation vials containing 8 ml of Triton-toluene scintillation fluid (recipe given in Section 3.2.6*i*) and count for ^{32}P and ^{3}H for 4 min on a suitable scintillation counter.

7. Calculations of the amount of $[\alpha-^{32}P]$ cGMP produced are as described for the $[\alpha-^{32}P]$-adenylate cyclase assay (Section 3.2.6).

As with the $[\alpha-^{32}P]$-adenylate cyclase assay extreme care should be taken to minimize the risks associated with the use of ^{32}P-label. Particular attention is

necessary with washes and eluates from the column steps, some of which will have
very high levels of radioactivity.

5. Assay of cyclic nucleotide phosphodiesterase activity

5.1 Introduction

The intracellular level of cyclic nucleotides can be decreased by several
mechanisms which include their degradation, the inhibition of synthesis, and
release from the cell. However, the only known enzymatic reaction for the
degradation of cyclic nucleotide is its conversion to 5'-nucleotide.

The enzyme responsible for this action is cyclic nucleotide phosphodiesterase
(E.C. 3.1.4.17), first identified by Butcher and Sutherland (34). The specific
reaction catalysed by the enzyme is the hydrolysis of the 3',5'-phosphodiester
bond of cyclic nucleotides, with cAMP and cGMP being considered the most
physiologically relevant of these. A schematic representation of the hydrolysis of
cAMP is presented in *Figure 5*.

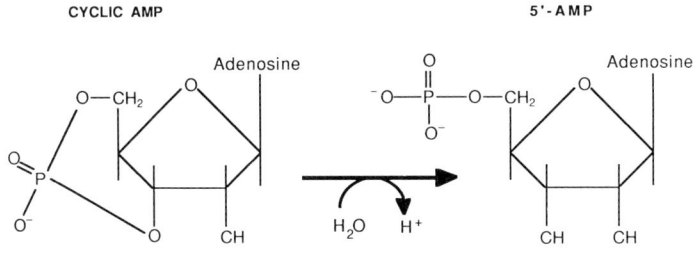

Figure 5.

The development of a sensitive radioactive assay for cyclic nucleotide
phosphodiesterase activity (35) has allowed an extensive study of cyclic
nucleotide phosphodiesterase activity. One of the most important findings to
emerge from these studies is that multiple forms of these enzymes exist in many
tissues (36). Some basic principles involved in the isolation and characterization
of such isoenzymes will be discussed in Section 5.6.

5.2 Overview of assay

The radioactive assay of cyclic nucleotide phosphodiesterase activity was first
described by Thompson and Appleman in 1971 (35) and is now the most widely
used method; it is equally applicable to cAMP or cGMP hydrolysing activity.
Use of radioactive tracers allows enzyme activity to be assayed at subsaturating
concentrations of substrate. This is a distinct advantage over other assay
methods, which are often insensitive and therefore inadequate for assay of

low-K_m (high affinity) phosphodiesterases. Other advantages include simplicity and lack of dependence on coupling enzymes.

The most commonly used substrate for this assay is tritiated cAMP or cGMP. Measurement of enzyme activity involves quantification of reaction product formed. A second incubation step is included to convert the product of the first incubation (5'-nucleotide monophosphate) to 5'-nucleotide, which is catalysed by 5'-nucleotidase present in snake venom. This allows the reaction product to be separated more efficiently from unreacted substrate.

The two incubation steps can be represented thus:

(i) 3',5'[8-³H]-cyclic NMP→5'[8-³H]-NMP

(ii) 5'[8-³H]-NMP→[8-³H]-nucleoside

Step (i) catalysed by cyclic NMP (cNMP) phosphodiesterase; Step (ii) catalysed by 5'-nucleotidase; NMP = nucleotide monophosphate.

Separation is by a 'batch' procedure, using Dowex-1-chloride anion exchange resin which selectively binds cyclic nucleotide. Quantification is by liquid scintillation counting of the isolated production (tritiated 5'-nucleotide).

5.3 Assay procedure

5.3.1 Reagents

First incubation

- 3',5' cNMP

- 3',5'[8-³H]-cNMP

- Buffer system, e.g. Tris–HCl, HEPES or TREACL: 10–100 mM, pH 7.0–8.0

- Magnesium chloride 1–10 mM

- Enzyme preparation.

Second incubation

- *Hannah ophiophagus* snake venom, 1 mg/ml in distilled H_2O

Isolation and quantification of reaction product

- Dowex 1-chloride anion exchange resin

- Scintillation fluid, e.g. Liquiscint, Ecoscint

The concentration of cyclic nucleotide in the assay can be varied from 0.1 μM to 1000 μM, with approximately 100 000 d.p.m. being the amount of tritiated label required per tube. Suitable buffering systems include Tris–HCl (Tris(hydroxymethyl) aminomethane/HCl), HEPES (N-2-hydroxy- ethyl piperazine-N'-2-ethanesulphonic acid) and TREACL (Triethanol-amine hydrochloride). The pH used will generally be in the 7.0–8.0 range; the optimum value will vary between enzyme preparations and should be determined through characterization studies. Mg^{2+} ion (in the millimolar concentration range) is required as an essential co-factor for phosphodiesterase activity.

5.3.2 Preparation of Dowex 1-chloride resin

Initially, 100 g of Dowex 1-Cl resin is washed in 1 M sodium hydroxide for 15 min after which it is extensively washed with deionized water until the pH of the eluate falls to 7.0; the Dowex 1-Cl resin is then collected and washed further in 1 M hydrochloric acid for 15 min followed by washing with deionized water until the pH of the eluate reaches 4–5.

5.3.3 Incubation procedure

First incubation

The first stage incubation is performed by combining the enzyme preparation with [8-^{3}H]-cNMP in the presence of Mg^{2+} and a suitable buffer. The total assay volume is 100 μl, which is an economical modification of the original method described by Thompson and Appleman (35) where an assay volume of 400 μl was used. Total counts should be retained and suitable blanks should be included in every assay protocol. All assays are performed such that less than 10% of the available substrate is utilized in the reaction. Linearity of product formation with the enzyme concentration is an important requirement in order that accurate kinetic analysis can be made.

Incubations are performed at 30–37°C and reactions are terminated by boiling the samples for 2 min in order to completely inactivate the phosphodiesterase activity. After this boiling step the samples should be cooled by placing them on ice. Termination of the phosphodiesterase assay by dilution, pH variance, or metal poisoning are less efficient processes than boiling. Likewise, TCA and ethanol precipitations can alter the catalytic efficiency of the 5'-nucleotidase incubation.

Second incubation

It is essential to separate the 5'[8-^{3}H]-NMP formed during the assay from unused 3',5'[8-^{3}H]-NMP, and this is achieved by adding an excess of 5'-nucleotidase, which enzymatically converts the 5'[8-^{3}H]-NMP to [8-^{3}H]-nucleotide. The most common source of 5'-nucleotidase used for this assay step is the *H. ophiophagus* snake venom.

Usually 25 μl of snake venom (1 mg/ml) is used per assay tube and the incubation is performed for 10 min at 30°C. Labelled nucleoside thus formed can be easily separated from charged cyclic nucleotide monophosphate by adsorption of the latter on to an anionic exchange resin (Dowex 1-chloride).

5.3.4 Isolation and quantification of reaction product

It is important to keep the resin well-stirred in order to maintain an homogenous suspension. Usually 400 μl of Dowex 1-Cl resin (1 part Dowex:2 parts water) is added to each assay tube. After a 15 min incubation at 4°C the labelled nucleotide is collected by sedimentation of the Dowex 1-Cl resin by centrifugation of the sample for 2 min at 14 000g using a bench-top microcentrifuge (e.g. Joblong or

MSE MicroCentaur). A 150 μl aliquot of the resulting supernatant is added to 4 ml of scintillation fluid (e.g. Ecoscint A) for liquid scintillation counting. One minute of counting, using a channel set for tritium, is generally sufficient.

5.3.5 Calculation of enzyme activity

For samples that require Dowex 1-Cl resin (1:2 water):

$$\text{Enzyme activity} = \frac{\text{c.p.m. of sample} \times \text{assay vol.} \times \text{NBF} \times \text{pmol cNMP}}{\text{Total c.p.m.} \times 150 \times \text{time} \times \text{enz. vol.}}$$

Correct 'c.p.m. of sample' by subtracting the appropriate blank value (see Section 5.3.6*ii*) before performing calculation of activity. 'Assay volume' ($= 100\ \mu$l) and 'enzyme volume' ($=$ volume of enzyme preparation added per assay tube) should be in millilitres. The 'pmol cNMP' term is the amount of unlabelled cNMP monophosphate present per assay tube, in picomoles. NBF is the 'nucleoside binding factor' (see Section 5.3.6*iii*). When 'Ethanol-Dowex' is used, omit the 'NBF' term. The '150' term is included since only 150 μl of the total assay volume is taken for scintillation counting.

Enzyme activity calculated in this way will have the units picomoles of cyclic nucleotide hydrolysed per minute per millilitre. If specific activity is required, divide activity by the protein concentration of the enzyme sample. Units for specific activity will be picomoles per minute per milligram of protein.

5.3.6 General assay considerations

i Validity of the phosphodiesterase assay

The radioactive assay of cyclic nucleotide phosphodiesterase is dependent upon the following assumptions:

(a) That all the cNMP remaining after the phosphodiesterase step of the reaction is bound to Dowex 1-Cl. The extent to which this is true will be reflected in the 'blank' value.

(b) That all the 5'-NMP generated in the assay is converted to nucleoside by the action of 5'-nucleotidase.

(c) That all the nucleoside remains free in solution and is not precipitated by the Dowex 1-Cl resin, if necessary ethanol should be included with the Dowex 1-Cl resin.

ii Binding of NMP to Dowex

Usually 2–5% of the 3',5'[8-^{3}H]-NMP does not bind to the Dowex 1-Cl resin. This leads to an apparent 'blank' value which has to be deducted from the test-sample value. This 'blank' value does not vary significantly with the concentration of cyclic nucleotide monophosphate used, although it is affected by the amount of labelled substrate added to the incubation.

Various agents do affect the ability of the NMP to bind to Dowex 1-Cl resin. For instance, the ionic strength of the buffer used for the enzyme preparation can markedly alter this binding phenomenon. Where the interference by contaminating salt is relatively small it should be sufficient to construct 'boiled blanks' which can simply be subtracted from the apparent activity measured in the presence of salt. When the effect of salt is high it is necessary to remove or significantly reduce its concentration in the enzyme sample. This may be achieved by several means, such as gel filtration on Sephadex G-25 columns or use of ultrafiltration cones.

iii Binding of adenosine to Dowex
It has been shown that labelled adenosine can bind non-specifically to Dowex 1-Cl resin (37, 38). This process is dependent upon the concentration of NMP used in the assay. For a range of substrate concentrations used in kinetic analysis (0.05 μM–1000 μM) the fraction of labelled adenosine binding to Dowex 1-Cl can be as high as 40–50% of the total adenosine generated in the assay. This value can be determined by assessing the amount of [^{3}H]-nucleoside (obtained, for example, from Amersham) bound to the Dowex 1-Cl resin in the presence of various concentration of substrate. Inclusion of ethanol can prevent this non-specific binding of labelled adenosine to Dowex 1-Cl resin (39).

iv Catabolism of adenosine
Sample preparation such as homogenates can contain adenosine deaminase (37). Adenosine generated in the assay can be converted thus into inosine. Inosine binds to Dowex 1-Cl resin leading to a serious underestimation of phosphodiesterase activity. However, inclusion of ethanol can effectively prevent this occurring. Usually the Dowex resin is prepared as a 1:1:1 mix of Dowex:H$_2$O:ethanol.

5.4 Preparation of the enzyme sample

5.4.1 Multiple isoenzyme forms
The multiple forms of cyclic nucleotide phosphodiesterase have been classified into four classes based upon their kinetic properties, inhibitor sensitivities and immunological characteristics. These classes are:

(a) *Calcium and calmodulin-activated cyclic nucleotide phosphodiesterase.* Micro-molar concentrations of calcium together with calmodulin can stimulate both cAMP and cGMP phosphodiesterase activity between two- and tenfold (36, 40).

(b) *Cyclic-GMP-activated cyclic AMP phosphodiesterase.* Micromolar concentrations of cGMP can activate cAMP phosphodiesterase activity by approximately fivefold (36, 41).

(c) *Cyclic-AMP-specific phosphodiesterase.* This enzyme class hydrolyses cAMP with high affinity and is further divided into two subclasses based upon ability to be inhibited by either cGMP (42, 43) or the phosphodiesterase inhibitor Ro-20-1724 (42).

(d) *Retinal cyclic GMP phosphodiesterase*. This enzyme hydrolyses cGMP with high-affinity and is regulated by light-bleaching of rhodopsin (44).

5.4.2 Sources of enzyme

Cyclic nucleotide phosphodiesterase activity can be assayed from a variety of sources such as:

- tissue homogenates, subcellular fractions of tissue, such as purified plasma membranes, endoplasmic-reticulum, cytosol;
- partially-purified enzyme preparations; for instance, active fractions isolated using DE-52 ion-exchange chromatography;
- purified phosphodiesterase.

Homogenates are usually prepared in an isotonic buffer of pH 7 to 8 that contains a cocktail of protease inhibitors. Some workers have included reducing agents in the homogenization buffers, and it is clear that although such agents can stabilize the enzyme activity, they can also activate sulphydryl protease which can act upon the phosphodiesterase. Proteolysis of the phosphodiesterase often leads to the generation of catalytically active fragments which exhibit different sensitivities to activator or inhibitor agents when compared to the native form (45, 46). This can lead to the identification of apparent novel forms which are simply proteolytic fragments of a native phosphodiesterase.

Phosphodiesterases appear to associate with the membrane environment by three distinct mechanisms: they are either transmembrane proteins, attach to the membrane via a short peptide stalk or are peripheral proteins. Removal of such membrane-bound proteins can be achieved by either detergent solubilization, hypotonic-shock treatment or high-ionic strength extraction (usually achieved using 0.5 M sodium chloride).

5.4.3 Purification of cyclic nucleotide phosphodiesterase

The purification of cyclic nucleotide phosphodiesterase can be achieved by manipulating the conditions to take advantage of the kinetic, physical, and immunological properties of each individual isoenzymic forms.

The purification methods developed have involved the use of several chromatographic procedures. These include:

- anionic-exchange chromatography, e.g. DEAE-52, Mono-Q (on FPLC system)
- hydrophobic chromatography, e.g. amino-pentyl agarose
- affinity chromatography, e.g. cGMP–Sepharose, cilostamide- agarose, cal-modulin–Sepharose
- dye-ligand chromatography, e.g. Affi-Gel blue agarose
- monoclonal antibodies linked to Ultra-Gel.

Using these techniques, most of the distinct isoenzymic forms have been successfully purified to apparent homogeniety. These include the cytosolic and particulate forms of the cGMP-activated cAMP phosphodiesterase (41), the calmodulin-stimulated cyclic nucleotide phosphodiesterase (47, 40), the cAMP-specific phosphodiesterase, which includes the Ro-20-1724 inhibited form (39, 38, 42), and the cGMP-inhibited enzyme (43, 42).

As an example of a purification scheme, the protocol used for the purification of the peripheral insulin-activated cAMP phosphodiesterase is represented diagrammatically in *Figure 6*.

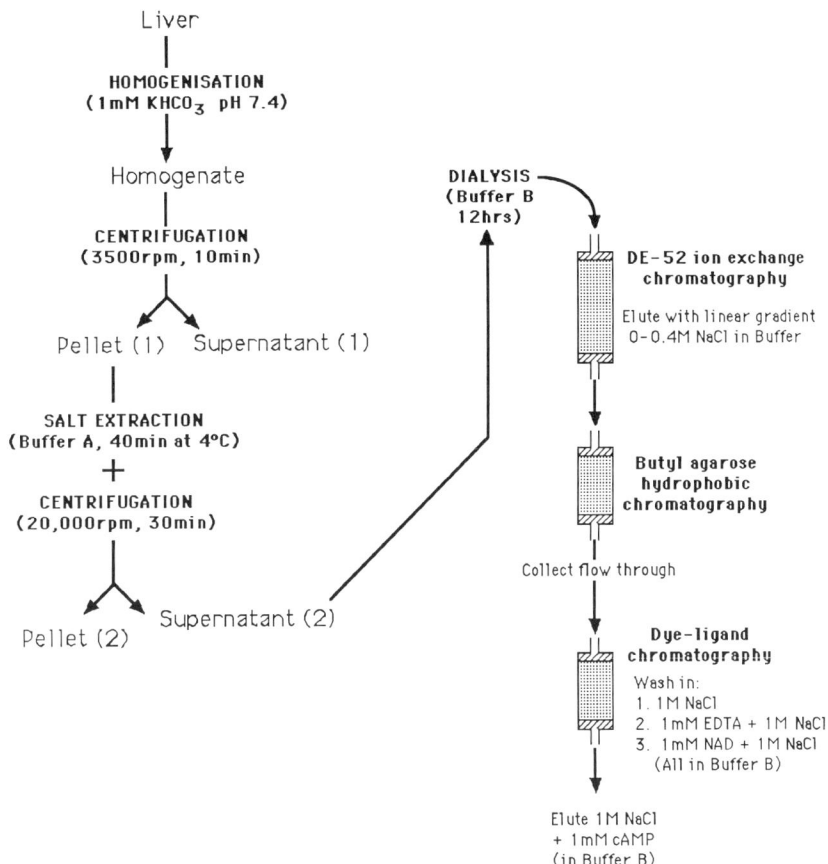

Figure 6. Purification scheme of rat liver plasma membrane cAMP phosphodiesterase. Chromatography steps are: (a) Diethylaminoethylcellulose-52; (b) Butyl agarose; (c) Affi-Gel blue agarose. Buffer A consists of 20 mM Tris–HCl, pH 7.4, 0.5 M NaCl, 5 mM $MgCl_2$ and 2 mM mercaptoethanol. Buffer B consists of 20 mM Tris–HCl, pH 7.4, 5 mM $MgCl_2$, 2 mM mercaptoethanol. This purification procedure allows a 2500-fold purification and a 15% yield of the phosphodiesterase.

5.5 Further techniques for investigation of phosphodiesterase

5.5.1 Inhibitor sensitivities

Most individual phosphodiesterases studied thus far exhibit distinct inhibitor sensitivities. The inhibitors used are usually derivatives of the substrates such as xanthine (e.g. 3-isobutyl-1-methylxanthine, a good non-specific inhibitor). They are often insoluble at high concentrations in normal buffering solutions and it is often necessary to firstly dissolve them either in water at pH 2 to 3 or in an organic solvent such as dimethylsulphoxide, ethanol or acetone prior to serial dilution into an appropriate buffer solution. The results obtained from these studies alalow one to use selective inhibitors to identify and assess the role of individual forms from crude extracts.

The markedly distinct inhibitor sensitivities for three purified rat liver phosphodiesterases is shown in *Table 4*.

Table 4. Phosphodiesterase inhibitors

IC_{50} values (μM)

Agent	P.M. PDE	D.V. PDE	cGS-PDE
Cyclic GMP	1000	2	N/A
Milrinone	7.3	1	1000
Amrinone	1000	71	1000
Ro 20-1724	7.2	50	210
ICI 118233	1000	3	1000
Carbazeran	9	4.5	43
IBMX	6	6	40

Inhibitor sensitivities of various purified cAMP phosphodiesterases. Assays were performed at 1 μm cAMP. IC50 is the concentration of inhibitor that elicits a 50% inhibition of phosphodiesterase activity. Purified enzyme preparation were obtained using purification protocols described previously (38, 41, 42). The 'P.M.' PDE is the plasma membrane, 'high-affinity' cyclic AMP phosphodiesterase; the 'D.V.' PDE is the 'dense-vesicle' high-affinity cAMP phosphodiesterase; the 'cGS-PDE' is the cyclic GMP-stimulated cAMP phosphodiesterase. Inhibitors were dissolved in dimethylsulphoxide such that the concentration of solvent in the assay was less than 0.5%, which has no effect upon the activity of these enzymes.

5.5.2 Identification of hormone-activated phosphodiesterase

One technique which can be used to identify and characterize hormonal activation of phosphodiesterase activity is to assay subcellular fractions isolated from cells pre-treated with submaximal concentration of hormone. Percoll fractionation of hepatocytes allows rapid isolation of the subcellular fractions. This method usually takes 2 h, and employs a gentle homogenization procedure such that latent protease and phosphatase activity is not released.

5.5.3 Immunological analysis

A definitive method used for the identification of novel forms of cyclic nucleotide phosphodiesterase has been the development of antisera that specifically identify the individual forms of phosphodiesterase (36). This has allowed rapid isolation of native active forms of these individual enzymes. This method obviates problems of differential extraction and proteolysis and the characterization of such isolated forms probably reflects more accurately their properties *in vivo*.

Antibodies can be covalently attached to Sepharose and used to immunopurify the phosphodiesterase and can also be used to isolate hormonally activated forms of the enzyme.

Protocol 9. Immunopurification of phosphodiesterase

1. Incubate 100 μl of antibody-linked Sepharose with 500 μl of protein extract (1 mg/ml) at 30°C.
2. Centrifuge sample for 2 min at 14 000*g*.
3. Wash Sepharose twice with Tris–HCl (20 mM, pH 7.4) plus 0.5 M NaCl.
4. Centrifuge samples for 2 min at 14 000*g* and resuspend pelleted Sepharose in appropriate buffer for assay of cAMP phosphodiesterase activity.

To date, the various antisera that have been generated to purified phospho-diesterases are: monoclonal antibodies to the cGMP-activated PDE, monoclonal antibodies to the calmodulin-stimulated PDE (40), monoclonal/polyclonal antibodies to the cGMP-inhibited cAMP PDE (43, 48), and polyclonal antibodies to the insulin-activated cAMP PDE (42).

5.5.4 Summary of phospodiesterase assay

The assay of cyclic nucleotide phosphodiesterase activity using tritiated cyclic nucleotide as the radioactive tracer has been described in this section. It is the most widely used method for assessing enzyme activity, and is both rapid (the entire assay can be performed within an hour) and highly sensitive, being useful for the assessment of both high- and low-affinity cyclic nucleotide phospho-diesterase activity.

This assay allows one to follow a particular phosphodiesterase activity through a purification scheme to apparent homogeneity. In turn, the purification of phosphodiesterases has allowed the development of highly selective inhibitors and specific antisera and it is the use of these agents which has confirmed the existence of multiple forms and which should allow an extensive study of the mechanisms through which these enzymes are regulated *in vivo*. One of the most exciting recent developments is the ability to isolate the distinct phosphodiesterases using antibody-coupled chromatography—a procedure which can obviate problems of differential extraction and proteolysis and which can allow isolation of hormone-induced covalently modified forms of the phosphodiesterase.

6. Measurement of cAMP-dependent protein kinase activity

6.1 Introduction

Many different protein kinases have been characterized from a variety of cells and tissues. They can be classified into the following groups: cyclic nucleotide-dependent kinases; calcium/calmodulin-dependent kinases; and cyclic nucleotide and calcium independent kinases. These enzymes catalyse the incorporation of phosphate into the seryl and/or threonyl residues of their target proteins. More recently, a distinct group of enzymes specific for tyrosyl-phosphorylation have also been described (reviewed in Hunter and Cooper (49)), and these include growth factor receptors and certain retroviral/oncogene products (see Chapter 7).

A detailed description of the purification and characterization of the cAMP-dependent protein kinase will not be attempted here (see Reimann & Beham, reference 50, for a review). Rather, this section will describe the most commonly-used assay methods for assay of cAMP-dependent protein kinase, obtained from column fractions or extracts of cells or tissues. In general, assay of cAMP-dependent protein kinases in crude homogenates can be complicated by the presence of ATPase activity and heat-stable protein kinase inhibitor protein. These problems are most easily overcome by using ion-exchange chromatography on DEAE–cellulose in order to remove these proteins and enrich the kinase activity. Also, cells and tissues are usually homogenized in the presence of phosphatase inhibitors (commonly 15–150 mM NaF, 10 mM glycerophosphate) and proteinase inhibitors (commonly 1 mM phenylmethylsulphonyl fluoride and 0.001% leupeptin).

These protein kinases, on activation by the appropriate cyclic nucleotide, catalyse the transfer of the γ-phosphoryl group of ATP to either a serine or threonine residue on the substrate protein. A convenient measure of their activity is to follow the incorporation of $[^{32}P]$ from $[\gamma\text{-}^{32}P]$ATP into a suitable substrate, followed by removal of excess ATP. The two most widely-used procedures to achieve this involve either protein precipitation or SDS-polyacrylamide gel electrophoresis (SDS-PAGE), by the method of Laemmli (24). In the first method, the protein is precipitated by trichloracetic acid, and absorbed either on to filter paper, or glass-fibre or nitrocellulose filters, followed by extensive washing. The amount of radioactivity incorporated in the protein may then be measured by scintillation counting. The assay protocol for both methods is described below.

6.2 Assay methodology

6.2.1 Assay incubation

The substrate is added to a mixture of 20 mM Tris–Cl, pH 7.4, 50 mM KCl, 10 mM β-mercaptoethanol and water. To this a 50 μl aliquot of the enzyme is added (this can either be a crude subfraction, partially purified enzyme, or a purified preparation of the kinase). The assay is initiated by combining this mixture with 10 mM MgCl, 0.1 mM ATP (0.5 μCi of $[\gamma^{-32}P]$ ATP). The total assay volume is usually 250 μl and the samples are incubated at 30°C for 30 min. The concentrations given above represent final concentrations.

6.2.2 Filter assay

After the incubation has been completed, a 100 μl aliquot of the assay sample is immediately spotted on to Whatman 3MM filter discs (2.5 cm, numbered in pencil), allowed to soak and then dropped into 10% TCA. Each Whatman disc is washed twice in 10% TCA (15 min each), then washed four times in 5% TCA (15 min each), and finally washed in ethanol (for a further 15 min). The individual discs are then dried and placed into scintillation fluid prior to liquid scintillation counting. The object of such an extensive washing procedure is to allow the complete removal of any excess labelled nucleotide, whilst unaltering the interaction of phosphorylated substrate with the Whatman disc.

6.2.3 SDS-PAGE assay

With the SDS-PAGE method the phosphoprotein can be conveniently resolved from lower-molecular-weight, contaminants, and the amount of $[^{32}P]$ incorporated estimated by autoradiography.

   Here, the assay is terminated after 30 min by addition of an equal volume of Laemmli buffer; after which it is boiled for 5 min and then subjected to SDS-polyacrylamide electrophoresis.

6.2.4 Choice of substrate

Although many proteins or peptides may be used as substrates, the most frequently preferred is a histone mixture, since it is readily available at low cost, stable, lacks endogenous protein kinase activities, allows a high level of ^{32}P-incorporation, and is easily precipitated.

   We use the following substrates to assess the kinase activity of a given enzyme preparation:

● Histone Type II A (Sigma P4020) or Type II AS (Sigma H7755).

● Casein (Sigma C4765), which can be obtained either partially dephosphorylated or hydrolysed.

● Protamine sulphate, grade X (Sigma P4020).

● Peptide substrates.

   With the exception of synthetic peptides, substrates listed are at a final concentration of 800 μg/ml in the incubation.

6.3 Assay of the inhibitor protein

The inhibitor protein can be assayed by making use of its specific interaction with, and inhibition of, the free catalytic unit of cAMP-dependent protein kinase.

Details of the purification, assay and characterization of the inhibitor may be found in the review of Whitehouse and Walsh (51). In summary, the inhibitor fraction to be assayed is diluted in 10 mM $MgCl_2$, pH 6.8, containing 15 mM β-mercaptoethanol and 0.5 mg/ml BSA, so that a 30 μl aliquot added to the kinase assay will give between 10 and 50% inhibition. The extent of inhibition observed will depend upon the concentration of inhibitor and of catalytic unit present; both of these must be carefully controlled.

Acknowledgements

The work described here was supported by MRC programme and British Heart Foundation grants to Professor M. D. Houslay, and Asthma Research Council, Scottish Home and Health, and Royal Society grants to N. J. Pyne. We thank Professor M. D. Houslay for his support and comments during this work.

References

1. Cornell, N. W. (1980). *Anal. Biochem.*, **102**, 326.
2. Khym, J. X. (1975). *Clin. Chem.*, **21**, 1245.
3. Sharps, E. S. and McCarl, R. L. (1982). *Anal. Biochem.*, **124**, 421.
4. Steiner, A. L., Wehmann, R. E., Parker, C. W., and Kipnis, D. M. (1972). *Adv. Cyc. Nucleotide Res.*, **2**, 51.
5. Gilman, A. G. (1970). *Proc. Natl. Acad. Sci. USA*, **67**, 305.
6. Brown, B. L., Ekins, R. P., and Albano, J. D. M. (1972). *Adv. Cyc. Nucleotide Res.*, **2**, 25.
7. Kuo, J. F. and Greengard, P. (1970). *J. Biol. Chem.*, **245**, 2493.
8. Rubin, C. S., Erlichman, J., and Rosen, O. M. (1974). *Meth. Enzymol.*, **38**, 308.
9. Birnbaumer, L., Codina, J., Mattera, R., Cerione, R. A., Hildebrandt, J. D., Sunyer, T., Rojas, F. J., Caron, M. J., Lefkowitz, R. J., and Iyengar, R. (1985). *Mol. Aspects Cell Regul.*, **4**, 131.
10. Gilman, A. G. (1987). *Ann. Rev. Biochem.*, **56**, 615.
11. Krishna, G., Weiss, B., and Brodie, B. B. (1968). *J. Pharmacol. Exp. Ther.*, **163**, 379.
12. White, A. A. and Zenser, T. V. (1971). *Anal. Biochem.*, **41**, 372.
13. Ramanchandran, J. (1971). *Anal Biochem.*, **43**, 227.
14. Salomon, Y., Londos, C., and Rodbell, M. (1974). *Anal. Biochem.*, **58**, 541.
15. Hunt, N. H., Martin, T. J., Michelangeli, V. P., and Eisman, J. A. (1976). *J. Endocrinol.*, **69**, 401.
16. Mac Neil, S., Crawford, A., Amirrasooli, H., Johnson, S., Pollock, A., Ollis, C., and Tomlinson, S. (1980). *Biochem. J.*, **188**, 393.
17. Lakey, T., MacNeil, S., Humphries, H., Walker, S. W., Munro, D. S., and Tomlinson, S. (1985). *Biochem. J.*, **225**, 581.
18. Houslay, M. D. and Elliott, K. R. F. (1979). *FEBS Lett.*, **104**, 359.
19. Heyworth, C. M. and Houslay, M. D. (1983). *Biochem. J.*, **214**, 93.
20. Marchmont, R. J., Ayad, S., and Houslay, M. D. (1981). *Biochem. J.*, **195**, 645.
21. Johnson, G. L. and Bourne, H. F. (1977). *Biochem. Biophys. Res. Commun.*, **78**, 792.
22. Hildebrandt, J. D., Sekura, R. D., Codina, J., Iyengar, R., Manclark, C. R., and Birnbaumer, L. (1983). *Nature, Lond.*, **302**, 706.

23. Heyworth, C. M., Whetton, A. D., Wong, S., Martin, B. R., and Houslay, M. D. (1985). *Biochem. J.*, **228**, 593.
24. Laemmli, U. K. (1970). *Nature, Lond.*, **227**, 680.
25. Seamon, K. B. and Daly, J. W. (1986). *Adv. Cyc. Nucleotides and Prot. Phos. Res.*, **20**, 1–150.
26. Heyworth, C. M., Hanski, E., and Houslay, M. D. (1984). *Biochem. J.*, **222**, 189.
27. Gawler, D., Milligan, G., Spiegel, A. M., Unson, C. G., and Houslay, M. D. (1987). *Nature, Lond.*, **327**, 229.
28. Katada, T. and Ui, M. (1982). *J. Biol. Chem.*, **257**, 7210.
29. Lowry, O. H., Rosenburg, N. J., Farr, A. L., and Randall, R. J. (1951). *J. Biol. Chem.*, **193**, 265.
30. MacNeil, S. (1987). *Biochem. J.*, **242**, 607.
31. Waldman, S. A. and Murad, F. (1988). *Pharmacol. Rev.*, **39**, 163.
32. Amirrasooli, H. (1981). Studies of human platelet adenylate cyclase and guanylate cyclase activity. Thesis, University of Sheffield.
33. Karczewski, P. and Krause, E. G. (1978). *Acta Biol. Med. Ger.*, **37**, 961.
34. Butcher, R. W. and Sutherland, E. W. (1962). *J. Biol. Chem.*, **237**, 1244.
35. Thompson, W. J. and Appleman, M. M. (1971). *Biochemistry*, **10**, 311.
36. Beavo, J. A., Hanson, R. S., Harrison, S. A., Hurwitz, R. L., Martins, T. J., and Mumby, M. C. (1982). *Mol. Cell. Endocrinol.*, **28**, 387.
37. Rutten, W. J., School, B. M., and Dupont, J. S. H. (1973). *Biochim. Biophys. Acta*, **315**, 378.
38. Marchmont, R. J., Ayad, S. R., and Houslay, M. D. (1981). *Biochem. J.*, **195**, 645.
39. Thompson, W. J., Epstein, P. M., and Strada, S. J. (1979). *Biochemistry*, **18**, 5228.
40. Mumby, M. C., Martins, T. J., Chang, M. S., and Beavo, J. A. (1982). *J. Biol. Chem.*, **250**, 6320.
41. Pyne, N. J., Cooper, M. E., and Houslay, M. D. (1986). *Biochem. J.* **234**, 325.
42. Pyne, N. J., Cooper, M. E., and Houslay, M. D. (1987). *Biochem. J.*, **242**, 33.
43. Harrison, S. A., Chang, M. L., and Beavo, J. A. (1986). *Circulation*, **73**, 109.
44. Bitensky, M. W., Wheeler, M. A., Rasenick, M. M., Yamazaki, A., Stein, P. J., Halliday, K. R., and Wheeler, G. L. (1982). *Proc. Natl. Acad. Sci. USA*, **79**, 3408.
45. Tucker, M. M., Robinson, J. B., and Stellwagen, E. (1981). *J. Biol. Chem.*, **256**, 9051.
46. Price, B., Pyne, N. J., and Houslay, M. D. (1987). *Biochem. Pharmacol.*, **36**, 4047.
47. Morrill, M. E., Thompson, S. T., and Stellwagen, E. (1979). *J. Biol. Chem.*, **254**, 4371.
48. Pyne, N. J., Anderson, N., Lavan, B. E., Milligan, G., Nimmo, H. G., and Houslay, M. D. (1987). *Biochem. J.*, **248**, 897.
49. Hunter, T. and Cooper, J. A. (1985). *Ann. Rev. Biochem.*, **54**, 897.
50. Reimann, E. M. and Beham, R. A. (1983). *Methods in Enzymology* (ed. J. D. Corbin and J. G. Hardman), Vol. 99, pp. 51–4. Academic Press, London and New York.
51. Whitehouse, S. and Walsh, D. A. (1983). *Methods in Enzymology* (ed. J. D. Corbin and J. G. Hardman), Vol. 99, pp. 80–93. Academic Press, London and New York.

4

Cytoplasmic free calcium: measurement and manipulation in living cells

MAURICE B. HALLETT, ROBERT L. DORMER, and ANTHONY K. CAMPBELL

1. A role for cytosolic Ca^{2+} in cell activation and injury

The survival, specialized functions, and reproduction of eukaryotic animal and plant cells, and many prokaryotes, requires Ca^{2+} in the medium bathing the cells (1, 2, 3). In 1883, Sydney Ringer showed that the normal beating frog heart requires external Ca^{2+}. The problem facing cell physiologists over the century since these pioneering experiments is how can one distinguish a 'passive' structural role for Ca^{2+} in a particular cell type from an 'active' regulatory one? In the latter case it is a change in free Ca^{2+} at a particular site within the cell which is responsible for initiating a response, for example cell movement, secretion or division.

The phenomena we are dealing with therefore involve a change of state in the cell, initiated by a primary stimulus and capable of modification by a secondary regulator. The primary stimulus may be physical (for example, an action potential releasing neurotransmitter at a nerve terminal), or chemical (for example, a hormone activating intermediary metabolism). Many of these phenomena involve individual cells crossing thresholds at different times and at different concentrations of stimulus. The free Ca^{2+} in the cytosol varies in a healthy cell from about 0.1 μM (resting) to 1–5 μM (activated). At higher concentrations Ca^{2+} can damage macromolecules and organelles, sometimes irreversibly, although mild rises in cytosolic Ca^{2+} can help cells protect themselves from attack (4, 5).

Thus the central question concerning Ca^{2+} in cell physiology or pathology is how to demonstrate definitively that a change in Ca^{2+} within the cell is responsible for initiating a physiological event in response to a stimulus such as a hormone. Additionally, in the case of cell injury, one can ask whether a rise in intracellular Ca^{2+} is a cause or consequence of cell injury, and whether intracellular Ca^{2+} can protect against attack or even cell death.

Indirect experiments based on manipulation of extracellular Ca^{2+} or pharmacological substances of dubious specificity may provide some clues that Ca^{2+} is involved in the molecular mechanisms underlying cell activation or injury. However, three experimental strategies are necessary to establish whether a primary agonist or pathogen acts via intracellular Ca^{2+}, and whether a secondary regulator works by altering the Ca^{2+} transient or by a mechanism independent of Ca^{2+}.

(a) Direct measurement of free Ca^{2+} within the intact cell, and correlation of the time course of any Ca^{2+} change with the physiological or pathological event.

(b) Measurement of Ca^{2+} fluxes across the cell and organelle membranes to ascertain the source of any cytosolic Ca^{2+} rise, together with the mechanism of release into the cytosol, e.g. voltage sensitive or receptor-mediated channels and $InsP_3$ induced release.

(c) Manipulation of intracellular Ca^{2+}, coupled with measurement, to show that the Ca^{2+} change is *necessary* for the event to occur.

2. Principles of measurement of cytosolic Ca^{2+}

The first attempt to measure cytosolic Ca^{2+} was carried out in 1928 by Pollack, who injected the red dye alizarin sulphate into an amoeba. However, it was not until the late 1960s that the work of Ridgway and Ashley, and Blinks, established in giant cells that the bioluminescent protein, or photoprotein, aequorin from the jellyfish *Aequorea* could measure changes in free Ca^{2+} in living cells (1).

The ideal indicator of cytosolic free Ca^{2+} should satisfy eight criteria:

(a) Specificity.
Ca^{2+} should be the only ion generating a signal from the indicator, particularly in the presence of Mg^{2+}, K^+, and Na^+ (1–100 mM).

(b) Sensitivity.
To cover the entire range of free Ca^{2+} in resting, stimulated and injured cells, the indicator should be capable of quantifying free Ca^{2+} in the range of 10 nM–100 μM. Furthermore the signal from the indicator should be readily detectable over any background signal arising from the cell itself, or from the apparatus.

(c) Speed of response.
Ca^{2+} transients occurring within 10 msec, and lasting many minutes, should be detectable.

(d) Incorporation into cells.
It should be possible to incorporate the Ca^{2+} indicator into the cytoplasm of the living cell without significant disturbance to cell structure and function. Once there, it should ideally be stable for hours, or even days in a cell culture, and should be non-toxic.

(e) Effect on cell Ca^{2+} balance.
The indicator should not significantly disturb the Ca^{2+} balance of the cell.

(f) Diffusion.

The diffusion of the indicator should not be a factor affecting the signal from the Ca^{2+} transient.

(g) Distribution of free Ca^{2+} within the cell.

All cells exhibit some polarity. To establish Ca^{2+} as the intracellular mediator of a primary stimulus, the free Ca^{2+} increase must ultimately be localized within the cell, and this localization correlated with chemical and structural changes associated with cell activation or injury. The ideal indicator will permit investigation of changes in Ca^{2+} in different regions of the cell.

(h) Availability.

The Ca^{2+} indicator should be readily available and inexpensive.

Three groups of indicators satisfy several, if not all, of these criteria and have been used inside cells.

(a) Ca^{2+}-activated photoproteins (aequorin, obelin)

(b) Indicator dyes.

  ● spectrophotometric (murexide, antipyrl azo III, arsenazo III

  ● fluorescent (quin2, fura-2, indo-1)

  ● nuclear magnetic resonance (FBAPTA)

(c) Ca^{2+} sensitive microelectrodes.

Two have found wide application in both small and large cells from the plant and animal kingdom, the Ca^{2+}-activated photoproteins discovered by Shimomura and colleagues in 1962 and the fluorescent indicators developed by Tsien.

3 Fluorescent chelator indicators

There are three main groups of fluorescent chelators, all introduced by Tsein (6, 7, 8) which have revolutionized the study of cytoplasmic free Ca^{2+} in small mammalian cells. These indicators display changes in their spectral characteristics on binding Ca^{2+} (see *Table 1*). The first indicator, quin2, is relatively weakly fluorescent and displays significant fluorescence only in the Ca^{2+}-bound form. This presented a major problem since it was difficult to definitively attribute fluorescence intensity changes to the indicator and not to autofluorescence, light scatter etc. The second group of indicators, fura-2 and indo-1, are more strongly fluorescent and display significant fluorescence in both the Ca^{2+} bound and free forms. This enables dual wavelength investigations to be carried out which elimiantes the possible misinterpretation of artefactual fluorescence changes. (See section 3.3.2.) The ratio of intensities of fura-2 and indo-1 at two wavelengths is independent of the absolute concentration (or amount) of the indicator. This therefore enables Ca^{2+} to be measured within individual cells where the thickness of the cell (and hence the amount of indicator) may not be constant. The third group of indicators, fluo-3 and rhod-2, are excitable at longer

Table 1. Ca^{2+} chelators useful in cell biology

	K_d* (nM)	Mode	Wavelength commonly used	
			Excitation	Emission
Quin2	114	Single wavelength	340	490
Fura-2	224	Dual excitation	340/380	505
Indo-1	250	Dual emission	340	410/485
Fluo-3	450	Single wavelength	506	526
Rhod-2	1000	Single wavelength	553	576
BAPTA	100	Non-fluorescent chelator	—	—
Dimethyl BAPTA	40	Non-fluorescent chelator	—	—
Nitr-5	145→6000	'Caged Ca^{2+}'	—	—
5-^{19}F-BAPTA	700	^{19}F NMR Probe	—	—
4-^{19}F-BAPTA	2500	^{19}F NMR Probe	—	—

* K_d in approximately physiological conditions, 37°C.

Note: all these chelators can be introduced in the cytoplasm of cell populations as their membrane permeant acetoxymethyl esters.

wavelengths and so many have less interference from autofluorescence. All these indicators are commercially available quin2, fura-2, and indo-1 from a range of sources including Molecular Probes, Calbiochem and Boehringer Manheim, and fluo-3 and rhod-1 from Molecular Probes.

3.1 Incorporation of indicators into cells

As with other Ca^{2+} indicators, it is important that the fluorescent chelators are introduced into the cytosol without subsequent leakage into the extracellular medium or into intracellular organelles. In large cells such as the sea-urchin egg or barnacle muscle direct micro-injection of the indicator is possible. Fura-2 has also been introduced into the smaller mast cell ($d = 10\ \mu m$) via a patch pipette. The major advantage of the fluorescent chelator indicators however is their ability to be 'loaded' into populations of small cells without the need for micro-injection. This can be achieved by using the acetoxymethyl esters of the indicators, which are sufficiently hydrophobic to cross biological membranes (*Protocol 1*). In the presence of cellular esterases, the ester bonds are cleaved and the hydrophilic Ca^{2+}-indicator molecule is released and entrapped within the cell. The efficiency of loading by this mechanism depends upon the conditions used but often more than 50% of the ester added to the cell suspension is converted to the acid in the cytoplasm.

The major potential problem with this approach arises from its dependence on the presence and location of cellular esterases. In some cells, there are significant amounts of esterase in cytoplasmic granules, e.g. in mast cells and neutrophils, which may lead to loading of the indicator into this or other non-cytoplasmic compartments (9, 10). With the lower intracellular concentrations of fura-2 needed for detection of the signal, the conversion of the ester to the acid may be sufficiently rapid in the cytosol to prevent significant accumulation into intracellular organelles. Another problem may be encountered associated with the hydrophobicity of the esters, which must be dissolved initially in DMSO. Dilution of the stock solution in an aqueous medium can lead to the formation of microcrystals which may be endocytosed by the cells. Addition of the dispersing agent, Pluronic F127 (Molecular Probes) to the stock (see *Table 2*) and inclusion of protein in the medium reduces the formation of precipitate. This is particularly important with monolayers, where there is a tendency for precipitate to settle on the cells and consequently become endocytosed.

Protocol 1. Loading fluorescent Ca^{2+} indicators using their acetoxymethyl esters. Steps indicated by asterisk (*) may be omitted

1. Dissolve ester in dry DMSO to give stock (1–5 mg/ml). Store aliquots at $-20°C$.

2*. Mix 2.5 µl Pluronic F127 (25% w/v in DMSO) with 5 µl ester stock.

3. Add 1 µl of ester solution (1–5 mg/ml) to 1 ml cell suspension† (approx. 10^6 -5×10^7 cells/ml) in Ca^{2+}-containing medium with protein (e.g. 0.1% BSA).

4. Incubate cells 20–60 min (either at 0°C or room temperature).

5. Resuspend cells in fresh medium.

6*. Store cells at 0°C to reduce fura-2 leakage rate.

7. Record excitation or emission fluorescence spectrum (quartz cuvettes). Check peak fluorescence (e.g. near 360 nm emission at 505 nm excitation for fura-2) is significantly higher than non-loaded cells, and spectrum is unimodal.

8. Treat cells with either digitonin (150 µM) or non-fluorescent Ca^{2+}-ionophore (e.g. ionomycin or Br-A23187). Check spectral change consistent with indicator (e.g. peak at 340 nm emission 505 nm excitation for fura-2).

9. Add EGTA (50 mM). Check spectral change consistent with indicator (e.g. peak approx 350 nm at 505 nm excitation for fura-2 with spectra crossing at 360 nm).

10. Observe with fluorescence microscope using appropriate filters (e.g. Zeiss LP420/G365) to check that the fluorescence is evenly distributed throughout the cell population and within individual cells.

11*. Fractionate cells to provide evidence for cytoplasmic location of indicator.

† Cell monolayers can be loaded by adding the indicator stock to the minimum volume of medium required to cover the cells. The cell monolayer may be held in the fluorimeter beam (e.g. by clamping a cover-slip on which the monolayer is grown at 45–60 degrees to the incident beam within the sample cuvette). If this is not possible, treating the monolayer with digitonin (150 µM) may release sufficient fluorophore to enable it to be detected in a standard cuvette. Its Ca^{2+} sensitivity can then be determined by recording spectra before and after the addition of EGTA as described above.

Endocytosis can be prevented by low temeprature and loading is often performed at 4°C. This low temperature also reduces pinocytosis and thus the possibility of cell-associated fluorescence originating from unconverted ester within pinosomes. It is therefore important for each different cell type to establish optimal conditions for loading, and to demonstrate the cellular location of the indicator. Cellular location can be determined by fluorescence microscopic examination, preferably at a Ca^{2+}-independent wavelength (e.g. around 360 nm excitation for fura-2) and by subcellular fractionation of the cells. Attempts have also been made to identify a cytosolic location for the indicator by determining the steady-state anisotropy of polarized fluorescence (11).

A second problem which has been reported for fura-2 is incomplete de-esterification leading to the production of an entrapped fluorescent molecule insensitive to Ca^{2+} in the physiological range (9, 12, 13). This problem was related to the batch of fura-2-AM and may be due to a contaminant in the commercially supplied ester (12). Problems of this kind will be revealed by observing the excitation spectrum of loaded cells and by comparing the spectral properties (see *Figure 1*, and *Protocol 1*) of the product released from lysed cells

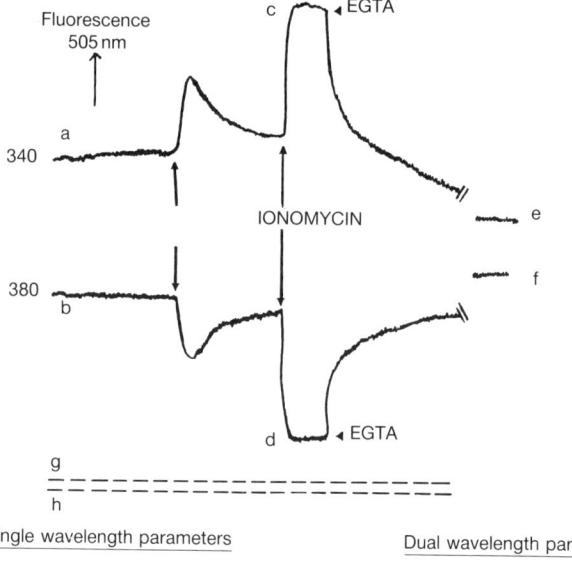

Single wavelength parameters

$F = a$

$F_{max} = c$

$F_{min} = e$

$Ca^{2+} = K_d (F - F_{min})/(F_{max} - F)$

Dual wavelength parameters

$R = (a - g)/(b - h)$

$R_{max} = (c - g)/(d - h)$

$R_{min} = (e - g)/(f - h)$

$\beta = (f - h)/(d - h)$

$Ca^{2+} = K_d \beta (R - R_{min})/(R_{max} - R)$

Figure 1. The measurement of parameters for Ca^{2+} measurement by fluorescent chelators. The example show the fura-2 response from rat neutrophils to the peptide f-met-leu-phe (see reference 11). The autofluorescence signals g.h. are from non-loaded cells. These values could also be obtained by the addition of Mn^{2+} (see Section 3.2).

with the standard fura-2 acid. If not too large, a correction can be made for the presence of Ca^{2+}-insensitive fluorescent material within the cells (see Section 3.3.1).

Hydrolysis of each ester bond during loading will liberate one formaldehyde, one acetate molecule and two hydrogen ions. It is desirable to resuspend the cells in fresh medium after loading in order to remove these possibly harmful by-products from the medium. It is also important to determine the effect of loading on cell viability and function. We (and others) consistently find a 50% decrease in ATP content in neutrophils although this seems to have little effect on the responsiveness of these cells.

3.1.1 Leakage of cytoplasmic indicator

Fluorescent chelators may not be permanently entrapped within the cytoplasm but may leak out. We have found in neutrophils that at $0°C$ this leakage is practically absent, but at $37°C$ is detectable over hours. In experiments lasting minutes, however, it is rarely necessary to correct for leakage. If possible, loaded cells should be kept at $0°C$ before use, but if not leaked indicator should be removed by resuspension in fresh medium. Extracellular indicator will obviously report the high Ca^{2+} there and so distort the intracellular signal (see Section 3.3.4). A simple test for the presence of leaked indicator is the addition of extracellular EGTA or Mn^{2+}. Extracellular indicator will respond immediately to the change in Ca^{2+} or to the quenching agent Mn^{2+}, whereas intracellular indicator will not.

3.2 Calculation of cytoplasmic Ca^{2+} from fluorescent signals

From knowledge of the dissociation constant for the Ca^{2+}-chelator complex the free Ca^{2+} concentration can be calculated from the measurement of fluorescence intensity. In practice the autofluorescence background from the cells and the maximum possible signal must also be defined before such a calculation is possible. With single wavelength measurement, the maximum fluorescence F_{max} achieved by saturating the indicator with Ca^{2+} may be determined by addition of the ionophore ionomycin or by cell lysis. Subsequent addition of a Ca^{2+} chelator such as EGTA produces the signal from Ca^{2+}-free dye, F_{min} (*Figure 1*). From consideration of law of mass action, it can be shown that

$$Ca^{2+} = K_d(F - F_{min})/(F_{max} - F). \qquad \text{[eqn 1]}$$

F_{min} can alternatively be calculated from knowledge of the fold increase (f) in fluorescence upon maximum Ca^{2+} binding relative to the Ca^{2+} free indicator, f. Using Mn^{2+} to quench the indicator fluorescence (F_q), F_{min} can thus be calculated.

$$F_{min} = F_q + 1/f(F_{max} - F_q) \qquad \text{[eqn 2]}$$

(for quin2 at 340 nm f is approximately 6).

For dual wavelength measurements, the situation is slightly more complex. For example, with fura-2 the signals at 340 nm and 380 nm defined as F_1 and F_2 respectively will be given by

$$F_1 = S_{f_1} \times c_f + S_{b_1} \times c_b \qquad \text{[eqn 3]}$$

$$F_2 = S_{f_2} \times c_f + S_{b_2} \times c_b \qquad \text{[eqn 4]}$$

where c_f and c_b are the concentrations of free and Ca^{2+} bound indicator and S is the constant of proportionality with the suffixes f and b being free and Ca^{2+} bound, and 1 and 2 being 340 and 380 nm respectively. The ratio of these two signals (R) can be used to calculate the free Ca^{2+} from the equation

$$Ca^{2+} = K_d\beta(R - R_{min})/R_{max} - R) \qquad \text{[eqn 5]}$$

where as before the subscripts min and max refer to the signal with minimum and maximum Ca^{2+} saturation and $\beta = S_{f_2}/S_{b_2}$ (*Figure 1*). The coefficient β is determined experimentally as $F_{2,min}/F_{2,max}$. In equation 5 it is important to note that $R = (F_1 - F_{1,auto})/F_2 - F_{2,auto})$, where F_{auto} is the autofluorescence from the cells. Accurate determination of the autofluorescence is therefore of prime importance (see Section 3.3.1). The use of Mn^{2+} as outlined above can provide an estimate of the autofluorescence (where autofluorescence $= F_q$) after determination of R_{min}. An alternative approach is to calculate the autofluorescence from the F_1 and F_2 maximum and minimum values as described later (Section 3.3.1).

However, if the signal is small relative to the autofluorescence, the subtraction of the large autofluorescence signal can result in gross errors in calculation of R. For this reason the fluorescence signal should be as large as possible without introducing significant Ca^{2+} buffering effects (see Section 3.3.3).

An important parameter in calculating the free Ca^{2+} is obviously the dissociation constant K_d. The values often used are based on Tsein's determinations (see *Table 1*). However, the K_d is sensitive to temperature and ionic strength as well as viscosity (14). In practice calibration of the signals to define K_d or $K_d \times \beta$ for a particular cell system is recommended. For fura-2 the experimentally determined values for the K_d at 37°C range from 168 nM to 998 nM, although most reports give a K_d of around 200 nM. It is suggested that the indicator generated in the cell should be released into the medium and the relationship between fluorescence and Ca^{2+} be at least partially characterized using a high, low, and intermediary (around the expected K_d) Ca^{2+} concentration to check each batch of indicator.

Another relationship which may sometimes be useful is the approximation, which holds good over the Ca^{2+} range 100–1000 nM, that the fluorescence intensity (proportional to fractional saturation) is linearly related to log (Ca^{2+}). The change in saturation is therefore proportional to log x, where x is the fold increase in Ca^{2+} concentration.

3.3 Problems in interpreting fluorescent signals

The principles for calculating the cytosolic free Ca^{2+} concentration outlined in Section 3.2 may be subject to problems arising particularly from the autofluorescence of the cells. This is a problem which must always be borne in mind and is particularly relevant at shorter wavelengths (e.g. 340 nm). The introduction of fluorophore in sufficient concentration to give a good signal to background ratio is often the only practical solution to this problem. However, even in this situation there is also the possibility for radiationless energy-transfer occurring between an endogenous molecule and the indicator. A further problem is raised by the possibility that incomplete de-esterification of the indicator can lead to additional background fluorescence. In this section an attempt is made to introduce analysis of dual wavelength fluorescence data as a means of calculating and correcting for the autofluorescence and other Ca^{2+}-sensitive fluorescence within indicator-loaded cells. The calculations are simple but would be laborious if the fluorescence data is not stored in digital form for computer calculations.

3.3.1 Calcium-insensitive fluorescence and autofluorescence

The presence of constant autofluorescence from the cells, probably due to NADPH and flavin nucleotides, presents no problems if a single wavelength method is employed for Ca^{2+} monitoring. However, it must be subtracted from the fluorescent signals before the ratio of signals produced at two wavelengths can be calculated. One approach often adopted is to measure autofluorescence on a non-loaded population and use the value to correct the signals from the loaded cells. Potential problems exist in performing this type of correction. For example, the value obtained may be different from that in the loaded cell population, e.g. caused by chelator loading reducing ATP content. One solution to this problem is the use of Mn^{2+} to quench the indicator fluorescence and leave only autofluorescence (see Section 3.2). The problem can also be resolved by calculating the Ca^{2+} insensitive fluorescence intensity from the signals produced in the fura-2 loaded cells. This latter approach also overcomes the potential problem of partial hydrolysis of fura-2 AM (see Section 3.1) to produce a Ca^{2+} insensitive but fluorescent product (which may also be quenched by Mn^{2+}). The method employed for calculating the magnitude of the Ca^{2+}-insensitive fluorescence and autofluorescence depends upon determining a parameter $\alpha = S_0/S_{Ca}$, where S is the proportionality constant as before, the subscript 0 or Ca referring to unsaturated and Ca^{2+}-saturated signal. The parameter of α is equal to the ratio F_0/F_{Ca} where the $F =$ fluorescence of fura-2 in the absence (0) or presence (Ca) of saturating Ca^{2+}.

The signal at 340 (or 380) at any Ca^{2+} conc. is given by

$$F = S_{Ca}(c_b) + S_0(c_f) + F_I \qquad \text{[eqn 6]}$$

where S_{Ca}, S_0, c_b, and c_f have been previously defined, and F_I is the fluorescence insensitive to Ca^{2+} plus autofluorescence. At high Ca^{2+},

$$F_{Ca} = S_{Ca}C_T + F_I, \qquad \text{(eqn 7]}$$

and at zero Ca^{2+}

$$F_0 = S_0C_T + F_I, \qquad \text{(eqn 8]}$$

from equation 7 $C_T = \dfrac{F_{Ca} - F_1}{S_{Ca}}$, and by substitution into equation 8

$$F_0 = \frac{S_0}{S_{Ca}}(F_{Ca} - F_1) + F_1, \qquad \text{(eqn 9]}$$

or

$$F_1 = \frac{F_{Ca} - F_0}{\alpha - 1}. \qquad \text{[eqn 10]}$$

At the end of the experiment, addition of ionomycin to induce fura-2 Ca^{2+}-saturation (F_{Ca}) and EGTA to produce free fura-2 (F_0), the autofluorescence plus other Ca^{2+} insensitive fluorescence can thus be calculated from equation 10. This value is subsequently subtracted from all fluorescence values before calcualtion of the ratio R. It should be noted that equation 10 is applicable to all wavelengths which are Ca^{2+} sensitive. At Ca^{2+} insensitive wavelengths such as the isobestic point $\alpha = 1$ and the equation fails. At 380 nm the value of α is given by β in equation 5.

3.3.2 Changes in Ca^{2+} insensitive fluorescence during experiment

The method above provides a means for establishing the Ca^{2+}-insensitive fluorescence at the end of the experiment. Nevertheless, the possibility may also exist that the autofluorescence changes during cell stimulation. Such an event may be difficult to detect by subjective inspection of the signals, since Ca^{2+} may also be changing at the same time. With single wavelength measurements this problem is insoluble. However, with dual wavelength measurements, this possibility can be eliminated. One strategy is to simultaneously monitor the emission by the excitation at the isobestic point. However, in order to measure Ca^{2+} in the dual wavelength manner, three wavelengths must be examined. Furthermore, an assumption must be made that changes at the Ca^{2+}-insensitive wavelength (e.g. 360 nm excitation) occur equally at the Ca^{2+} sensitive wavelengths (e.g. 340 nm and 380 nm). An alternative approach is to use the information contained within the 340 nm and 380 nm signals. Since the amount of fura-2 under observation is fixed, a change in Ca^{2+} will merely shift the 340 nm and 380 nm signals by amounts which, when the differences in the constants of proportionality are taken into account, will be oppositely balanced. This balanced signal can be calculated and provide a signal which is independent of Ca^{2+}. From equations 3, and 4 by subtracting $c_f = C_T - c_b$ where $C_T = $ the total concentration of fura-2.

$$c_b = (S_{f_1} \times C_T - F_1)/(S_{f_1} - S_{b_1}) \qquad \text{[eqn 11]}$$

and

$$F_2 = S_{f_2}C_T - c_b(S_{f_2} - S_{b_2}). \qquad \text{[eqn 12]}$$

By substituting equation 11 into equation 12 and defining a coefficient

$$A = (S_{f_2} - S_{b_2})/(S_{b_1} - S_{f_1})$$

it can be shown that

$$AF_1 + F_2 = C_T(S_{f_2} + A \times S_{f_1}),$$ [eqn 13]

i.e. $AF_1 + F_2$ will be proportional to C_T at all times, provided the constants do not change during the experiment and provided that other non-fura-2 or Ca^{2+} insensitive fluorescence does not change. The coefficient A can be calculated from standard calibrations and also at the end of the experiment from the ionomycin and EGTA procedure. This method therefore also provides a check for changes in autofluorescence during a Ca^{2+} change. Furthermore a check is also made that the constants of proportionality which are crucial for the quantitative interpretation of the fluorescence signals (see equations 3 and 4) remain constant throughout the course of a Ca^{2+} transient.

3.3.3 Distortion of Ca^{2+} change due to buffering by indicator

Ca^{2+} buffering is an important event which must inevitably accompany attempts to measure cytoplasmic Ca^{2+} by a Ca^{2+} binding indicator. It is important that the buffering effect is kept to a minimum while also providing sufficient specific signal for the correction due to autofluorescence to be small. In practice, we have found that in rat neutrophils it is difficult to use an intracellular fura-2 concentration of less than 25 μM, as the correction required becomes too high. It is also important to quantify the effect of the indicator in the Ca^{2+} buffering, capacity of the cell. The buffering capacity of the indicator is defined as L/K_d, where L is the concentration of free indicator (15). In order to determine L, the total concentration of indicator (L_T) in the cytoplasm must be determined. If all the fura-2 is cytoplasmic (see Section 3.1) this is simply a matter of dividing the total indicator taken up by the cells by the total cytoplasmic volume. It may however be necessary to construct a calibration curve to convert fluorescence intensity (e.g. at the isobestic point) into molar quantity of fluorescent indicator. The concentration of free ligand can then be calculated from the measured free Ca^{2+} concentration using the equation

$$L = L_T[1 - 1/(k_d/Ca + 1)].$$ [eqn 14]

The buffering capacity can then be calculated. At 100 μM total fura-2, ($K_d = 224$ nM), in the resting cell ($Ca^{2+} = 100$ nM), the buffering capacity of the indicator will be approx 300:1. (The buffering effect will lessen as the Ca^{2+} concentration rises.) The 'fast' Ca^{2+} buffering capacity of some cells is about this same order of magnitude, although the 'slow' buffering capacity (in seconds time-scale) in axons and in neutrophils is about tenfold higher (11, 15). This means that in a slow Ca^{2+} transient, 10 times as much Ca^{2+} will be bound to cytoplasmic buffers as to the indicator and the additional buffering effect of the indicator will be minimal. In a situation where a stimulus releases a limited pool of Ca^{2+} the effect of increasing the intracellular concentration of the indicator on the measured free Ca^{2+} rise can be used to estimate the Ca^{2+} buffering capacity of the cell. Increasing the intracellular concentration of indicator by a factor I_f will increase the total Ca^{2+} buffering capacity of the cells by a factor C_f (measured by the decrease in the free Ca^{2+} from the pool) then

$$I_f \times C_i + C_c = C_f(C_c + C_i) \qquad \text{[eqn 15]}$$

where C is the Ca^{2+} buffer capacity and the subscript i that for the indicator and c that for the cell. It can be shown therefore that the buffer capacity of the cell is given by

$$C_c = C_i(I_f - C_f)/C_f - 1). \qquad \text{[eqn 16]}$$

In a cell with cytoplasmic volume 200 fl and buffering capacity of 3000 : 1, a $1\mu M$ Ca^{2+} rise will result from 0.6 fmol Ca^{2+}/cell released from a limited cell store. In the presence of the Ca^{2+} chelating indicator this same amount of Ca^{2+} may result in a lower free Ca^{2+} concentration. However, an unlimited source of Ca^{2+}, e.g. an extracellular source, will provide sufficient Ca^{2+} to bind to the chelator and ultimately may provide the same free Ca^{2+} concentration. In this case, the presence of the chelator will slow the approach to the maximum free Ca^{2+} level. (It is also this slow influx of Ca^{2+} from the medium into the cell, during loading with the chelator, which compensates for the presence of the chelator.) This stresses the importance of extracellular Ca^{2+} for maintaining the normal Ca^{2+} homeostasis in the cells during the loading step (see Section 7.2).

3.3.4 Effect of heterogeneity in population response

When measuring a signal from a population, the possibility that individual cells do not behave uniformly must always be considered (1, 17).

The simplest situation to consider is the existence of two populations. If the Ca^{2+} concentration in a fraction of the cells (f) is x-fold higher than the remainder, then from the logarithmic approximation (see Section 3.2), it can be shown that the apparent Ca^{2+}-fold change in the population will be x^f. Since the arithmetic mean fold increase in Ca^{2+} concentration in the population will be $1 - f + fx$. It can be seen that the actual local and mean Ca^{2+} changes must be underestimated. This underestimating effect, however, reduces the bias caused by fura-2 in the extracellular medium or in a subpopulation of cells totally permeable to extracellular medium. For example, if 10% of the fura-2 were extracellular (i.e. fully Ca^{2+} saturated), and the remainder were within intact cells at 100 nM (31% saturated), the apparent saturation of the indicator would be approx. 38%. The calculated intracellular Ca^{2+} of the population will therefore only be 136 nM (whereas the population mean is 0.1 mM). The presence of the extracellular fura-2 has therefore only produced a 36% overestimation of the true intracellular Ca^{2+} concentration.

In comparing cell population measurements made using fura-2 and Ca^{2+}-activated photoprotein, quantitative discrepancies will arise if the population does not respond homogeneously. Whereas fura-2 will underestimate the mean response, the photoprotein because of the power relationship will overestimate the mean Ca^{2+} concentration (see reference 17).

3.3.5 Interference from other ions

All the fluorescent chelator indicators bind other divalent cations, often more strongly than Ca^{2+}; for example, Mn^{2+}, Fe^{2+}, Zn^{2+}, Co^{2+}, and Ni^{2+}. The important physiological ions which may interfere with the signal are Mg^{2+} and Zn^{2+}. Quin 2 has selectivity for Ca^{2+} over Mg^{2+} of 10^4, approximately equivalent to the anticipated intracellular concentration difference between these two ions. It must therefore be assumed for quin2 measurement that Mg^{2+} remains constant at around millimolar. In contrast, the K_d for Mg^{2+} binding to fura-2 is around 10 mM, consequently reducing the possibility of interference.

Zn^{2+} binds to quin2 with high affinity, the K_d being about 26 pM (16) and to fura-2 with K_d of 1.6–2 nM (14). Little is established about intracellular Zn^{2+}, but from ^{19}FNMR measurements of 5-fluoro BAPTA in Erhich cells (see Section 5.3) the free Zn^{2+} concentration may be surprisingly high, about 1 nM. The possibility of interference from Zn^{2+} must therefore exist. Zn^{2+} quenches quin2 fluorescence and may thus lead to an underestimation of the Ca^{2+} saturation of the indicator. The reaction of Zn^{2+} with fura-2 is, however, similar to that of Ca^{2+}, causing similar spectral shifts. The use of TPEN (N,N,N',N'-tetrakis-(2pyridyl-methyl)-ethylenediamine), a lipid soluble chelator of Zn^{2+}, with low affinity for Ca^{2+} and Mg^{2+}, may be useful in reducing cytoplasmic Zn^{2+} concentrations, and hence minimizing interference with the Ca^{2+}-indicator.

3.4 Instrumentation

3.4.1 Cell populations

The measurement of fluorescence from cell suspensions usually requires a cuvette housing which can be maintained at 37°C and provides a stirred solution to keep the cells suspended and in the light path. With single wavelength measurement no special optical set up is necessary, although obviously with wavelengths around 340 nm quartz cuvettes will improve the signal strength. For dual emission studies, e.g. with indo-1, two monochromators and detectors set at right angles to the sample will monitor dual emission simultaneously. For dual excitation studies with fura-2, the light at two wavelengths must be alternatively 'flashed' into the cuvette. This can be achieved by a spinning filter wheel or an optical chopper intercepting the light output from two monochromators. Obviously, with dual wavelength measurements the signals at the two wavelengths must remain separate. This can be achieved using digital data acquisition and the appropriate software (see Section 3.4.3), which will also enable the cytosolic free Ca^{2+} concentration to be calculated from the new data.

With monolayers of cells, a cover-slip on which the cells are grown can be held in position at an angle in the light path. Obviously the smaller the angle of incidence the greater the area of the cover-slip that will be illuminated. Light scatter from the cover-slip can be reduced if angles other than 45°C are used. Perhaps the simplest approaches are to use a cap for a standard cuvette with a recess to allow the cover-slip to be held in the medium or to wedge the cover-slip diagonally across a cuvette.

3.4.2 Individual cells

Unlike photoproteins, fluorescent molecules can emit several thousand photons before eventual destruction (e.g. fluorescein is estimated to emit 10^4–10^5 photons/molecule). This property means in practice that the intensity of emitted fluorescence is proportional to the intensity of the excitation light. The laser used for flow cytometry, or a 50 W UV microscope lamp if focused on to a small area, will provide fluorescent signals from individual loaded cells which are easily detectable by standard photomultiplier tubes. However, as the concentration of indicator may vary between individual cells, only those methods which are independent of this parameter are applicable, i.e. dual wavelength measurements with fura-2 or indo-1.

Indo-1 has the most suitable spectral characteristics for flow cytometry, where a single laser is required for excitation and emission can be monitored at two wavelengths simultaneously using two detectors. Flow cytometry, however, provides a means of determining the population distribution of cytoplasmic Ca^{2+} but cannot follow an individual cell during stimulation and thus is of limited application.

Microfluorometry, on the other hand, offers more possibilities. The fluorescence from an individual cell can be monitored during stimulation and the Ca^{2+} concentration calculated by the method described in Section 3.2. Dual excitation is again possible using a spinning filter wheel or the optical chopper system with two monochromators. Dual emission would be possible by conducting the emitted light to two photomultiplier tubes. The 340/380 pair of excitation wavelength for fura-2 is difficult to use without quartz optical components in the microscope. However, sufficient signal at the shorter wavelength can be achieved by using 350 nm. In this case the 350:380 ratio is merely substituted for the 340:380 ratio in equation 5, the parameter β being dependent only upon the 380 signals. The cell can be held in position during microscope examination either by adherence to a solid surface or by the use of low temperature melting point agarose, enabling solution stimuli to be added to the cells and maintained at the physiological temperature using an appropriate microscope stage heater. The output from individual cells can thus be monitored one by one. An attractive alternative is digital imaging, which enables a field of cells to be analysed as individuals.

By coupling a video camera to an epi-fluorescence microscope, the digital image of a cell field can be captured. This enables pixel-by-pixel analysis of the dual excitation signal. The Ca^{2+} within individual cells in the field or within a single cell can therefore be quantified. Obviously, image processing hardware and software capable of digitizing and manipulating the paired images is required which can handle the data with appropriate speed. This is now becoming available commercially (see Section 3.4.3). A discussion of problems and solutions of digital imaging is given by Tsein and Poenie (18).

3.4.3 Commercially available instruments

Provided that warming and stirring of the sample chamber can be achieved, a number of standard commercially-available fluorimeters can be used for single wavelength measurements. For single excitation, dual emission measurement, a T-format arrangement is employed with the detectors at right angles to the incident excitation light. Commercial suppliers of this type of system include (i) Applied Photophysics, London; (ii) Spex, New Jersey, USA, and Glen Spectra, Stanmore, UK, (iii) S6M Instruments, Illinois, USA, and (iv) Photon Technology International Inc. (P.T.I.), New Jersey, USA. The latter four sources also produce systems with dual beam chopped excitation light, and digital photon counting detection suitable for dual excitation measurement and data manipulation. Ratio imaging (dual wavelength excitation) hardware and software is available as CM/IM Series from Spex, Imagescan from P.T.I. and also from Meridian, Michigan, USA, and Joyce-Loebl Ltd, Gateshead, UK. Videos cameras for imaging must have sufficient light sensitivity, and silicon intensified target cameras (e.g. Dage MTI) or image-intensified charge-coupled-device cameras (e.g. Photonic Science, Tunbridge) can be used.

4. Photoproteins

Luminous coelenterates (Cnidarian and Ctenophores), apart from anthozoans, and radiolarians contain proteins of molecular weight approximately 20 000 which emit blue light (max 440–490 nm) when they bind Ca^{2+}. The energy for light emission comes from a chemiluminescent reaction (*Figure 2*). Coelenterazine is the prosthetic group on these proteins and is an imidazololopyrazine,(2-*p*-hydroxybenzyl)-6-(*p*-hydroxyphenyl)-8 benzyl-3-7-dihydroimidazo[1, 2-a] pyr-

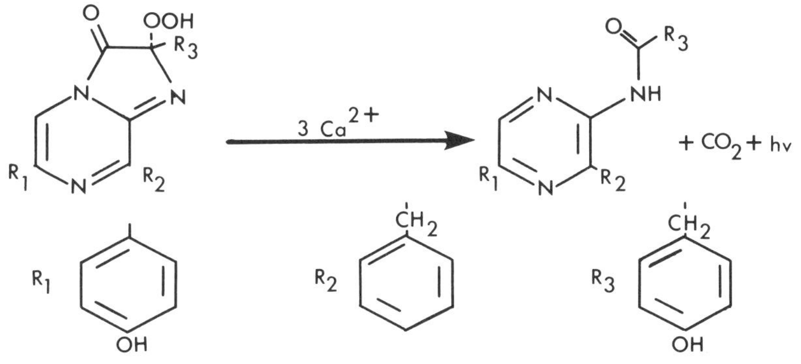

Figure 2. Reaction of Ca^{2+} ions with photoprotein.

azin-3-one) which when oxidized produces coelenteramide in an excited state which then emits light. The quantum yield on the protein is 0.1–0.2. At saturating Ca^{2+} the light emission decays exponentially with the constant of 1.4–$4\,s^{-1}$, depending on the photoprotein (19). As the Ca^{2+} concentration decreases the apparent rate constant decreases allowing free Ca^{2+} to be quantified over the range 10 nM–100 μM.

Coelenterazine is also the luciferin in luminous decapods, copepods, several squid and fish, but only in the coelenterates and radiolarians are Ca^{2+} activated photoproteins found. Here the prosthetic group and oxygen are firmly attached to the protein enabling the whole complex to be isolated (*Protocol 2*) and stored in a stable form. Once the chemiluminescent reaction has taken place, the Ca^{2+} has been removed, the reacted prosthetic group dissociates from the protein. The apoprotein can then be reactivated in the presence of coelenterazine and O_2 to form Ca^{2+}-activated photoprotein again (*Protocol 3*).

Protocol 2. Extraction and purification of photoproteins

1. Cut off outer rings from umbrella of *Aequorea* and shake out luminous cells, or scrape *Obelia* from *Laminaria* seaweed.

2. Rinse in 0.5 M NaCl (large volume) to remove Ca^{2+} in sea water.

3. Homogenize (polytron or blender) in 1 l of 50 mM EDTA and saturated $(NH_4)_2SO_4$ pH 6–6.5 for Aequorin or 1 l (2 : 1 w/v) 100 mM tris, 0.5 M NaCl, 50 mM EDTA pH 7 for obelin. Aequorin can be stored at -70°C, this crude Obelin cannot.
NB $(NH_4)_2SO_4$ can cause large losses in crude extracts.

4. Remove particulate material by muslin and filtration on a Buckner funnel, and centrifugation for 30 min at 35 000g.

5. Dilute obelin 1 : 10 and load on to DEAE–cellulose in a Buckner funnel. Wash 100 ml 5 mM EDTA/50 mM NaCl pH 7, then 100 ml of same medium $+0.5$ M NaCl to elute the obelin. Obelin can be stored in this form for months at -70°C or as a $(NH_4)_2SO_4$ precipitate.

6. Purify aequorin by:
  (a) Ammonium sulphate fractionation 25–75% cut.
  (b) Gell filtration on Sephadex G-50.
  (c) Ion-exchange on QAE-Sephadex A-50 (or Sepharose for faster column), equilibrated in 10 mM EDTA pH 5.5, elute green fluorescent protein with 5 mM EDTA, 5 mM Na acetate pH 4.75 and aequorin with a linear NaCl gradient, 500 ml 0–1 M NaCl in 10 mM EDTA pH 5.5).
  (d) Sephadex G-50 fine.
  (e) Ion-exchange chromatography on DEAE-Sephadex A50 with pH step and salt elution.

Protocol 2. *Continued*

7. Purify obelin by:

(a) $(NH_4)_2SO_4$ fractionation (50%-saturated cut removes pink protein).

(b) DEAE–cellulose (50 mM tris, 1 mM EDTA pH 7.5, 0–1 M NaCl, 500 ml gradient).
   This can be desalted and freeze-dried for use at this stage, though it is not pure.

(c) Further purify as for aequorin, step 6.

8. Desalt and concentrate using ultrafiltration against chelex-treated water or buffer to reduce EDTA concentration to $< 10\ \mu$M. Freeze dry in aliquots with Calex resin (polyacrylamide-parvalbumin from Sigma) ATP or chelex-100 beads, or 0.1% gelatin, to protect the photoprotein. Large losses can occur otherwise. Store dessicated at $-70°$C. Overall yield 20–30%.

Protocol 3. Preparation of apo photoprotein and reactivation

1. Add a few microlitres of stock photoprotein ($c.\ 10^4$–10^{10} luminescent counts) to 50 μl 10 mM Tris, 1 mM EDTA, 0.5 M NaCl, 5 mM mercaptoethanol or dithiothreitol, 0.1% gelatin pH 7.5 plus 50 μl 200 mM Tris, 0.5 mM EDTA pH 7.5. Assay for Ca^{2+}-activated photoprotein.

2. Add 4 μl 50 mM $CaCl_2$ in dark and observe the bright blue flash. Leave 10–15 min. at room temperature. Reassay photoprotein to show $> 99\%$ consumption.

3. Add 5 μl 200 mM EDTA pH 7.5 and remove coelenterazine on a 0.7×8 cm Sephadex G-25 (superfine) column. This gel filtration is not absolutely necessary. Apo photoprotein should be used fresh, freeze-dried, or can be stored at $-70°$C though $> 75\%$ loss sometimes occurs on freeze-drying or storage at $-70°$C.

4. Reactivate photoprotein by adding a few μl of apo photoprotein to 100 μl 10 mM Tris, 1 mM EDTA, 0.5 M NaCl, 5 mM mercaptoethanol, 0.1% gelatin pH 7.5, plus 1–5 μl 100 μM coelenterazine (TLC or HPLC pure) in methanol. $> 50\%$ reactivation within 1 h maximal by 3 h (require 24 h with impure coelenterazine). Coelenterazine (available from London Diagnostics, Minnesota, USA) is very unstable in solution and is light sensitive. Store dry in the dark.

4.1 Preparation of reagents

Two luminous hydrozoans have provided sufficient photoprotein for physiological experiments and mRNA for cloning: the jellyfish *Aequorea* (species *aequorea=forskalea=?victoria*) and the hydroid *Obelia geniculata*.

Aequorea medusae occur in large numbers in certain parts off the East and

West coasts of USA, and also in European waters, in the summer. *Obelia* is of world-wide distribution and occurs plentifully off the British coasts, particularly near Plymouth, growing on brown seaweeds, e.g. *Laminaria*. *Aequorea* medusae can be collected with a net, and a ring at the outer edge of the bell dissected, containing the luminous cells, which can be shaken out (20). Sufficient *Obelia* on *Laminaria* can be collected by scuba diver within an hour or so. The hydroids are scraped off and the photoprotein obelin extracted (*Protocol 2*). The photoprotein is then purified by ammonium sulphate fractionation and ion-exchange chromatography. The purified photoprotein can be stored for several months or years as a freeze-dried powder, at $-70°C$ or dessicated. 0.1% gelatin or chelex 100 beads are then required to prevent large losses of photoprotein during freeze-drying. Ca^{2+} contamination must be minimized throughout. In the latter stages when the EDTA concentration is reduced or zero, Ca^{2+} is removed using a chelex 100 column or acrylamide-parvalbumin column (*Protocol 4*) Photoprotein assay for quantifying yields is shown in *Protocol 5*.

Protocol 4. Preparation of Ca^{2+}-free medium by Chelex treatment

1. Suspend Chelex 100 (200–400 mesh Na form) in 0.1 M HCl, allow to settle and pour off supernatant.
2. Resuspend in 0.1 M NaOH, allow to settle and pour off supernatant.
3. Resuspend in 0.1 M HCl, neutralize with 0.1 M NaOH (do not dip pH electrode in suspension, but measure pH in samples of supernatant).
4. Pour column and elute with double distilled water until conductivity and pH of eluate is identical to starting material.
5. Elute buffer (10 mM Tris, pH 7.4). Check pH and conductivity and collect.

Protocol 5. Photoprotein assay

1. Add few microlitres of 1/100–1/1000 extract to 0.5 ml 200 mM Tris, 0.5 mM EDTA pH 7.5.
2. Place in front of a sensitive photomultiplier tube, e.g. in a commercial or home-built chemiluminometer (2).
3. Add 0.5 ml 80 mM $CaCl_2$ or $CaAc_2$ and record luminescent counts or analogue light signal for 5–10 sec (6 half-times = 98% total consumption).

This assay can detect down to 1–10 tipomol (10^{-21}–10^{-20} mol) of obelin.

For incorporation into cells two- to threefold higher yields within the cells can sometimes be achieved by starting with apo photoprotein prepared as described in *Protocol 2*. More than 50% reaction is achieved by adding coelenterazine for 1–2 h.

Aequorin is also commercially available from Dr J. R. Blinks, Rochester, and from Baxter-Travenol, Tokyo, Japan.

4.2 mRNA isolation and cloning

To clone photoprotein 'genes' mRNA is extracted and purified as described in *Protocol 6*. A cDNA library is then constructed (21) using a plasmid such as pBR322 or an expression vector such as pUC18. A cDNA library from *Aequorea victoria* has been constructed (i) using *E. coli* SK 1592 homopolymer dC-tailed cDNA inserted into the dG-tailed PstI site of pBR322 (44), and (ii) in *E.coli* DH1 using the method of Okayama and Berg (45). High levels of expression of aequorin were obtained by inserting the *PstI/EcoRI* or *HindIII/EcoRI* fragments of aequorin cDNA into pUC9, incorporating both lac and tac promotors (46). Clones with plasmid combining photoprotein can be detected using either an oligonucleotide probe:

G

3'-ACCATATGGACCTAGG-5'

C

or assay of apoprotein (*Protocol 3*) or its mRNA (*Protocol 6*). Since as little as 10 tipomol (10^{-20} mol) of photoprotein can be detected, with a good expression as few as 1–10 bacteria should be detectable.

Protocol 6. Isolation and assay of mRNA

1. Extract *Aequorea* luminous cells or *Obelia* hydroids in 4M guanidine isothiocyanate (1:1 w/v) containing 5 g/litre Na lauryl sarcosine, 25 mM citrate pH 7, 0.1 M mercaptoethanol and 0.1% antifoam A.

2. Remove particulate material on nylon mesh and by centrifugation.

3. Layer supernatant on 4 ml 5.7 M CsCl, 25 mM Na citrate pH 5 (previously autoclaved with 0.2% diethyl pyrocarbamate), and centrifuge 18 h at 100 000g.

4. Dissolve RNA pellet in 1 ml 10 mM Tris 1 mM EDTA pH 7.4, extracted in neutralized phenol/Sevag (1:1 v/v water saturated phenol/Sevag; Sevag= $CHCl_3$:pentan-2-o-1) (secondary amyl alcohol) 24:1 v/v.

5. Precipitate RNA in the upper aqueous phase with a final concentration of 0.4 M NaCl, 70% v/v ethanol and store at $-70°C$.

6. Centrifuge and redissolve pellet in 10 mM Tris, 1 mM EDTA pH 7.4 (A_{260}/A_{280} for obelin RNA = 1.95, A_{260} standard = calf liver RNA), purify on poly A^+ mRNA on 0.7 × 2 cm column of oligo (dT)-cellulose equilibrated in 10 mM tris pH 7.4, 1 mM EDTA, 0.4 m NaCl, 0.5% w/v SDS and eluted with 3 × 1 ml 10 mM Tris pH 7.4, 1 mM EDTA, 0.5% SDS. Precipitate RNA as step 5 for storage.

Protocol 6. *Continued*

7. Assay for photoprotein RNA by adding 2 μl redissolved RNA to 10 μl rabbit reticulocyte lysate, incubate 30°C 1 h, add 50 μl 10 mM tris, 1 mM EDTA, 5 mM mercaptoethanol, 0.1% gelatin pH 7.4 + 1 μl, 100 μM coelenterazine 2 h (or 24 h). Add 100 μl 200 mM Tris, 0.5 mM EDTA pH 7.5 and assay for Ca^{2+}-activatable photoprotein.

4.3 Incorporation into cells

There are now five methods available for incorporating Ca^{2+}-activated photoproteins into cells.

- Micro-injection (8, 47).
- Cell-permeation including mechanical treatment, reversible osmotic swelling, and specialized media (22).
- Cell fusion with an erythrocyte ghost or liposome (23, 25).
- Release from internalized micropinocytotic residues (24).
- Intracellular biosynthesis from mRNA or cDNA (25).

We have found that higher yields of intracellular entrapment can be achieved by using initially the apo-photoprotein or mRNA, followed by reactivation to the photoprotein by adding 1–10 μM coelenterazine to the cells for 1 h (25). Originally, we failed to incorporate photoprotein into the cytoplasm of cells using lipsomes, but recently using pH- or temperature-sensitive lipsomes a method which is generally applicable for populations of phagocytic or non-phagocytic cells now seems available (25). pH-sensitive lipsomes are composed and made by sonication of dioleoyl phosphatidyl ethanolamine with palmitomyl homocysteine (molar ratio 4:1), with photoprotein or mRNA, with or without palmitoyl-antibody for cell targetting (*Protocol 7*). These are then incubated with the cells. Temperature sensitive lipsomes are composed of dipalmitoyl phosphatidyl dioline, sonicated with photoprotein or apo-protein. Fusion with cells occurs at 41°C following binding at 0°C (*Protocol 7*).

Protocol 7. A universal method for incorporating bioluminescent proteins into cells using heat-sensitive immunoliposomes (52)

A. *Coupling palmitic acid to cell specific antibody*

1. React palmitic acid (30 mmol) with *N*-hydroxysuccinimide (30 mmol) and dicyclohexylcarbodiimide (30 mmol) in dry ethyl acetate (140 ml) overnight at room temperature.

2. Remove precipitated dicyclohexylurea by filtration, and rotary evaporate the filtrate to yield white crystals of the *N*-hydroxysuccinimide ester of palmitic acid (NHSP).

Protocol 7. *Continued*

3. Recrystallize from hot ethanol.

4. React NHSP with cell specific antibody (molar ratio 10:1) in 137 mM NaCl, 2.7 mM KCl, 1.5 mM KH_2PO_4, 1.0 mM Na_2PO_4, 2% (w/v) deoxycholate pH 7.4 at 37°C for 12–18 h.

5. Separate coupled antibody from free NHSP on Sephadex G-75 (30 × 1.2 cm). Concentrate palmitoyl-antibody to approx. 15 mg/ml by ultrafiltration.

B. *Immunoliposome preparation*

1. Dry dipalmitoyl phosphatidyl choline (2.5 mg in chloroform) under nitrogen to give a film of lipid on glass container.

2. Add material to be entrapped (200 μl in saline containing EDTA 0.5 mM pH 7.4). Sonicate in bath for 20 min. Sonicate on ice with probe sonicator (30 sec) to give stable emulsion.

3. Dilute liposome suspension to 2.5 mg lipid/ml with saline containing EDTA 0.5 mM pH 8.0 and incubate at 41°C for 10 min.

4. Inject 0.25 mg of palmitoyl-antibody (approx. 17 μl) into the liposome suspension at a rate of 1 μl/min at 41°C. Incubate at 43–44°C for 20 min then cool slowly to room temperature (approx. 30 min).

5. Dialyse overnight against saline with EDTA 0.5 mM pH 7.4.

C. *Fusion of immunoliposomes with cells*

1. Cool cells and immunoliposomes to 4°C for 10 min.

2. Pellet cells and add immunoliposomes, mix and incubate at 4°C for 60 min.

3. Wash cells and incubate in fresh medium at 18°C for 30 min.

4. Incubate cells at 37°C for 30 min. Wash cells.

It is essential to quantify how much photoprotein is cytosolic, rather than entrapped within vesicles on the cell membrane or internalized vesicles. We routinely use four criteria to check this:

(a) Careful disruption of the cell, followed by isolation of $10^5 g$ supernatant = cytosol and broken organelles.

(b) Co-elution with a fluorescent indicator also entrapped within liposomes, e.g. fluorescein-albumin, or propidium iodide to label the nucleus.

(c) Specific permeabilization of the cell membrane to Ca^{2+} using activation of the classical complement pathway. Add an antiserum specific for a cell surface antigen, then human or guinea-pig serum (1/10, v/v) as a source of complement. Within 10–30 sec the membrane attack complex ($C5b6789_n$) will cause a μM rise in cytosolic Ca^{2+}, producing photoprotein chemiluminescence.

(d) Check the viability of the cells by determining ATP content, lactate dehydrogenase release and trypan blue exclusion (53) and also check functional activity of the cells.

4.4 Application and calibration in cells

Cells which have been loaded with aequorin or obelin include *E. coli*, several plant cells, protozoa, invertebrate and vertebrate eggs, nerve fibres, and muscle cells, and cells as small as human neutrophils (1, 2, 8, 20). The photoprotein chemiluminescent signal from cells can be converted to free Ca^{2+} manually (*Figure 3*) or ideally by interfacing the chemiluminometer to a computer.

At time t:

$$\text{fractional photoprotein consumption} = k_{app} = I_t/P_t$$

where I_t = light intensity at time $t = dh\upsilon/dt$
P_t = total potential photoprotein light yield at time t

when the Ca^{2+} transient is fast and low photoprotein consumtpion is $<10\%$. Thus, by exposing all intracellular photoprotein to high Ca^{2+} using a detergent, e.g. Nonidet P40, at the end of the experiment P_t can be estimated. However, if the cytosolic free Ca^{2+} remains 1 μM for a few minutes or more significant photoprotein consumption occurs, but P_t can be estimated at the end by scanning the total light yield between t and the end of the experiment, together with the detergent exposable light yield at this time (see reference 17).

1. Record photoprotein signal from cells stimulated (S) on chart recorder.
2. Add detergent (D), record signal.

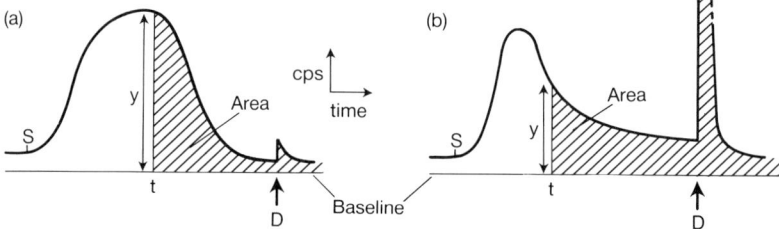

3. Measure height, y, of trace above baseline at time point, t. (units=mm).
4. Calculate area, A, to right of ordinate by counting squares on chart paper (units=mm. sec). If detergent spike is large compared to physiological signal, as in example b, use scalar to give total counts in spike and add to area after conversion of millimetres to counts/sec.
5. Calculate y/A to give rate constant, k, at time t. (units=sec^{-1}).
6. Convert k to Ca^{2+} using calibration curve.
7. Repeat at new time point.

Example: a, y=95 mm, A=1805 mm.sec, k=5.3×10^{-2} sec^{-1}, Ca^{2+}=7.9 μM.
b, y=500 c.p.s., A=8.8×10^6 cts, k=5.7×10^{-5} sec^{-1}, Ca^{2+}=190 nM.

Figure 3. Manual method for calculating Ca^{2+} concentration from photoprotein signal.

K_{app} is related to free Ca^{2+} using a calibration curve in the assumed intracellular ionic composition. This is important since K_{app} is affected by pH, Mg^{2+}, and high Na^+ or K^+.

At saturating Ca^{2+} (i.e. Ca^{2+} 0.1 mM)

$$K_{app} = K_{sat} = 1.4 \text{ sec}^{-1} \text{ for aequorin}$$
$$\text{or } 4 \text{ sec}^{-1} \text{ for obelin}$$

At low Ca^{2+} (c. 0.1 μM)

$$K_{app} = K_{sat} K_1 K_2 K_3 Ca^3$$

where $K_1 K_2 K_3$ = dissociation constant for the three Ca^{2+} sites (all K = c. 10–20 μM).

4.4.1 Single-cell analysis

Individual cells isolated with a micropipette or cell sorter can be investigated by placing the cell in a special cup very close to the photomultiplier tube (8, 47). Alternatively, individual cells in a population can be observed using an image intensifier (2) provided each cell contains sufficient photoprotein.

Single-cell analysis enables oscillations in cytosolic free Ca^{2+} in individual cells to be examined, as well as waves of free Ca^{2+} moving across the cell.

4.5 Problems

Photoproteins are capable of quantifying free Ca^{2+} in the range 10 nM–100 μM. However, the practical range inside cells is limited by four factors:

(a) The Ca^{2+}-independent light signal from the photoprotein ($c.10^{-6}–10^{-7} \text{ sec}^{-1}$), and from oxidative reactions in many cells (ultraweak chemiluminescence).

(b) The amount of photoprotein incorporated into the cells.

(c) The consumption rate.

(d) The time course of the Ca^{2+} transient.

For example, when cells are attacked by membrane pore forming proteins such as complement, the cytosolic free Ca^{2+} can be $> 10 \mu$M for several minutes resulting in $>95\%$ photoprotein consumption. However, by entrapping apo obelin or mRNA into the cells the free Ca^{2+} can still be monitored following addition of coelenterazine at a time when no signal from the original photoprotein is detectable. Formation from mRNA will also enable sufficient photoprotein to be entrapped to detect single cell signals, thereby circumventing the problem of observing the hetereogeneity and oscillations in small cell populations.

Photoproteins and the fluors fura-2 and indo-1 are therefore complementary and both viable techniques. Photoproteins have the advantage of a wide range for free Ca^{2+}, can be used at concentrations which do not disturb the cell Ca^{2+} balance, and are much less susceptible to leakage or intracellular redistribution than the fluors in long-term studies or cell injury. Photoproteins are the only way at present of measuring free Ca^{2+} in bacteria. Transgenic organisms and cells offer exciting prospects for the future.

Important discoveries arising from the use of photoproteins in both large and small cells include identification of primary stimuli (23) and secondary regulators (48) which work through Ca^{2+}, four ranges for free Ca^{2+} in cells (resting, activated, reversible injury, death) (49), localization of Ca^{2+} transients within the cell (50), the source of free Ca^{2+} either extra- or intracellular (51), and the role of oscillation frequency as a possible mechanism for generating thresholds in cells (47).

4.6 Instrumentation

A chemiluminometer suitable for the detection of Ca^{2+}-activated photoprotein signals needs to be able to quantify chemiluminescent reactions emitting as little as 1 photon per second over the 400–600 nm spectral range, over a temperature range (usually 0–37°C), and be capable of detecting flashes of just a few milliseconds in duration to glows lasting many minutes or even hours. This can be achieved using highly sensitive bialkali (or multialkali) photomultiplier tube with a stabilized, but variable high voltage supply set up in digital mode (43). Analogue devices cannot detect low pulse rates, their nose is very variable, and unlike the discriminator in digital systems, cannot distinguish low energy noise pulses emanating from the dynode chain from those originating at the photocathode. The photomultiplier will require some cooling if the reaction is above room temperature. In summary, the main components will therefore be:

(a) A thermostatable light tight housing, with an injector port, having the flexibility to cope with tubes containing cell suspensions and individual cells, preferably with the necessary apparatus to measure the cellular end response. The best way to achieve this flexibility is to construct the housing in modular form, if necessary using fibre optics as light guides.

(b) A specially selected bialkali photomultiplier tube with a dark current <0.1 nA if 11-stage, or <1 nA if 13-stage, placed as close to the light source as possible. It should be cooled to at least +4°C, or even to −20°C.

(c) A stabilized high voltage 500–2000 V setting, the photocathode highly negative (some people argue that the noise is lower with the anode highly positive and the photocathode at earth potential).

(d) A pre-amplifier to provide the discriminator with the pulse in an appropriate form.

(e) An eight-digit scalar to display the accumulated photon counts, a chart recorder output, and a computer interface to process the data, print, and plot it on a printer and VDU.

These requirements can best be met by building the equipment specifically for Ca^{2+} measurements. Photomultiplier tubes are available from RCA, Thorn EMI, Centronics, and Hamamatsu. Modular electronics for pulse counting, discriminators and high voltage supplies are available from Thorn EMI, Applied

Photophysics and EG and G (Princeton Electronics). Fully assembled apparatus providing digital counting but lacking flexibility of sample housing is also available commercially from Berthold (Biolumat), CLEAR, LKB Wallac, and Lumac.

5. Other methods of measuring cytosolic Ca^{2+}

5.1 Ca^{2+}-sensitive micro-electrodes

In cells large enough to impale with micro-electrodes without inflicting major damage, Ca^{2+}-sensitive micro-electrodes can be used to monitor the cytoplasmic Ca^{2+} concentration or activity. As with other ion-sensitive electrodes, the method depends upon the establishment of Nernst equilibrium across the ion sensitive region of the electrode, the generated voltage giving a measure of the ion concentration. Two important factors to be optimized in the construction of such a micro-electrode are miniaturization and speed of response. Detailed discussion of the method, of the theory and of construction are given elsewhere (15) but the Ca^{2+} exchanger in the electrode tip of crucial importance. A neutral carrier such as ETH1001 incorporated into the tip as a PVC-containing gel can give electrodes with sufficient sensitivity and response times to be useful in some cellular experiments.

5.2 Metallochromic indicators

The metallochromic indicators, murexide, arsenazo III, and antipyrylazo III have been used in giant non-mammalian cells. These indicators increase absorbance on binding Ca^{2+}, and have K_d in the 10–100 μM range. Since absorbance changes cannot be as easily detected as luminescence (chemiluminescence or fluorescence), the indicator must be used in high concentrations, where Ca^{2+} buffering may present problems (see Section 3.3.3). Perhaps the major reason for the decline in the use of such compounds as Ca^{2+} indicators is the problem of introducing the indicator into the cytoplasm of small cells. Although micro-injection can be used with large cells, it is unlikely that the techniques available for introduction of membrane-impermeant molecules into the cytoplasm of small cells (see Section 4.3) would be able to produce the high intracellular concentrations of dye required. An excellent and detailed discussion of the indicators is given by Thomas (15).

5.3 ^{19}F NMR probe for free Ca^{2+}

Fluorinated Ca^{2+} chelators have been synthesized (26–29) which can be loaded into small cell populations in the same manner as the fluorescent chelators, and provide a ^{19}F NMR spectrum from which the free Ca^{2+} concentration can be calculated. Analogues of BAPTA (see Section 7.2) have been used; e.g. the 4-fluoro and 5-fluoro derivatives (see *Table 1*). The ^{19}F NMR signal distinguishes free BAPTA from the Ca^{2+} complexed form. A major advantage of the method is that the Ca^{2+}-complex signal is also distinguished from the signals generated by Zn^{2+}, Pb^{2+}, and Fe^{2+} (see Section 3.3.4). The time required to accumulate a spectrum is, however, long and the time-resolution would not enable rapid Ca^{2+} transients to be accurately measured. However, an increase in free Ca^{2+} in mouse thymocytes associated with stimulation by succinyl concanavalin A and ionophore A23187 was detected (27, 29), and in human red cells treated with A23187 (29). It is interesting to note that in both Ehrlich Ascites cells (29) and mouse thymocytes (27) the resting Ca^{2+} was calculated to be 250 nM (by two independent groups), a value significantly higher than the 100 nM reported by fluorescent chelators. The reason for this discrepancy is not clear, but a Zn-BAPTA complex was detected in the ^{19}NMR spectrum from Ehrlich cells equivalent to 1 nM free Zn^{2+}. This would be sufficient to bind to and partially quench quin2 fluorescence (see Section 3.3.4).

6. Ca^{2+} flux measurement

The majority of measurements of Ca^{2+} fluxes in whole cells have been carried out by using $^{45}Ca^{2+}$, which is readily obtainable as $^{45}CaCl_2$ in aqueous solutions from Amersham or NEN-Dupont. Although $^{45}Ca^{2+}$ fluxes provide little information about Ca^{2+} within the cell, their measurement allows information to be obtained about mechanisms of transport into and out of the cell; for example, whether the processes are active or passive and what are the effects of pharmacological channel blockers. In polarized cells, $^{45}Ca^{2+}$ fluxes may allow determination of which face of the cell is involved in Ca^{2+} movements. By analysis of the kinetics of $^{45}Ca^{2+}$ fluxes an estimate of the number and nature of intracellular compartments involved in Ca^{2+} movements can be obtained. Additional insights can also be derived by comparing with measurements at the subcellular level.

Historically, the measurements of $^{45}Ca^{2+}$ movements within giant cells indicated that Ca^{2+} was not freely diffusible within cells and that a variety of hormones affect unidirectional Ca^{2+} movements into and out of cells. In many cases such studies were able to predict the effect of hormones in increasing the cytoplasmic free Ca^{2+} concentration and the involvement of release of Ca^{2+} from intracellular stores in this process.

6.1 Whole cell preparations

$^{45}Ca^{2+}$ fluxes can be studied using whole tissues either as an isolated perfused organ, an intact tissue epithelium mounted in an Ussing chamber configuration, allowing transcellular fluxes to be determined, or as isolated tissue slices or fragments. The advantages of these preparations are that they are closer to the *in vivo* situation, retaining cell–cell contacts, normal cellular configurations and possible physiological presentation of hormones. However, the practical disadvantages are great: in all cases there is a lack of sensitivity of measurement of the effects of hormones on $^{45}Ca^{2+}$ fluxes due to the extracellular space which constitutes a large, non-regulated compartment which can include binding to surface glycoproteins of the glycocalyx (30). In addition, since in whole organ and tissue systems it is usually necessary to measure loss or gain of $^{45}Ca^{2+}$ in the perfusate or incubation medium to obtain kinetic measurements of influx or efflux by the cells, sensitivity is reduced compared to measurements of cell content of $^{45}Ca^{2+}$ at various times. Another serious problem may be to distinguish the cell type contributing to the cellular $^{45}Ca^{2+}$ fluxes: even organs which are predominantly of a single cell type will contain endothelial and smooth muscle cells of the vasculature which actively transport $^{45}Ca^{2+}$ and respond to a number of hormones and neurotransmitters.

These disadvantages can be reduced by dissociating tissues into single cells or small groups of cells in which the extracellular space may be negligible and binding to the glycocalyx can be reduced by up to 99.9%. Dissociated cells can be exposed to a variety of agents without problems of diffusion to the cell surface. Rapid sampling of aliquots of cells also allows kinetics of changes in $^{45}Ca^{2+}$ content to be measured. Obviously, these advantages have to be balanced against the possible effects of loss of tissue structure, cell polarity and cell-cell contacts on any responses observed.

6.2 Procedures

$^{45}Ca^{2+}$ fluxes in whole cells can be measured using experimental protocols which can vary in detail but may be reduced to two basic procedures. Representative protocols for the measurement of fluxes are shown in *Protocol 8*.

Protocol 8. Measurement of $^{45}Ca^{2+}$ fluxes

1. Pre-equilibrate isolated cells (approx. 10^6/ml) at 37°C.

INFLUX

2. Add 2–4 μCi/ml $^{45}Ca^{2+}$ ±effector

3. Sample* cells at various times.

4. At suitable time sample medium, determine total radioactivity added.

EFFLUX

2. Add 2–4 μCi/ml $^{45}Ca^{2+}$.

3. Incubate for 1–2 h.

4. Centrifuge cells—keep aliquot of supernatant to determine total radioactivity added.

5. Wash cells once in ice-cold non-radioactive medium.

6. Incubate at 37°C.

7. Sample* cells at various times.

* SAMPLING

1. Remove aliquot of cells (approx 10^5 cells).

2. Add to 20 vol. ice-cold 0.9% (w/v) NaCl in filtration tower.

3. Filter and wash with 10 vol. ice-cold 0.9% NaCl.

4. Transfer filter to 2 ml distilled water and sonicate.

5. Determine $^{45}Ca^{2+}$ content in 1 ml by liquid scintillation counting. Assay aliquot for protein.

6.2.1 Influx

Influx is measured in the unstimulated cells simply by adding ^{45}CaCl$_2$ as a tracer to the extracellular medium which would normally contain 1.3 or 2.6 mM CaCl$_2$. The amount of radioactivity required will depend upon the number of cells available for sampling, but in commonly used procedures where cells are used at approximately 10^6/ml or tissues at 15–30 mg wet weight/ml. ^{45}CaCl$_2$ can be added at 2–4 μCi/ml. ^{45}Ca^{2+} influx is then determined at a number of time points from 1 min or less to up to 3–4 h, either by its disappearance from the medium or its appearance in the cells or tissues (see sampling procedures below).

Particularly in experiments using whole-organ perfusion or tissue fragment incubation a radioactively-labelled extracellular marker must be included to correct for ^{45}Ca^{2+} in the extracellular space. ^{3}H-labelled markers are preferable as they can be used concomitantly with ^{45}Ca^{2+}; the most commonly used are inulin, sucrose, or mannitol and need to be included at approximately 10 times the radioactivity as ^{45}Ca^{2+} to allow accurate dual-isotope liquid scintillation counting.

The effects of a hormone on ^{45}Ca^{2+} influx can be measured either as an initial, unidirectional influx by adding it at the same time as the ^{45}Ca^{2+} (to measure initial rates at < 1 min, add cells to mixture of ^{45}Ca^{2+} and hormone) or by pre-incubating with hormone for various times before adding ^{45}Ca^{2+}.

6.2.2 Efflux

Efflux is measured by first incubating unstimulated cells with ^{45}Ca^{2+} as described for influx. In theory, it is preferable to label all exchangeable Ca^{2+} pools to equilibrium but in practice, with most *in vitro* preparations, except for cells in culture, this may take longer than the time during which the cells remain viable. It is common to label for 1–2 h in which time the ^{45}Ca^{2+} influx (which may be determined separately) approaches a plateau. Efflux is initiated by removing tracer ^{45}CaCl$_2$ and sampling to determine reappearance of ^{45}Ca^{2+} in the medium from the cells or loss of ^{45}Ca^{2+} from the cells themselves.

The ^{45}Ca^{2+} can be removed by rapid washing: this is easily affected in perfusion systems by changing to non-radioactive perfusate; tissue fragments or dissociated cells can be washed using cooled solutions before returning to medium at 37°C to commence sampling for efflux. These latter protocols take several minutes and therefore risk losing rapid components of cellular efflux. Ca^{2+}EGTA (equimolar, 20 mM CaCl$_2$ and EGTA (31) or 6.5 mM CaCl$_2$ + 5 mM EGTA (32)) has also been used by adding it to the medium after loading to chelate ^{45}Ca^{2+} and initiate efflux.

As for influx, the effect of hormones may be determined either by addition at the beginning of the sampling period or at later times during efflux. The latter protocol may be important to uncover effects on intracellular Ca^{2+} release which may be observed during rapid loss of ^{45}Ca^{2+} from the extracellular space or dissorption from cell surface sites.

142

6.2.3 Sampling

Whole organs or tissue fragments can be studied either by vascular perfusion or perifusion systems respectively. Influx or efflux of $^{45}Ca^{2+}$ can be determined by loss or gain of isotope in the effluent. However, as discussed above, this has the disadvantage of lack of sensitivity in detecting changes in cellular fluxes in response to hormones. The use of small tissue fragments allows measurement of $^{45}Ca^{2+}$ content of cells by sampling at various times of incubation, washing the tissue and determining the radioactive content as follows:

(a) Blot tissue lightly on filter paper and weigh.

(b) Dry tissue overnight, either in a vacuum desicator or in an oven at approx. 60°C and weigh to constant weight.

(c) Solubilize tissue; the simplest method is to use a commercial solubilizer which is compatible with liquid scintillation cocktails.

(d) Determine radioactivity by standard dual isotope liquid scintillation counting method.

The following calculations can then be made:

(i) Wet wt (a) − dry wt (b) = Total tissue water volume.

(ii) (d.p.m. of extracellular marker) ÷
   (concentration of radioactivity of extracellular marker in incubation medium)
   = extracellular water volume.

(iii) (i)–(ii) = intracellular water volume.

(iv) (Concentration of $^{45}Ca^{2+}$ radioactivity in incubation medium)
   × (extracellular water volume)
   = $^{45}Ca^{2+}$ in extracellular space.

(v) $\dfrac{^{45}Ca^{2+} \text{ content of tissue} - (iv)}{(iii)} = {}^{45}Ca^{2+}$ space of cells.

Various methods of washing tissues have been developed to minimize the amount of $^{45}Ca^{2+}$ in the extracellular space. In particular, Van Breemen (33) characterized a method used for smooth muscle in which tissue was washed in 10 mM La^{3+} at 4°C for at least 30 min, with the aim of displacing extracellular and surface-bound $^{45}Ca^{2+}$ whilst inhibiting loss of intracellular $^{45}Ca^{2+}$.

However, secondary changes in cell $^{45}Ca^{2+}$ content or cell function cannot be ruled out and it may be preferable to rely on the use of extracellular marker as described above or use a dissociated cell preparation. Dissociated cells can be rapidly (within 10 sec) separated from the incubation medium either by filtration on polycarbonate filters or centrifugation through an inert oil. In either case, although the extracellular space is normally very low, this should at least be confirmed in preliminary experiments. Our experience (34) is that retention of extracellular markers, when aliquots of cells are filtered on 3 μm pore-size filters, is negligible. This allows a simple and rapid method for determining cell $^{45}Ca^{2+}$ content under a variety of conditions, aliquots of cells are diluted into a large volume of ice-cold 0.9% (w/v) NaCl on a filtration apparatus, rapidly filtered and washed on the filter. The filters are then placed in 2 ml of water, sonicated, and aliquots taken for liquid scintillation counting. This method also allows sampling errors to be normalized by determining the protein content of each sample.

6.3 Analysis of data

The most commonly used method is to apply the model described by Borle (35, 36) in which parallel cellular compartments are designated as being in equilibrium with the medium. This allows data to be fitted to exponential curves where each compartment is expected to possess its own rate constant for Ca^{2+} flux and to be of a defined pool size. Data can be analysed using curve-fitting programmes which utilize non-linear regression analysis to determine the number of exponential terms giving the best fit. An equation of the form $A_i(1 - e^{k_i t})$ describes Ca^{2+} influx into a pool (i) where A is the size of that pool and k_i the rate constant for exchange. Data can also be analysed by graphical means in which the rates of uptake at various times are plotted against time (*Figure 4*). The best fit can then be determined such that the number of straight lines in such a plot is equivalent to the number of exponential components. The flux rate (ϕ) for each pool can then be obtained by extrapolation of each straight line to time 0 and can be expressed either as nanomoles per milligram protein min^{-1} or per unit cell surface area min^{-1}. The $t_{\frac{1}{2}}$ (time taken for flux rate to decrease by 50%) can also be obtained from the graph allowing the rate constant ($t_{\frac{1}{2}} = \ln 2/k$) and pool size ($\phi/k$) to be calculated. Efflux data can be analysed in similar ways using equations of the form $A_i e^{-k_i t}$ to describe Ca^{2+} efflux from each pool.

In most cells, the fastest phase of $^{45}Ca^{2+}$ flux is thought to be equilibration of extracellular pools and/or the outer cell surface. Intracellular compartments typically show a fast pool exchanging with a rate constant of the order of 10^{-2}/min, and slower compartments may be observed. Metabolic inhibitors may be used to suggest whether an active intracellular process is involved in the exchange but further interpretation of the nature of the compartments involved requires subcellular fractionation approaches.

7. Manipulation of intracellular Ca^{2+}

Both measurement and manipulation of cytosolic free Ca^{2+} play equally important parts in establishing a role for the ion in the regulation of a cellular event. In this section, experimental agents and procedures which can alter the intracellular Ca^{2+} concentration in cells or modify the Ca^{2+} change triggered by a physiological agent are discussed (see *Table 2*).

1. Draw graph of $^{45}Ca^{2+}$ content against time.

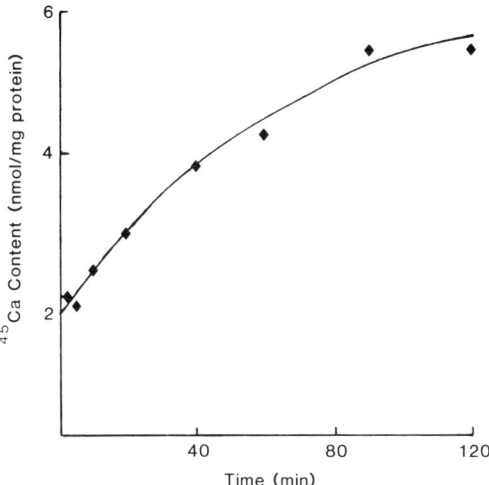

2. Draw tangents to curve at various times. Determine $^{45}Ca^{2+}$ influx rate at each time from slopes of tangents.
3. Draw graph of log ($^{45}Ca^{2+}$ influx) against time.

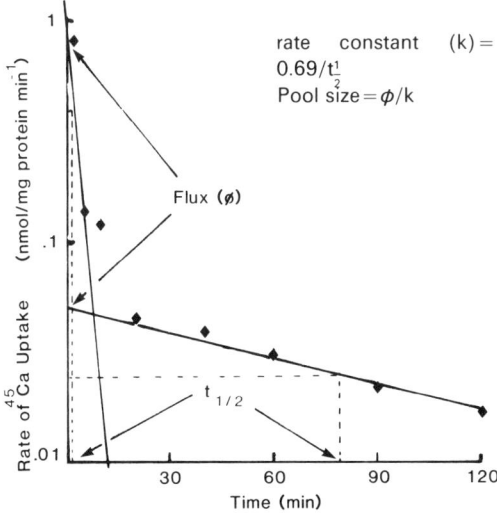

rate constant $(k) = 0.69/t_\frac{1}{2}$

Pool size $= \phi/k$

4. Determine best fit of points to straight line(s) using linear regression analysis.
5. Determine intercepts with y axis (flux rate ϕ) and time to 50% max flux $(t_\frac{1}{2})$.
6. Calculate rate constant (k) and pool size.

Figure 4. Simplified graphical analysis of $^{45}Ca^{2+}$ influx data. The raw data is from dissociated rat submandibular acini published in reference 42. For the data shown, the following parameters were calculated:

	Faster phase	Slower phase
Intercept (nmole/mg protein/min)	0.81	0.050
Half time (min)	1.8	78.0
Rate constant (min^{-1})	0.39	0.0089
Pool size (nmol/mg protein)	2.10	5.62

145

Table 2. Some agents which are useful in manipulating cytoplasmic Ca^{2+}

Agent	Solvent	Concentration range
Ionophores:		
Ionomycin	DMSO	1–5 μM
A23187	DMSO	1–10 μM
Br-A23187	DMSO	1–10 μM
Chelators:		
BAPTA-AM	DMSO/Pluronic	1–10 μM
methyl-BAPTA-AM	DMSO/Pluronic	1–10 μM
Ca^{2+} channel blockers:		
Ni	H_2O	mM
CO	H_2O	mM
D600	DMSO	10–100 nM
nifedipine	DMSO	10–100 nM
verapamil	methanol, DMF	μM
cromoglycate	H_2O	1–10 μM
Other agents:		
TMB-8	methanol, DMF, H_2O	10–100 μM
dantrolene sodium	H_2O	10–100 μM
Nitr 5-AM	DMSO/Pluronic	μM
BAYK 8644	DMSO	1–100 μM

7.1 Ca^{2+} inophores

In small mammalian cells or cell populations, where micro-injection of Ca^{2+} or Ca^{2+}/EGTA buffers is not possible, the intracellular Ca^{2+} concentration can be rapidly elevated by increasing the permeability of the palsma membrane to Ca^{2+}. In untreated cells, the enormous electrochemical gradient driving Ca^{2+} into the cell is maintained by outwardly directed Ca^{2+} pumps and by a highly impermeant plasma membrane. Ca^{2+}-ionophores complex Ca^{2+} in a lipid soluble form which enable Ca^{2+} to cross the membranes and consequently Ca^{2+} can enter the cell down its electrochemical gradient. In the absence of millimolar concentrations of extracellular Ca^{2+}, ionophores can also release Ca^{2+} from membrane bound stores within the cell. Under these conditions, the released Ca^{2+} will ultimately be removed from the cytoplasm by the plasma membrane pumps.

Two Ca^{2+} ionophores have been widely used, A23187 and ionomycin. The latter is preferable since it is reported to have greater selectively for Ca^{2+} over Mg^{2+} and monovalent ions than the former (37). Ionomycin is also useful in experiments with fluorescent chelator indicators, since unlike A23187, it is non-fluorescent. Recently, a non-fluorescent bromide analogue of A23187 (38) has become available (Calbiochem) which may also be usable in fluorescent experiments.

The concentration of ionophores used in experimental procedures is often in the range of 0.1–10 μM. The 'effective' concentration per cell is dependent upon cell number and may also be lowered by non-cellular ionophore binding sites in the cell suspension, such as proteins.

7.2 Intracellular Ca^{2+} buffers

Rises in intracellular Ca^{2+} can be suppressed by increasing the Ca^{2+} buffering capacity of the cell. The fluorescent chelators, quin2, fura-2 and indo-1 as mentioned previously (Section 3.3.3) increase the buffering capacity of the cytoplasm and can therefore also be used to manipulate the cytoplasmic concentration of Ca^{2+}. The capacity of Ca^{2+} buffer at any free Ca^{2+} concentration is defined as the ratio of buffered to free Ca^{2+}.

Since

$$K_d = (Ca^{2+})(L)/(CaL)$$

where (Ca^{2+}) is the concentration of free Ca^{2+}, (L), the concentration of free ligand, and (CaL) the concentration of buffered Ca^{2+}, the buffering capacity will be given by $(L)/K_d$. The effectiveness of the buffer therefore depends upon the dissociation constant of the chelator and its intracellular concentration. Quin2 has both a high affinity for Ca^{2+} and low fluorescence, necessitating high intracellular concentrations, i.e. conditions which favour buffering. In contrast, fura-2 has both a lower affinity (K_d 200 nM) for Ca^{2+} (factor of 2) and is more fluorescent (factor of 30). Under conditions where quin2 and fura-2 give comparable fluorescence signals, the quin2 loaded cells will therefore have an increased buffering capacity 60 times that of the fura-2 loaded cell. By increasing the intracellular concentration of chelator, the buffering effect of the chelator will become apparent. This approach can be used progressively to suppress a physiological transient in order to determine its relationship to physiological response. Two non-fluorescent chelators (39) which can also be introduced into the cytoplasm as their permeant esters are BAPTA ($K_d = 100$ nM) and the very effective buffer dimethyl-BAPTA ($K_d = 40$ nM).

An important consideration in the use of buffers is the kinetics of Ca^{2+} binding in relation to the kinetics of the Ca^{2+} transient. For the buffers discussed here with 1:1 binding the time-constant for approach to equilibrium following a small change in Ca^{2+} concentration (reciprocal of relaxation time, t) is given by the following equation

$$t = 1/[(K_1(Ca^{2+} + L) + K_2)] \qquad \text{[eqn 17]}$$

where K_1 and K_2 are the association and dissociation binding constants and Ca^{2+} and L are the concentrations of free Ca^{2+} and free chelator. For fura-2 $K_1 = 5.95 \times 10^8 \ M^{-1} \ sec^{-1}$ and $K_2 = 97 \ sec^{-1}$ (at 20°C) (see reference 8). At a resting free Ca^{2+}, 100 nM and free fura-2 concentration 50 μM, the relaxation time will be 33 μsec. This would be fast enough to effectively buffer very fast transients in Ca^{2+}. EGTA, for example, with K_1 and K_2 values of $2 \times 10^6 \ M \ sec^{-1}$ and 0.4 sec^{-1} respectively, under similar conditions would have a relaxation time of 10 msec, slow enough to cause some problems in buffering rapid Ca^{2+} transients, in the microsecond time-scale.

When loading cells with Ca^{2+} chelators, it is usually important that extracellular Ca^{2+} be available to replace cytoplasmic Ca^{2+} as it binds to the intracellular chelator. In the absence of extracellular Ca^{2+}, the cytoplasmic Ca^{2+} concentration becomes extremely low and Ca^{2+} is even forced to leave 'stores' within the cell. Under these conditions 'Ca^{2+}-depleted' cells are produced, which may be useful for some experimental situations.

Recently, a Ca^{2+} chelator Nitr5 whose affinity for Ca^{2+} is altered by flash photolysis, has been developed. Nitr-5 is also available as its membrane permeant ester (Calbiochem) and can be introduced into the cytoplasm of small cells (40). Following flash photolysis, the affinity for Ca^{2+} is reduced from 145 nM to 6 μM in 0.3 msec and Ca^{2+} is thus released from the chelator. This therefore provides a form of 'caged Ca^{2+}' within the cell which can be released rapidly on demand. This may prove useful for kinetic studies in fast responding cells or by 'stepping up' the Ca^{2+} concentration synchronously in a cell population at a defined time.

7.3 Ca^{2+} channel blockers

The rise in intracellular Ca^{2+} accompanying a physiological stimulus may be also prevented by pharmacological interference of Ca^{2+} influx by 'blocking' the Ca^{2+} permeability channel (41). Ions such as nickel and cobalt have this property in some cells. Some voltage sensitive channels can be blocked by the pharmacological agents D600, nifedipine, and verapamil, and opened by BAYK 8644. Other agents such as cromoglycate can block Ca^{2+} entry via Ca^{2+} channels on specific cell types. There are also agents which have been reported to prevent release of Ca^{2+} from sites within the cell, such as TMB-8, dantrolene sodium, and ryanodine (*Table 2*).

8. Conclusions

The study of intracellular Ca^{2+} in cell activation and injury has come a long way since the pioneering concept of Heilbrunn some 50 years ago (see reference 1). The techniques described in this chapter are aimed at helping to answer four questions:

(a) Is a rise in intracellular Ca^{2+} directly responsible for cell activation?

(b) If so, is the source of the Ca^{2+} extra- or intracellular, and how is it released into the cytoplasm?

(c) How and where does the Ca^{2+} act?

(d) Do secondary regulators act by altering the Ca^{2+} transients, the Ca^{2+} mechanisms or independently of Ca^{2+}?

Exciting prospects lie ahead for the application of transgenic cells expressing Ca^{2+} indicators and proteins responsible for Ca^{2+} processing. Which approach one chooses depends not only on the importance of the discovery one is aiming at but also on our individual expression as creative thinkers. The significance of our contribution is perhaps well expressed in the quotation from Bertrand Russell (1917):

Maurice B. Hallett, Robert L. Dormer, and Anthony K. Campbell

In art nothing worth doing can be done without genius, in science even a very moderate capacity can contribute something to a supreme achievement.

Acknowledgements

The work described here has been supported by the MRC (UK), and the Arthritis and Rheumatism Council. We are also grateful to Enid Davies for accurate typing of a difficult text.

References

1. Campbell, A. K. (1983). *Intracellular Calcium: Its Universal Role as Regulator*. Wiley, Chichester.
2. Campbell, A. K. (1988). *Chemiluminescence: Principles and Applications in Biology and Medicine*. Ellis Horwood/VCH, Chichester.
3. Nordin, B. E. C. (ed.) (1988). *Calcium in Human Biology*. Springer-Verlag, London.
4. Campbell, A. K. and Morgan, B. P. (1985). *Nature*, **317**, 164, 166.
5. Morgan, B. P., Luzio, J. P., and Campbell, A. K. (1986). *Cell Calcium*, **7**, 399, 511.
6. Tsien, R. Y. (1983). *Ann. Rev. Biophys. Bioeng.*, **12**, 91.
7. Tsein, R. Y., Poenie, M., and Rink, T. J. (1985). *Cell Calcium*, **6**, 145.
8. Cobbold, P. H. and Rink, T. J. (1987). *Biochem. J.*, **248**, 313.
9. Almers, W. and Neher, E. (1985). *FEBS Lett.*, **192**, 13.
10. Steinberg, S. F., Bilezilhan, J. P., and Awqati, Q. (1987). *Am. J. Physiol.*, **253**, C744.
11. Al-Mohanna, F. A. and Hallett, M. B. (1988). *Cell Calcium*, **9**, 17.
12. Scanlon, M., Williams, P. A., and Fays, F. S. (1987). *J. Biol. Chem.* **262**, 6308.
13. Highsmith, S., Bloebaum, P., and Snowdowne, K. (1988). *Biochem. Biophys. Res. Commun.*, **138**, 1153.
14. Grynkiewicz, G., Poenie, M., and Tsien, R. Y. (1985). *J. Biol. Chem.*, **260**, 3440.
15. Thomas, M. V. (1982). *Techniques in Calcium Research*. Academic Press, London and New York.
16. Hesketh, T. R., Smith, G. A., Moore, J. P., Taylor, M. V., and Metcalf, J. C. (1983). *J. Biol. Chem.*, **258**, 4876.
17. Hallett, M. B. and Campbell, A. K. (1982). In *Chemical and Biochemical Luminescence* (ed. L. J. Kricka and T. J. N. Cater), pp. 89–133. Marcel Dekker, New York.
18. Tsien, R. Y. and Poenie, M. (1986). *Trends Biochem. Sci.*, **11**, 450.
19. Ashley, C. C. and Campbell, A. K. (ed.) (1979). *Detection and Measurement of Free Ca²⁺ in Cells*. Elsevier, Amsterdam.
20. Blinks, J. R., Mattingly, P. H., Jewell, B. R., van Leuwin, M., Harrer, G. C., and Allen, D. G. (1978). *Methods in Enzymology* (ed. M. A. De Luca), Vol. 57, pp. 292–328. Academic Press, New York, London
21. Prasher, D. C., McCann, R. O., and Cormier, M. J. (1987). *Methods in Enzymology* (ed. M. A. De Luca and W. D. McElroy), Vol. 133, pp. 288–307. Academic Press, New York, London.
22. Campbell, A. K., Dormer, R. L., and Hallett, M. B. (1985). *Cell Calcium*, **6**, 69.
23. Campbell, A. K. and Hallett, M. B. (1983). *J. Physiol.*, **338**, 537.
24. Hallett, M. B. and Campbell, A. K. (1983). *Immunology*, **50**, 487.

25. Campbell, A. K., Patel, A. K., Razavi, Z. S., and Capra, F. (1988). *Biochem. J.*, **252**, 143.
26. Metcalf, J. C., Hesketh, T. R., and Smith, G. A. (1985). *Cell Calcium*, **6**, 183.
27. Smith, G. A., Hesketh, R. T., Metcalf, J. C., Feeney, J., and Morris, P. G. (1983). *Proc. Natl. Acad. Sci. USA*, **80**, 7178.
28. Marban, E., Kitakaze, M., Kusuoka, H., Porterfield, J. K., Yue, P. T., and Chacko, V. P. (1987). *Proc. Natl. Acad. Sci.*, **84**, 6005.
29. Gupta, R. K. (ed.) (1987). *NMR Spectroscopy of Cals and Organisms*. CRC Press, Boca Raton, FL.
30. Borle, A. B. (1968). *J. Cell Ciol.*, **36**, 567.
31. Van Breemen, C. and Casteels, R. (1974). *Pflügers Arch.*, **348**, 239.
32. Aaronson, P., Van Breemen, C., Loutzenhiser, R., and Kolber, M. A. (1979). *Life Sci.*, **25**, 1781.
33. Van Breemen, C., Farinas, B. R., Casteels, R., Gerba, P., and Wuytack, F. (1973). *Phil. Trans. R. Soc. London*, Ser. B, **265**. 57.
34. Dormer, R. L., Poulsen, J. H., Licko, V., and Williams, J. A. (1981). *Am. J. Physiol.*, **240**, G38.
35. Borle, A. B. (1969). *J. Gen. Physiol.*, **53**, 43.
36. Borle, A. B. (1969). *J. Gen. Physiol.*, **53**, 57.
37. Liu, C. and Hermann, T. E. (1978). *J. Biol. Chem.*, **235**, 5892.
38. Deber, C. M., Tsu-Kun, J., Mack, E. and Grinstein, S. (1985). *Anal. Biochem.*, **146**, 349.
39. Tsien, R. Y. (1980). *Biochemistry*, **19**, 2396.
40. Adams, S. R., Kuo, J. P. Y., and Tsien, R. Y. (1986). Abstract No. 26, Society of General Physiol.
41. Towart, R. (1985). In *Control and Manipulation of Calcium Movement* (ed. J. R. Parrat), p. 169. Raven Press, New York.
42. McPherson, M. A. and Dormer, R. L. (1984). *Biochem. J.*, **224**, 473.
43. Stanley, P. E. (1982). In *Clinical and Biochemical Luminescence* (ed. L. J. Kricka and T. J. N. Carter), pp. 219–260. Marcel Dekker, New York.
44. Prasher, D. C., McCann, R. O., and Cormier, M. J. (1985). *Biochem. Biophys. Res. Commun.* **126**, 1259.
45. Okayama, H. and Berg, P. (1986). *Mol. Cell Biol.*, **2**, 167.
46. Inouye, S., Sakaki, Y., Goto, T., and Tsugi, F. I. (1986). *Biochemistry*, **25**, 8425.
47. Woods, N. M., Cuthbertson, K. S. R., and Cobbold, P. H. (1986). *Nature, Lond.*, **319**, 600.
48. Roberts, P. A., Newby, A. C., Hallett, M. B., and Campbell, A. K. (1985). *Biochem. J.*, **227**, 669.
49. Dormer, R. L., Hallett, M. B., and Campbell, A. K. (1984). In *Control and Manipulation of Calcium Movement* (ed. J. R. Parrat), pp. 1–27. Raven Press, New York.
50. Gilkey, J. C., Jaffe, L. F., Ridgway, E. B., and Reynolds, G. T. (1978). *J. Cell Biol.*, **76**, 468.
51. Hallett, M. B. and Campbell, A. K. (1984). *Cell Calcium*, **5**, 1.
52. Sullivan, S. M. and Huang, L. (1986). *Proc. Natl. Acad. Sci. USA*, **83**, 6117.
53. Elliott, K. R. F. (1979). *Techniques in Life Sciences*, B2/1, part B204. Elsevier/North Holland, Amsterdam.

5

Inositol phosphate second messengers

C. J. KIRK, A. J. MORRIS, and S. B. SHEARS

1. Introduction

The role of inositol lipid hydrolysis in receptor-mediated cell signalling was first suggested by Michell in the mid-1970s (1). However, the current growth in interest in this subject probably stems from the more recent realization that receptor-mediated hydrolysis of phosphatidylinositol 4,5-bisphosphate [PtdIns(4,5)P_2] generates two intracellular second messengers. Following receptor activation, this lipid is hydrolysed by a 'phosphoinositidase C' to yield diacylglycerol, which activates protein kinase C, and inositol 1,4,5-trisphosphate [Ins(1,4,5)P_3], which mobilizes Ca^{2+} from an intracellular pool (see references 2 and 3 for reviews).

The primary concern of this chapter is with the methods used for the study of Ins(1,4,5)P_3 and the ever-growing number of other inositol phosphates which are found in cells, some of which may have messenger roles in their own right. Since these intracellular messengers are derived from membrane phospholipids, we will also describe techniques for the extraction and separation of these parent compounds.

2. Labelling protocols for investigating inositol lipids and their degradation products

PtdIns(4,5)P_2 and PtdIns4P are quantitively minor phospholipids, comprising 0.1–1.0% of the total phospholipid complement of most cells. In the liver, for example, PtdIns(4,5)P_2 is found at a concentration of about 4 nmol/g dry wt and the estimation of this compound in biological samples requires very sensitive assay techniques (see Section 5). The recent identification of other minor inositol

---

Abbreviations used: Ins, Ins$P$, InsP_2, InsP_3, InsP_4, InsP_5, and InsP_6 are myoinositol and its mono-, bis-, tris-, tetrakis-, pentakis-, and hexakisphosphates; Gro$P$Ins, Gro$P$Ins$P$, and Gro$P$InsP_2 are glycerolphosphoinositol and its mono- and bisphosphates. PtdIns, phosphatidylinositol; PtdIns3P, phosphatidylinositol 3-phosphate; PtdIns4P, phosphatidylinositol 4-phosphate; PtdIns(4,5)P_2, phosphatidylinositol 4,5-bisphosphate. All locants are designated according to the D-nomenclature.

phospholipids (PtdIns3P and PtdInsP_3, see reference 4 and 5) add further to the complexity of assaying these compounds.

In stimulated cells, Ins(1,4,5)P_3 and related compounds are found in the micromolar concentration range and are difficult to separate from other organic phosphates; hence these compounds do not lend themselves to assay by routine methods based on the measurement of inorganic phosphate, and alternative approaches have been developed. These include an enzymatic method based upon Ins(1,4,5)P_3 3-kinase (6) and a binding assay (7). However, both these methods can prove expensive if applied to large numbers of samples and most studies of receptor-mediated inositol lipid hydrolysis continue to utilize cells which have been pre-labelled with a radioactive precursor of PtdIns(4,5)P_2. In this way, the degradation of inositol lipids and the generation of inositol phosphates may be followed by the change in the radioactivity of the individual compounds. Most early studies of stimulated inositol lipid hydrolysis utilized cells which had been pre-labelled with [^{32}P]-P$_i$ and this isotope, which is readily detected by liquid scintillation spectrometry, has advantages in studies which are confined to the inositol lipids themselves. However, routine methods for the analysis of the inositol phosphates are based upon ion-exchange chromatography which achieves very poor separation from the nucleotides of the cell (see Section 4). Some workers have described the separation of inositol phosphates from nucleotides with activated charcoal (8) but, in the authors' experience, this is not a quantitative procedure. Consequently, any studies which aim to characterize inositol phosphate accumulation or metabolism are best conducted with an alternative radiolabelled precursor which is not incorporated into cellular nucleotide pools. The precursor of choice is *myo*-inositol, which is available in ^{3}H- or ^{14}C-labelled forms. Most cells will only incorporate inositol into the inositol lipids and phosphates derived therefrom, so this greatly simplifies the interpretation of such experiments.

The precise design of labelling protocols will depend upon the nature of the experimental system and the question to be addressed. In freshly isolated tissues and cells, where experiments must be concluded within a few hours at the most, short-term labelling has been successfully employed to reveal receptor-mediated degradation of inositol lipids and accumulation of inositol phosphates (9,10). Short-term labelling has also been particularly useful in pulse-chase experiments, e.g. in investigating the possible existence of discrete, hormone-sensitive inositol lipid pools (11). The inositol lipids can also be labelled *in vivo*, by injecting experimental animals with appropriate radioactive precursors (10).

Experiments using short-term labelling techniques cannot yield quantitative measurements of inositol lipids and their derivatives unless steps are also taken to determine the specific radioactivity of the various compounds involved. For lipids which may constitute only 0.1% of the total lipid complement of many cells, this is no easy task and it requires sensitive separation and assay techniques (see Section 5).

Reliable estimates of the extent of inositol lipid hydrolysis or inositol phosphate accumulation may be made if cells or tissues have previously been labelled to equilibrium with an appropriate precursor. This is most easily achieved in culture cells, which may be grown through several cell divisions in a medium containing radioactive inositol of known specific radioactivity (12). Before making deductions about the cellular concentrations of inositol lipids or inositol phosphates, it is essential to demonstrate that equilibrium labelling has been achieved by assaying the specific activity of at least two of the intermediates involved.

A further consideration in the design of labelling protocols is the concentration of inositol in the medium in which cells and tissues are incubated. Some workers have sought to maximize the incorporation of radioactive precursor into lipid pools by using incubation media free of non-radioactive carrier (11). However, labelling protocols should be carefully evaluated for the particular system in use. Some cells may contain large intracellular pools of a particular precursor and cellular uptake of the radioactive compound may be limited by membrane transport mechanisms (13).

For the study of inositol phosphate production, 10 mM Li^+ is often added to incubations 10 min prior to stimulation. Li^+ enters the cell and inhibits at least two inositol phosphate phosphatases (see reference 2). In this way, Li^+ amplifies the observed inositol phosphate response to agonists. However, it is important to stress that Li^+ has differential effects upon the intracellular concentrations of the inositol phosphates. Thus, if it is the intention to study physiological changes in the inositol phosphates, this is best done in the absence of Li^+.

3. Quenching and extraction procedures for inositol phosphates and inositol lipids

Several quenching and extraction methods for the inositol phosphates are in routine use. Most utilize acidic conditions in which non-cyclic inositol phosphates are stable at temperatures of 0–4°C. Whichever method is chosen, it must be demonstrated that all the relevant inositol phosphates are extracted with good efficiency ($\simeq 90\%$, see reference 14). Incubations have frequently been quenched with trichloroacetic acid or perchloric acid. Both of these acids must then be removed, as they react with the ion-exchange resin and prevent inositol phosphate binding (see Sections 3.1 and 3.2). Both acids are suitable for the efficient extraction of non-cyclic inositol phosphates, provided a carrier is present; phytate hydrolysate has been used (18), but phytic acid works just as well. In the authors' experience, the perchlorate quench has proved the least arduous when processing multiple samples of inositol phosphates but, by the same criterion, both are preferred to the acidified chloroform/methanol method (14).

Sometimes substantial proportions of cyclic inositol phosphates accumulate in cells after several minutes' stimulation (19). As acidic procedures hydrolyse cyclic inositol phosphates, they should be extracted with neutral chloroform/methanol or with a phenol-based extraction procedure (20, 21, Section 3.3). The phenol method extracts >95% of Ins$P$, InsP_2, InsP_3 and InsP_4 (20). The neutral chloroform/methanol method is suitable for Ins$P$, InsP_2 and InsP_3 (>85% recovery), although the recovery of InsP_4 may sometimes be lower (e.g. 79%, see 18).

A further problem has recently been observed in [^{3}H]-inositol labelled platelets quenched with chloroform/methanol. Here, 1-monomethylphospho-inositol 4,5-bisphosphate (MeInsP_3) was detected (22). MeInsP_3 was not detected in perchlorate-quenched incubations. Moreover, once isolated from the tissue MeInsP_3 was stable in perchloric acid, so its appearance in the extract seems dependent upon the presence of chloroform/methanol. It is not yet known if MeInsP_3 occurs naturally in platelets. Until this problem is resolved, it may be useful to check the inositol phosphate profiles in tissue extracts using more than one quench technique.

The routine extraction of the polyphosphoinositides requires the use of acidified chloroform:methanol mixtures. These may either be used directly to quench incubations (23, 24) or for the extraction of the lipids from the precipitated protein generated by the acid-quench techniques described below (10, 12).

3.1 Perchloric acid quench

Incubations are quenched with a final concentration of 4% (v/v) ice-cold HClO$_4$ containing 1 mg/ml phytic acid carrier (see above) and then placed on ice for 10 min. The precipitated protein is centrifuged (>1000g) for 5 min. The supernatant is saved and the perchlorate therein is then removed by one of two methods. The first (17) is by addition of 5 μl Universal Indicator followed by an appropriate volume of 1.2 M KOH/75 mM Hepes/60 mM EDTA to achieve neutrality. The samples are left on ice for 1 h, during which time a KClO$_4$ precipitate forms. This is removed by centrifugation (>1000g) for 5 min, and the supernatant is chromatographed (see below).

Alternatively, the perchlorate is removed by a modification (25) of a method first described by Sharpes and McCarl (15). 1 vol. of the HClO$_4$-quenched supernatant is added to 0.08 vol. of 50 mM EDTA (pH 7.0 with NaOH) followed by 1.5 vol. of freshly mixed 1:1 (v/v) 1,1,2-trichlorotrifluoroethane:tri-*n*-octylamine, all of which must be ice-cold. The samples are mixed vigorously in a tightly sealed tube, and centrifuged at 1000g for 5 min at 0–4°C. The largest possible proportion of the neutralized upper phase is saved but the precise amount, as with all phase splits, will depend upon the volumes used and the internal diameter of the tubes. If necessary, ice-cold water may be added in a volume equal to that of the removed upper phase. After further mixing and centrifugation, the upper phase is again withdrawn and combined with the earlier extract. This technique does not demand the routine use of universal indicator, which is an advantage when samples are to be analysed by HPLC since this reagent saturates the guard column. Otherwise, the relative advantages of the two neutralization techniques rests upon the particular equipment available for the mixing and centrifugation of multiple samples.

3.2 Trichloroacetic acid quench

This method (26) involves quenching incubations with an equal volume of 1 M trichloroacetic acid (containing 1 mg/ml phytic acid). The tissue is sedimented by centrifugation. The supernatants are washed five times with 2 vol. of water-saturated diethyl ether, in which the acid is soluble (16). The pH of the final extract is adjusted to 7–8 with $NaHCO_3$.

3.3 Neutral quench techniques for extracting cyclic inositol phosphates

3.3.1 Phenol method

This procedure is a modification of a method for extracting ATP from cells (20) and it preserves both cyclic and non-cyclic inositol phosphates. The use of phenol in this method means that a bottle of methanol must be at hand to wash any spillage from the skin. A solution of 5 mM KH_2PO_4/10 mM EDTA (pH 6.3) is saturated for at least 2 h with chloroform and then mixed with 0.5 vol. phenol (prepared by melting crystalline phenol at 57°C in a water bath). A phase separation of this mixture occurs upon centrifugation. The upper phase is used as the quench buffer.

Incubations are terminated with twice their volume of the quench buffer. Samples are mixed, placed on ice, and 1.3 vol. of phenol:chloroform:isoamyl-alcohol (38:24:1) is added. The mixture is shaken vigorously for 30 sec, and then centrifuged at 3600*g* for 15 min at 0–4°C. The upper phase is withdrawn for analysis.

3.3.2 Neutral chloroform/methanol method

This method is essentially that described by Berridge *et al.* (14) and Hawkins *et al.* (21). Incubations are quenched with 3.8 vol. chloroform/methanol (1:2 v/v), which is pre-cooled on solid CO_2. The samples are warmed to 0–4°C and chloroform (1.2 vol.) and water (1.2 vol.) are added, then the mixture is shaken and centrifuged for 5 min at 1000*g*. The upper aqueous phase is removed, and the interfacial material is washed twice with 0.4 vol. of an upper phase prepared by mixing chloroform:methanol:0.1 M sodium cyclohexane-1,2-diaminotetra-acetic acid (16:8:5). The original aqueous extract is combined with the two washes.

3.4 Quenching and extraction of inositol lipids

3.4.1 Procedures utilizing acidified organic solvents

These methods have the advantage of utilizing a combined quenching and extraction medium. However, they are not recommended where the harvesting of inositol phosphates form part of the experimental design. Hence, they are appropriate where the inositol lipids are to be studied, but information on the behaviour of the inositol phosphates is not required.

The inositol lipids may be directly extracted from tissues and cells using acidified organic solvents. We have extracted packed cells with 3.75 vol. of chloroform:methanol:12 M HCl (200:400:5) for 30 min at room temperature (24). 1.25 vol. of chloroform and 1.25 vol. of 0.1 M HCl are added with mixing and the phases are separated by gentle centrifugation. The lower (chloroform) phase, containing the extracted lipids, is then removed. The remaining upper phase can be further extracted with 2.5 vol. of chloroform, and the lower phases combined and washed with a 'synthetic upper phase' if required.

An alternative method is particularly useful for the extraction of culture cell lines grown on plasticware, which is not attacked by the methanol in the initial quenching and extraction medium (23). Incubations are terminated by aspirating the culture medium and replacing it with one volume of methanol/12 M HCl (100:6, v/v). Cells are scraped free from the plastic substratum with a rubber policeman and the whole extract is transferred to a glass tube containing 2 vol. of chloroform and vortexed. The culture vessel is then washed with a further volume of methanol, which is added to the extraction mixture. This mixture is sonicated for 2 min to ensure complete extraction. Half a volume of 100 mM KCl is added, with further mixing, and the phases are separated by centrifuging at 3000g for 5 min. The lower organic phase is transferred to a new tube and the aqueous phase is washed with 2 vol. of chloroform. The lower phases are then combined and washed with 1 vol. of methanol/1 M KCl (4:3, v/v), followed by a further centrifugation at 3000g for 5 min. The upper aqueous phase is discarded and the lower, lipid-containing, phase may be evaporated to dryness under N_2.

3.4.2 Lipid extraction from protein precipitates of acid-quenched samples

After the extraction of inositol phosphates as described in Sections 3.1 and 3.2, the lipids in the precipitated material may be further extracted with acidified chloroform/methanol (9, 12). The original method called for trichloracetic acid precipitates to be first washed with twice the extraction volume of 5% (w/v) trichloroacetic acid, containing 1 mM EDTA, and once with a similar volume of water (9). We have found that perchloric acid precipitates may be adequately washed once, with 1 vol. of water. Further washing leads to excessive re-dissolving of the protein pellet. Washed precipitates derived from either of these methods are then extracted twice with an appropriate volume of chloroform/methanol/12 M HCl (100:100:1 by vol.) and once with 0.66 vol. of chloroform/methanol/12 M HCl (200:100:1 by vol.). The three extracts are combined and mixed with 1 vol. of chloroform and 0.733 vol. of 0.1 M HCl. After mixing and centrifugation at 3000g for 5 min, the lower phase (which contains the lipids) is removed, washed once with new upper phase and dried under N_2.

4. Ion-exchange chromatography of inositol phosphates

A number of inositol phosphate isomers are now known to occur *in vivo*, including at least two of $InsP_3$ [$Ins(1,4,5)P_3$ and $Ins(1,3,4)P_3$, (27), three of $InsP_4$ ($Ins(1,3,4,5)P_4$, $Ins(1,3,4,6)P_4$, $Ins(3,4,5,6)P_4$; (26, 28–30)], as well as several $InsP_2$ and $InsP$ isomers. The resolution of these different isomers requires HPLC. However, inositol phosphates can also be separated routinely on small gravity-fed ion-exchange columns according to their degree of phosphorylation.

4.1 Analysis by gravity-fed columns

The columns can be constructed by inserting a bolus of glass wool into a Pasteur pipette, followed by 0.6 ml (packed) resin (Bio-Rad AG 1-X8 200–400 mesh, formate form). They are linked to a 20-ml reservoir (a plastic syringe barrel works well) by a small length of silicon tubing (Fisons SRT/4). Alternatively, the resin may be added to reusable polypropylene columns (trade name Econo-Column) which are available from Bio-Rad. A custom-built rack will facilitate the analysis of multiple samples.

The neutralized samples are loaded on to the column in 10 ml water. Inositol, being uncharged, passes through the column. However, in incubations of tissue preparations with [^{3}H]-inositol, the amount of radioactivity in the free inositol frequently greatly exceeds that in the inositol phosphates. Thus, it is worth washing the column with sufficient water until no further inositol is eluted. A typical complete elution system is as follows: 10 ml 0.06 M ammonium formate to elute glycerophosphoinositol; 10 ml 0.18 M ammonium formate for $InsP$; 10 ml 0.4 M ammonium formate/0.1 M formic acid for $InsP_2$; 10 ml 0.8 M ammonium formate/0.1 M formic acid for $InsP_3$; 10 ml 1.05 M ammonium formate/0.1 M formic acid for $InsP_4$; 10 ml 2 M ammonium formate/0.1 M formic acid for $InsP_5$ and $InsP_6$ (26, 31, 32). Note that $Ins(1:2\text{-cyclic})P$ elutes in the glycerophosphoinositol eluant because it is significantly less polar than $InsP$ (N. S. Wong, unpublished data). However, $Ins(1:2\text{-cyclic, }4)P_2$ and $Ins(1:2\text{-cyclic, }4,5)P_3$ respectively elute in the $InsP_2$ and $InsP_3$ eluants (21). Appropriate aliquots of the eluants are mixed with scintillation fluid and counted for radioactivity.

The ion-exchange resin exhibits batch variation, so it is essential to check the elution profiles with inositol phosphate standards. In the authors' experience, >95% of an appropriate standard should elute in the appropriate eluant, but achieving this may necessitate changes to the elution volume. For example, $InsP_3$ elution has required between 9 and 12 ml in past experiments. In the authors' laboratory, columns are not generally regenerated after use, but this can be achieved for a limited number of times (<5) by washing the column with 20 ml 2 M ammonium formate/0.1 M formic acid, followed by 20 ml water.

Some studies of receptor-mediated inositol phosphate production may only require measurement of the accumulation of total inositol phosphates. In this case, inositol and glycerophosphoinositol are first eluted from the column. Then total inositol phosphates may be eluted with 10 ml 2 M ammonium formate/ 0.1 M formic acid. To minimize the effects of biological variation between tissue preparations, the radioactivity in the inositol phosphates can be expressed relative to the radioactivity incorporated into the lipid (determined as described in Section 3).

4.2 Analysis by HPLC

Several procedures are now available, but as no method will yet separate all known inositol phosphates, the experimental objectives should be considered carefully at the outset. There has been a recent proliferation of methodology in this area and restrictions on space mean that only a selection can be described below (for further methods, see references 6, 33, and 34). Whichever method is used, increasing column age shortens the retention times of the various inositol phosphates, and increases the broadness of the peaks, thereby decreasing the efficiency of separation. Typically, 50 useful runs may be obtained from the columns described below. However, reproducibility is improved if the column is pre-washed with water for 1 h, and then for a further 20 min between runs. The column should be protected with a guard column containing similar material. Some eluants such as ammonium formate may dissolve the silica support in the columns. This can be prevented by saturating the eluant with silicate ions, using a silica pre-column immediately before the loading loop.

An unknown inositol phosphate from a tissue extract cannot be characterized solely by its co-elution with a known standard unless the presence of all other inositol phosphates can be excluded. Additional methodology will aid definitive characterization (see Section 8). Nevertheless, analysis by HPLC has been extremely useful in studying inositol phosphate metabolism. In the figures (*Figure 1a–1d*), describing the methods detailed here, the elution times of various standards are plotted, and these have been corrected by omitting the 'dead' volume between the injection loop and the fraction collector. Inositol, which elutes in the loading volume, is omitted for clarity.

A popular method (reference 26; see *Figure 1a*) uses a Whatman Partisil 10 SAX column, which is eluted using a two-pump system. Pump A contains water, and pump B contains 1.7 M ammonium formate (pH 3.7 with phosphoric acid). The elution is as follows: 0–5 min, B=0%; 5–10 min, B increases to 44%; 10–12 min, B=44%; 12–18 min, B increases to 59%; 18–23 min, B=59%; 23–33 min, B increases to 100%; 33–39 min, B=100%; 39–40 min, B decreases to 0%. The usual flow rate is 1.25 ml/min, and 0.25-min fractions are generally collected and mixed with 1 ml methanol:water (1:1) before 4 ml of scintillant is added (e.g. Triton:xylene (1:2) plus fluors). The following isomers should achieve a base-line separation: $InsP$, $Ins(1,4)P_2$, $Ins(3,4)P_2$, $Ins(4,5)P_2$, $Ins(1,3,4)P_3$, $Ins(1,4,5)P_3$ and $InsP_4$. However, this system will not reliably separate isomers of $InsP$ or $InsP_4$, nor the (1,3), (1,4), and (1,5) isomers of $InsP_2$.

A more recent method (32) uses a Whatman Partisphere WAX column (*Figure*

1b), which contains a weaker ion-exchange resin than the Partisil column described above. Pump A contains water and pump B contains 0.5 M diammonium hydrogen phosphate (pH 3.2 with phosphoric acid). The elution is as follows: 0–12 min, B increases from 0 to 25%; 12–53 min, B = 25%; 53–54 min, B increases to 99%; 54–70 min, B = 99%; 70–71 min, B decreases to 0%. The flow rate is 1 ml/min and, depending upon the resolution required, 0.3–0.5-min fractions are collected. These are dissolved in scintillant as described above. This system is particularly useful for separating $InsP_6$, $InsP_5$, $Ins(1,3,4,5)P_4$, $Ins(1,3,4,6)P_4$, $Ins(3,4,5,6)P_4$, $Ins(1,4,5)P_3$, and $Ins(1,3,4)P_3$. The useful lifetime of the column can be extended by reducing the %B between 12 and 53 min (see reference 32 for details).

Some systems have been developed for separating a small number of inositol phosphates. For example, $Ins(1:2cyclic,4,5)P_3$ can be separated from $Ins(1,4,5)P_3$ and $Ins(1,3,4)P_3$ by using a Whatman Partisil 10 SAX column eluted with 2 M ammonium formate (pH 5.5 with phosphoric acid) (20, *Figure 1c*). The flow rate is 1.25 ml/min, and 0.5 min fractions are mixed with 0.5 ml methanol and 4 ml of scintillant (e.g. xylene:triton 1:1 plus fluors) for the determination of radioactivity.

Stephens *et al.* (30) have developed a method (*Figure 1d*) for separating $InsP$ isomers which is particularly useful for identifying the products of ammonical hydrolysis (Section 8.5). This system uses two 12.5-cm Partispere SAX (5-μm particle size) columns, connected in series with a low dead-volume adaptor (Waters 'Direct-Connect') and eluted at a flow rate of 1 ml/min. Pump A contains water and pump B contains 0.2 M sodium acetate (pH 3.75 with acetic acid). The elution is as follows: 0–10 min, B = 0%; 10–11 min, B increases to 48%; 11–80 min, B = 48%; 80–81 min, B increases to 100%; 81–85 min, B = 100%; 85–86 min, B decreases to 0%.

5. Separation techniques for the inositol lipids

Lipid samples, prepared as described in Section 3.4, may be separated directly by affinity chromatography on neomycin-linked beads, by thin layer chromatography or by HPLC. This latter method does not separate PtdIns from phosphatidic acid and will not be further discussed here; details may be found in reference 35. Thin layer chromatography yields intact lipids which may be subjected to further analysis if they are first eluted from the silica gel support medium. It should be borne in mind, however, that TLC methods which are designed to separate the highly polar polyphosphoinositides may not achieve an acceptable separation of other phospholipids. An alternative method for the separation of the polyphosphoinositides involves the deacylation of these lipids, followed by separation of their glycerophosphoinositol esters by anion-exchange chromatography or HPLC. This method will not yield intact lipids but it avoids the tedious process of lipid elution from silica gel and is particularly useful in

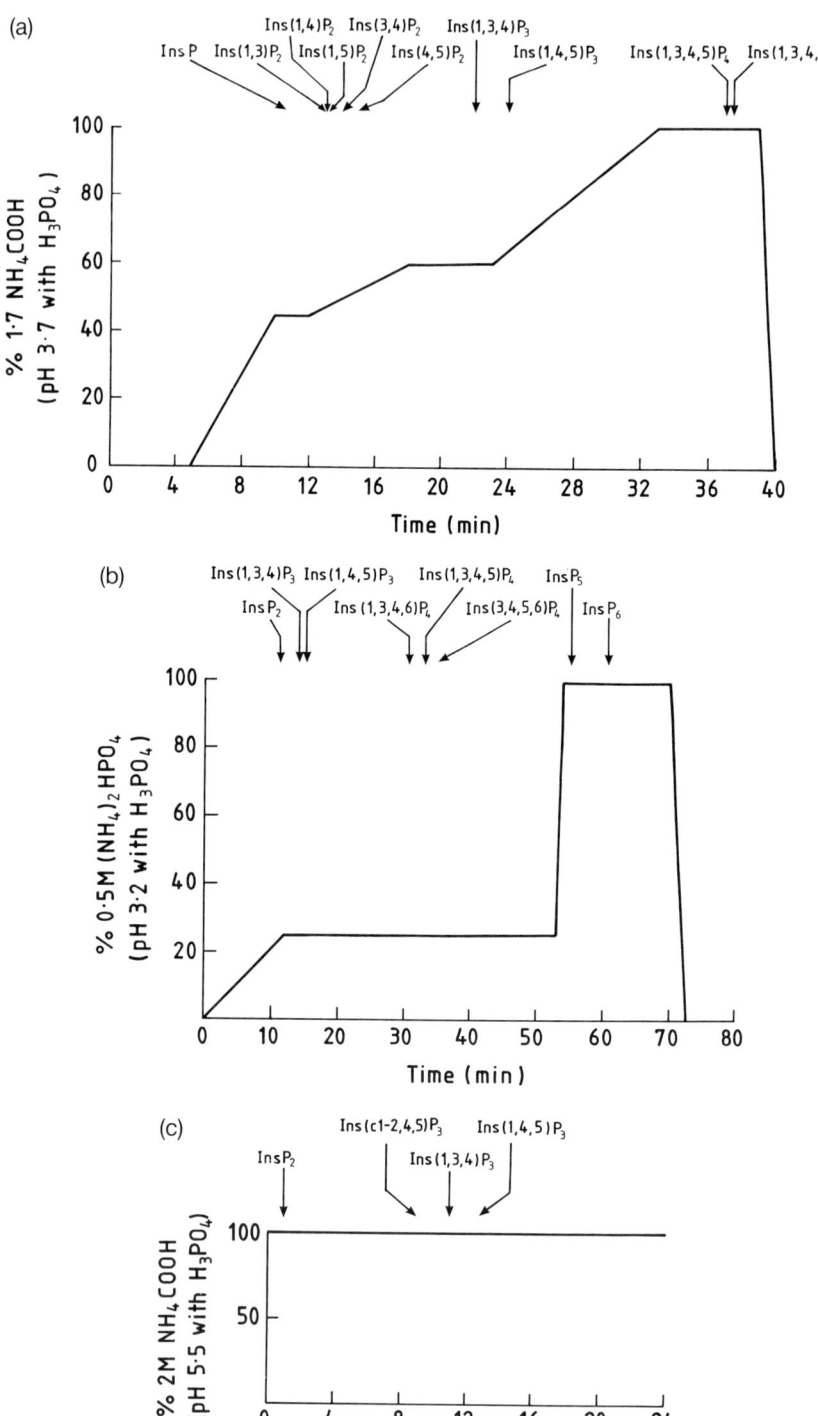

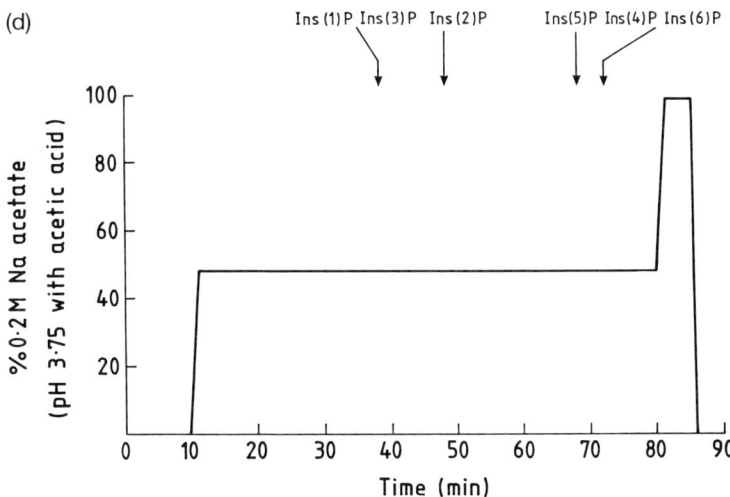

Figure 1a–d. Elution of inositol phosphates on HPLC. The point at which the peak of each inositol phosphate is known to elute in a typical run is arrowed. Inositol is omitted for clarity, but will elute in the load. Note that the extent to which individual isomers can be resolved is related to the broadness of the peaks, which is influenced by the peak size and the age of the column. See text for further details.

(a) Whatman Partisil 10 SAX eluted with 1.7 M ammonium formate (pH 3.7 with phosphoric acid.

(b) Whatman Partisphere WAX eluted with 0.5 M ammonium phosphate (pH 3.2 with phosphoric acid).

(c) Whatman Partisil 10 SAX eluted with 2 M ammonium formate (pH 5.5 with phosphoric acid).

(d) Whatman Partisphere 5 SAX eluted with 0.2 M sodium acetate (pH 3.75 with acetic acid).

labelling studies where interest centres on the phosphoinositol headgroups of the polyphosphoinositides.

5.1 Purification of phosphoinositides by affinity chromatography on neomycin-linked glass beads

The aminoglycoside antibiotic neomycin is known to interact specifically with the polyphosphoinositides. This property has been exploited in the design of an affinity column for the purification of these lipids (36), which has some advantages over TLC. Most notably, the affinity method is insensitive to non-lipid contaminants (e.g. haem), is able to yield chemically pure PtdIns4P and PtdIns(4,5)P_2 in a single step and has a far greater capacity than TLC.

The stationary phase is neomycin reductively coupled to reactive glass beads such as 'Glycophase CPG' (Pierce Chemical Co.) or 'Glyceryl Glass' (Sigma). Beads of a mesh size between 200 and 400 and a pore size of 240 Å have been found to be appropriate. The beads are coupled to neomycin by first oxidizing the glycerol groups on the glass to aldehydes by incubating 20 g of glass beads with

200 ml of 20 mM sodium periodate under vacuum and with constant stirring for 1 h at room temperature. After washing the beads three times with distilled water, they are added to 200 ml of 60 mM neomycin adjusted to pH 9.0 with NaOH. The mixture is stirred for a further hour under vacuum at room temperature, after which 200 mg of sodium borohydride is added. The beads are then washed sequentially with water, 1 mM HCl, 50% (v/v) methanol and, finally, chloroform:methanol:1.5 M ammonium acetate (3:6:1 by vol.). The neomycin-linked beads may be stored as a slurry in this final solvent mixture.

The capacity of the resin for the polyphosphoinositides has been reported to be around 2 μmol/ml and the resin may be used in gravity-fed columns. Lipids are applied to the column as the lower (organic) phase of a conventional Bligh and Dyer (37) extraction mixed with an equal volume of chloroform:methanol:1.5 M ammonium acetate (3:6:1 by vol.). Unbound lipids are eluted with 20 column volumes of this latter solvent mixture. PtdIns4P is eluted with 8 column volumes of chloroform:methanol:6 M ammonium acetate (3:6:1 by vol.) and PtdIns(4,5)P_2 is eluted with 8 vol. of chloroform:methanol:concentrated aqueous ammonia (3:6:1 by vol.). In order to prevent decomposition of the separated lipids, exposure to basic solvents should be kept as brief as possible. Hence the lipid fractions are eluted directly into tubes containing either: 2.4 M HCl (0.3 ml/ml eluant, for PtdIns4P), or 6 M HCl (0.3 ml/ml eluant, for PtdIns(4,5)P_2). After mixing and centrifugation, the lower (lipid containing) phases are recovered. Recovery is improved by re-extracting the upper phase with chloroform.

5.2 Thin-layer chromatography

A thin-layer chromatography method which can be adapted to separate either the polyphosphoinositides or phosphatidylinositol by itself may be successfully performed using HPTLC Kieselgel 60 plates (Merck) and a choice of solvent systems (38). A mixture of chloroform/methanol/aqueous ammonia/water (45:45:4:11 by vol.) will achieve a good separation of PtdIns4P and PtdIns(4,5)P_2 but will not reliably separate PtdIns3P from PtdIns4P (4) or PtdIns from phosphatidic acid and some other lipids. PtdIns may be separated from other lipids with chloroform/methanol/glacial acetic acid/water (50:32:11:3 by vol.), but the polyphosphoinositides remain at the origin in this solvent system.

A particularly effective method for the separation of all the inositol lipids and phosphatidic acid, and one that is in routine use in the authors' laboratory, was first described by Jolles and his colleagues (39). This may also be performed on commercially prepared plates, and we have found Polygram SIL N HR (Camlab) to be suitable. The plates are first impregnated with potassium oxalate by dipping them briefly in 40% methanol containing 2% potassium oxalate and 2 mM EDTA. After spotting, chromatograms are developed in chloroform/acetone/methanol/acetic acid/water (40:15:13:12:8). This method will also resolve

PtdInsP_3 (5) but we have noted that PtdIns4P may sometimes co-chromatograph with a phosphorus containing impurity (40). This problem may be overcome if samples are first chromatographed in a system designed to separate the polyphosphoinositides from other phospholipids (see 40 for details).

With either of these methods, lipids may be visualized with I_2 vapour, an appropriate lipid stain or by autoradiography. It is vital that individual spots are identified using authentic lipid standards of suitable purity.

5.3 Deacylation of inositol lipids and the separation of glycerophosphoinositol esters by anion-exchange chromatography

Since the headgroups of the polyphosphoinositides are so much more polar than those of other lipids, the glycerophosphoinositol esters derived from the deacylation of these lipids may be readily separated by anion-exchange chromatography (9). Deacylation may be performed either by alkali metal alcoholysis (41) or alkaline O→N transacylation using methylamine (42). The latter method is preferred due to the relative absence of side products (43), and excess reagent can be quickly removed by volatilization. We use 100-fold excess of methylamine over lipid phosphorous. For the deacylation of inositol lipids prepared from 6 ml of packed human erythrocytes (Section 6.1.1), we add the dried lipid extract to 2 ml of a mixture of 33% methylamine (in methanol, BDH Chemicals, Poole, Dorset, UK):H_2O:n-butanol:methanol, in the ratio 44:44:9:3 (42). An alternative mixture is 40% aqueous methylamine (Fluka):H_2O:n-butanol:methanol (36:8:9:47 by vol.). This is incubated at 50°C for 45 min in a tightly stoppered tube. The mixture is cooled and evaporated to dryness under vaccuum and then added to 2 ml of a mixture of n-butanol, petroleum ether, ethyl formate (20:40:1 by vol.), followed by 2 ml of water. The ethyl formate reacts with any remaining methylamine. After thorough mixing and gentle centrifugation, the lower aqueous phase, containing the water-soluble deacylation products, is carefully removed and the organic phase re-extracted with a further 2 ml of water.

The glycerophosphoinositol esters in the above extracts can be separated on the anion exchange columns (Bio-Rad AG1-X8 200–400 mesh, formate form) described in Section 4.1. If deacylation has been effected by alcohol metal alcoholysis, the extracts should first be neutralized with boric acid (41). Having loaded the extracts on to the columns, the deacylation products of glycerolipids other than the polyphosphoinositides are first eluted with 20 ml of 0.18 M ammonium formate/5 mM sodium tetraborate. Glycerophosphoinositol 4-phosphate (from PtdIns4P) is eluted with 10 ml of 0.3 M ammonium formate/0.1 M formic acid and glycerophosphoinositol 4,5-bisphosphate (from PtdIns(4,5)P_2) is eluted with 10 ml of 0.75 M ammonium formate/0.1 M formic acid. The radioactivity in these fractions may be determined by liquid scintillation spectrometry. Columns can be regenerated as described in Section 4.1, but should not be used to separate more than 4–5 extracts. It is important to check the elution characteristics of the anion exchange columns from time to time with appropriate standards.

5.4 Assay of lipid phosphorus

The mass of individual lipids, or their deacylated products, can be determined by assaying inorganic phosphate following complete digestion of the lipids. Glass tubes used in this assay should be pre-washed with 2 N H_2SO_4. Dried lipid extracts (or their deacylation products) can be digested in 0.06 ml 2 N H_2SO_4. Samples are heated for 1 h at 170°C in tightly sealed tubes, then cooled and neutralized with 0.3 ml 1.33% NH_4OH before being evaporated to dryness.

A variety of inorganic phosphate assays are available (see, e.g., reference 31), but a particularly useful procedure, sensitive to 2 ng phosphorus/tube, uses a malachite green reagent (44). This is prepared immediately prior to use by combining: 30 ml H_2O, 0.18 g malachite green carbinol base (Aldrich Chemical Co.), 20 ml 1.0 M HCl, and 10 ml 4.2% ammonium molybdate in 4.5 M HCl. When the above constituents have been mixed, 0.6 ml Triton X-100 is added and the reagent is diluted 1:1 with H_2O for use. 0.5 ml of this reagent is added to the dried lipid digest, followed by 0.5 ml H_2O or appropriate standards (2–24 ng phosphorus/tube). After 10 ± 2 min incubation at room temperature, absorbance is measured at 600 nm. A microspectrophotometer has been developed which increases the sensitivity of this assay 20-fold (44).

6. Preparation of radioactive inositol phosphates

It is now possible to chemically synthesize several different *myo*inositol phosphates including Ins(1,4,5)P_3 (45). The methods used are generally beyond the scope of most biochemical laboratories. Furthermore, they may not be simply adapted to allow the incorporation of radioactive precursors into the compounds concerned. For this purpose the most suitable starting materials are the inositol-containing phospholipids. The following sections describe methods for the isolation of radioactively-labelled inositol lipids from cell and tissue preparations and procedures by which these lipids may be selectively degraded to produce inositol phosphates.

6.1 Labelling strategies

6.1.1 Labelling the monoester phosphate groups of PtdIns4P and PtdIns(4,5)P_2 with [^{32}P]-PO$_4^{2-}$ in human erythrocytes

Mature human erythrocytes cannot synthesize PtdIns *de novo*, but they contain the kinases and phosphatases which catalyse the synthesis and degradation of the polyphosphoinositides. When incubated with [^{32}P]-PO$_4^{3-}$, these cells incorporate radioactivity exclusively into the monoesterified phosphate groups of PtdIns4P and PtdIns(4,5)P_2 and also into PtdOH (46). An extract of [^{32}P]-labelled erythrocyte lipids therefore provides a good starting point for the preparation of several [^{32}P]-labelled inositol phosphates including Ins4P, Ins(1,4)P_2 and Ins(1,4,5)P_3 (see *Figure 2*).

The erythrocyte lipids are labelled with [^{32}P]-PO$_4^{3-}$ by a slight modification of the method published by King *et al.* (40). Erythrocytes are sedimented from whole blood by gentle centrifugation (1000g, 10 min) and the plasma and buffy coat layer are removed. 6 ml of packed cells are next washed three times by resuspension and recentrifugation in ice-cold medium containing 188 mM Na$^+$ Hepes, (pH 7.2 with NaOH), 5 mM Na$^+$ pyruvate, 1.18 mM Mg^{2+} gluconate, 2 mM Ca^{2+} gluconate, 10 mM glucose, 1 mM inosine, 1 mM adenine, 1.8 mM KH$_2$PO$_4$, and 0.07% w/v bovine serum albumin (the Ca^{2+} gluconate and adenine have to be dissolved separately, the latter with warming). The cells are incubated at 37°C in 10 ml of this buffer with [^{32}P]-PO$_4^{3-}$ (generally 2–5 mCi). The incubation medium facilitates rapid entry of phosphate into the cells by omitting anions which would compete with PO$_4^{3-}$ uptake. In intact human erythrocytes, the rates of synthesis and degradation of PtdIns4P and PtdIns(4,5)P_2 are such that labelling of the 5-phosphate group of PtdIns(4,5)P_2 with [^{32}P]-PO$_4^{3-}$ is more rapid than that of the 4-phosphate group of PtdIns4P and hence of PtdIns(4,5)P_2 (see reference 40 for further explanation). After 7 h of labelling the monoester phosphate groups of PtdIns4P and PtdIns(4,5)P_2 reach a steady state at which their specific activities are 25–30% of the γ-phosphate of ATP (40). Shorter labelling times result in an asymmetric distribution of radioactivity between the 4- and 5-phosphate groups of PtdIns(4,5)P_2 with the

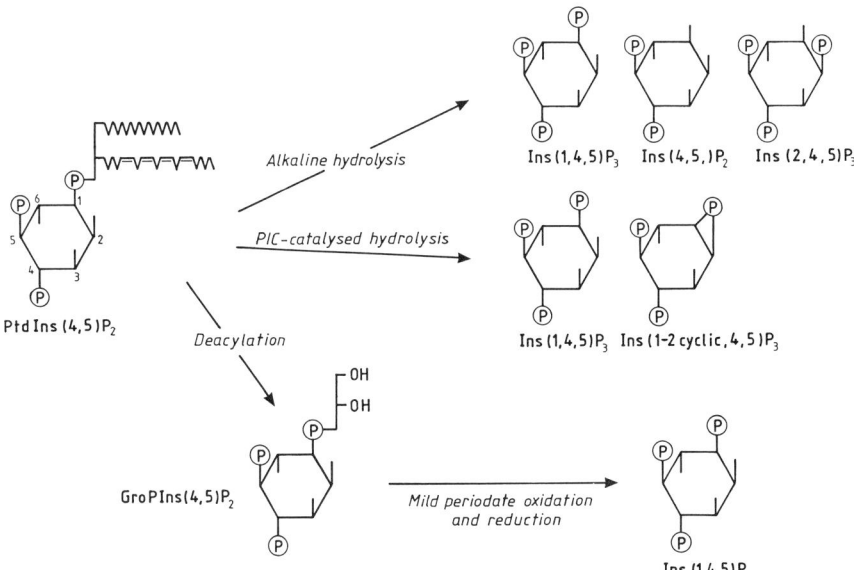

Figure 2. Preparation of inositol phosphates by enzymatic and chemical degradation of PtdIns(4,5)P_2.

5-phosphate predominant. This is useful for certain investigation of inositol phosphate metabolism (see Section 7).

The labelled erythrocytes are next lysed in 100 ml of ice-cold 20 mM Tris, (pH 7.3 at 25°C with HCl) plus 1 mM EDTA. Membranes are sedimented by centrifugation ($16\,000g$, 10 min) and washed free of haemoglobin by repeating this process six or seven times until they are white. The lipids are extracted by addition of 3.75 vol. of chloroform:methanol:12 M HCl (40:80:1) (37). The solubilized material is stood for 30 min before addition of 1.25 vol. of chloroform and 1.25 vol. of 0.1 M HCl. The mixture is vortexed and the resultant two phases separated by centrifugation. The lower phase containing extracted lipids is removed and dried under vacuum. The approximate yields of radioactivity in PtdIns4P and PtdIns(4,5)P_2 are 0.02% and 1% of added ^{32}P. Section 6.2 described methods by which the [^{32}P]-labelled PtdIns4P and PtdIns(4,5)P_2 can be quantitively converted to Ins(1,4)P_2 and Ins(1,4,5)P_3. The PtdIns(4,5)P_2 content of human erythrocytes is around 45 nmol/ml packed cells (47), so 6 ml of packed cells labelled to apparent isotopic equilibrium with 5 mCi of [^{32}P]-PO$_4^{3-}$ will generally yield 5–10 μCi of Ins(1,4,5)P_3 at an approximate specific activity of 0.02–0.04 μCi/nmol. If higher specific activities are required, the method of Spat *et al.* (48) is more appropriate. These workers employed [γ-^{32}P]-ATP to phosphorylate PtdIns and PtdIns4P in isolated human erythrocyte membranes.

6.1.2 Labelling the diester phosphate of the inositol-containing lipids with [^{32}P]-PO$_4^{3-}$ and the inositol ring with [^{14}C]- or [^{3}H]-Ins

Both these strategies require cell types which can synthesize PtdIns *de novo*. Section 2 describes the labelling of the inositol lipids of intact cells with [^{32}P]-PO$_4^{3-}$ and with [^{3}H]- or [^{14}C]-Ins. Of particular relevance to the efficient preparation of [^{3}H]- or [^{14}C]- labelled inositol phosphates are methods that maximize the incorporation of label into cell lipids. In the authors' laboratory, WRK-1 rat mammary tumour cells maintained in continuous culture are labelled for 5 days with 10 μCi/ml [^{3}H]-inositol (0.56 μM) (reference 20, see Section 2). About 10%, 2%, and 1% of added radioactivity was recovered in PtdIns, PtdIns$P$ and PtdIns(4,5)P_2 respectively.

6.2 Production of inositol phosphates by chemical and enzymic degradation of inositol lipids

The methods used for the degradation of radioactively-labelled inositol lipids to produce inositol phosphates all require hydrolysis of the phosphodiester phosphate by either specific enzymatic or nonspecific chemical means. Selective chemical degradation of inositol lipids by sequential deacylation, mild periodate oxidation and reduction can be used to yield inositol phosphates in which the original diester phosphate is preserved in monoester form. These methods are described below and depicted in *Figure 2*.

6.2.1 Phosphoinositoidase C-catalysed hydrolysis of inositol lipids

The Ca^{2+}-activated, PIC-catalysed hydrolysis of inositol lipids in erythrocytes is no longer recommended for the production of $Ins(1,4)P_2$ and $Ins(1,4,5)P_3$ because unknown contaminants are produced (49). However, the PIC-catalysed hydrolysis of inositol lipids is useful for the production of Ins1:2 cyclic phosphates (50, 51). The PIC-catalysed hydrolysis of $[^{32}P-4,5]$-$PtdIns(4,5)P_2$ in human erythrocyte membranes (Section 6.1.1) is achieved by incubating them at 37°C for 30 min with 1 mM Ca^{2+}. At neutral pH, this method yields only 1% $Ins(1:2 \text{ cyclic},4,5)P_3$ as a proportion of total $InsP_3$. The proportion of $Ins(1:2 \text{ cyclic},4,5)P_3$ increases to about 15% in 20 mM Tris (pH 5.5 with acetic acid) (N. S. Wong, unpublished data). The membranes are removed by centrifugation (16 000g for 30 min) then the cyclic and non-cyclic isomers are separated by HPLC (Section 4.2) and desalted (Section 6.5).

The preparation of radiolabelled $Ins(1:2 \text{ cyclic})P$ involves first labelling cells that can synthesize PtdIns *de novo*, (e.g. hepatocytes incubated with vasopressin, see Section 2). A dried lipid extract is made (Section 3.4) from which PtdIns is isolated by resuspending the extract in ethanol for 2 h at -20°C, after which time PtdIns is recovered by centrifugation (52). The PIC must then be presented to the substrate. PICs exist in both soluble and membrane-associated forms (see reference 2 for a short review). Where the PIC and its substrate are not present in the same phase, the mode of presentation of the enzyme to its substrate becomes important to ensure a high yield of the desired products. To this end, the procedures employed vary from simple emulsification of the lipid substrate with incubation medium to more complex preparations involving sonication and centrifugation to produce membrane vesicles (51). An effective method of the former type is used in the authors' laboratory; the PtdIns-enriched preparation is emulsified in an incubation buffer containing 5 mM Tris (pH 5.5 with maleate) by first dissolving it in diethyl ether followed by vigorous mixing with the aqueous buffer and, finally, gradual removal of the ether under vacuum.

A convenient source of PIC is a 100 000g cytosolic fraction of rat liver (Section 6.3). The cytosol is dialysed against the incubation buffer (5 mM Tris, pH 5.5 with maleate) and then incubated with the lipid emulsion in the presence of 1 mM Ca^{2+} at 37°C. The omission of Mg^{2+} from the buffer limits the degradation of the inositol phosphates by phosphatases present in the crude PIC preparation. Typically, 10 mg of cytosol protein contains sufficient activity to catalyse the hydrolysis of 2.5 mg of PtdIns in 60 mins. The reaction may be terminated by the addition of chloroform and methanol under neutral conditions (Section 3.3.2) and the inositol phosphates are then purified from the aqueous phase by high-voltage paper electrophoresis (53): samples are applied to Whatman No. 1 chromatography paper and run at 60 V/cm in pyridine:acetic acid:water (1:10:89) for between 60 and 90 min. The paper is dried and then the inositol phosphates are eluted from the paper in water.

6.2.2 Alkaline hydrolysis of inositol lipids and glycerophosphoinositol phosphates

Alkaline hydrolysis of the inositol lipids is a non-specific process producing mixtures of inositol phosphates which either lose their diester phosphate or retain it in the monoester form (54) (see *Figure 2*). Glycerophosphoinositol phosphates (GroPInsPs) are similarly susceptible to alkaline hydrolysis and, since these water-soluble deacylation products of inositol lipids are more amenable to prior purification (most simply by anion-exchange chromatography, Section 5.2), they often prove a better starting material. Base-catalysed phosphate migration between the D-1 and 2-hydroxyl groups is a minor side-reaction of this process (55) (see *Figure 3*). The procedure involves warming to 90°C in molar OH, either

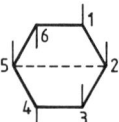

Figure 3. Haworth projection of *myo*-inositol showing the plane of symmetry that divides the molecule into chiral halves. C atoms are numbered according to the D-numbering system.

in a sealed tube placed over a boiling water bath or, for larger volumes, by refluxing. We use 1 M KOH (i.e. at least a 100-fold molar excess over the total lipid present) and heat for 30 min before the sample is cooled on ice and neutralized by addition of an appropriate volume of 11 M perchloric acid. This procedure precipitates K^+ perchlorate, which may be removed by brief centrifugation leaving the sample salt-free for purification by anion-exchange chromatography (Section 4). Alkaline hydrolysis of GroPIns[^{32}P]4P isolated from ^{32}P-labelled human erythrocyte lipids is a convenient method for the preparation of [^{32}P]-Ins4P.

6.2.3 Deacylation of inositol lipids followed by mild periodate oxidation and reduction of the glycerophosphoinositol phosphates formed

This procedure quantitatively converts inositol lipids into their corresponding inositol phosphates and may be performed by alkaline O→N transacylation using methylamine as described in Section 5.3 (42). For the deacylation of [^{32}P]-labelled inositol lipids prepared as in Section 6.1, we use 100-fold excess of methylamine over lipid phosphorus.

The use of periodate selectively to oxidize the glycerol moiety of GroPInsPs relies on the relative resistance of the inositol ring to attack by this reagent (56) (see also Section 8.1 for further explanation). We use a tenfold molar excess of 10 mM Na$^+$ periodate over total lipid phosphorus (1.6 ml in the [^{32}P]-labelled preparation described above) and incubate for 30 min in the dark at room temperature. The incubation is then quenched with a large excess (20 μl) of ethylene glycol. Periodate oxidation forms glycolaldehyde inositol phosphate esters which are reduced to the desired inositol phosphates with 200 μl of 1% aqueous 1,1-dimethylhydrazine (pH 4.0 with formic acid; see reference 57). The reaction is allowed to proceed at room temperature for 4 h. The inositol phosphates thus formed are then purified using a Bio-Rad anion-exchange column (see Section 4.1). A red contaminant (57) remains in the eluant.

6.3 Preparation of metabolites of Ins(1,4,5)P_3

Ins(1,3,4,5)P_4 may be prepared by the phosphorylation of Ins(1,4,5)P_3 using rat liver cytosol as a source of Ins(1,4,5)P_3 3-kinase, incubated at pH 9.0 to minimize 5-phosphatase activities (58). We prepare cytosol from a single rat liver from which the blood has been removed by perfusion with an ice-cold solution of 0.25 M sucrose, 5 mM Hepes (pH 7.2 with NaOH). The liver is homogenized, made up to a 30% (w/v) suspension, and centrifuged at 100 000g for 1 h. A 60-μl portion of the supernatant is incubated at 37°C with 1 ml of medium containing 40 mM glycine (pH 9.0 with KOH), 10 mM ATP, 20 mM MgSO$_4$ and the radiolabelled Ins(1,4,5)P_3. The incubation time necessary for optimum phosphorylation to Ins(1,3,4,5)P_4 will depend upon the chemical quantity of Ins(1,4,5)P_3. For 0.1 μCi of [^{32}P-4,5]-Ins(1,4,5)P_3 prepared as in Section 6.2.3, incubations of 30 min duration give a 75% yield of Ins(1,3,4,5)P_4. The incubations are quenched with perchloric acid, neutralized, and the Ins(1,3,4,5)P_4 is purified by HPLC (Section 4.2).

Ins(1,3,4)P_3 is prepared by dephosphorylation of Ins(1,3,4,5)P_4 using human erythrocyte plasma membranes (26). In the authors' laboratory, these are prepared from about 500 ml of human blood which is less than 2 weeks old. The erythrocytes are washed twice in ice-cold 154 mM NaCl (pH 7.5 with Hepes/NaOH) and divided equally into 6 × 250 ml centrifuge tubes. The cells are lysed by the addition of 5 vol. of ice-cold 20 mM Tris (pH 7.5 with HCl) and centrifuged at 17 000g (av.) for 30 min. The supernatant is aspirated to waste and the pellet gently resuspended in a similar volume of ice-cold lysis buffer. It is essential to use transparent centrifuge tubes to assist visualization of the pellet. The centrifugation and washing procedures are repeated and, after 2–3 washes, it is possible to discern a dark red 'button', beneath the looser, lighter colour ghosts. This is thought to be unlysed cells and it can be exposed and aspirated to waste if the tubes are gently tilted to one side. The fourth or fifth wash should consist of 154 mM NaCl (pH 7.5 with Hepes/NaOH), which assists the removal of haemoglobin and some non-specific phosphatase activity (31). This should be followed by further washes in 20 mM Tris (pH 7.5 with HCl) until the ghosts are completely free of haemoglobin. The washing procedure may be performed over two days by storing the pellets at 0–4°C overnight. The final pellets can be stored at −20°C for at least three months.

The optimum incubation conditions for the preparation of Ins(1,3,4)P_3 again depend upon the chemical concentration of substrate: a typical incubation in our laboratory consists of 1 μCi [^{3}H]-Ins(1,3,4,5)P_4 (50 μmol) plus 2.4 ml of packed ghosts added to 1.2 ml of medium containing 20 mM Hepes pH (7.2 with NaOH), 15 mM Mg^{2+} acetate, 0.6 mg/ml saponin. Over 85% of the Ins(1,3,4,5)P_4 is converted to Ins(1,3,4)P_3 after 30-min incubations, with no other detectable InsP_3 contaminants (see also reference 32). The incubations are quenched with 1.4 ml 2 M HClO$_4$, neutralized and purified by HPLC (Section 4.2).

6.4 Removal of salt from inositol phosphates following anion-exchange chromatography

Samples purified by HPLC should first be neutralized with 75 mM Hepes plus 50 mM EDTA, diluted tenfold and applied to a small Bio-Rad anion exchange column (Bio-Rad AG 1-X8 200–400 mesh, formate form; Section 4.1). Most HPLC eluants contain considerable amounts of phosphate (Section 4.2) and this can be eluted with 5 × 5 ml aliquots of 180 mM ammonium formate. The inositol phosphate is then eluted with the appropriate concentration of ammonium formate/formic acid (Section 4.1). A good freeze-drier may be used to lyophilize the samples directly. However, in the authors' experience this is a rather lengthy and unreliable procedure. Thus it is usual to remove the ammonia before attempting to freeze dry.

A long-standing method (31) for desalting the preparation is to apply the sample to an appropriately-sized cation-exchange column (Bio-Rad AG 50W-X8, 200–400 mesh, H$^+$ form). It is possible to improve the recoveries and obtain a much cleaner end-product if the H$^+$ resin is washed before use (two washes in 2 M NaOH, followed by washing to neutrality with water, and then two washes in 2 M HCl, before finally washing again with water until neutrality is reached). The column is washed with an equal volume of water, and the eluants are combined and lyophilized. Some workers include a few milligrams of mannitol before adding the sample to the column to improve recovery during desalting and lyophilization. However, this is to be avoided if the structure of the sample is to be determined with periodate (Section 8), since mannitol will quench the periodate oxidation reaction. In any case, if the sample if lyophilized in plastic tubes, the mannitol may be omitted. Some batches of resin give recoveries of over 85%, but this is variable and the authors have obtained yields as low as 50%, so it is wise to check each batch of new resin.

The variability in yields described above has led to other desalting techniques being introduced. For example, InsP_3, InsP_4 and InsP_5 can be desalted without the use of H$^+$ resin. Here, the samples are applied to a 100 μl formate column (Bio-Rad AG 1-X8, 200–400 mesh). For InsP_3 and InsP_4, the formate ions are eluted with 5 × 0.5 ml aliquots of 0.1 M ammonium bicarbonate then the inositol phosphates are eluted with 5 × 0.5 ml aliquots of 0.5 M ammonium bicarbonate (freshly made, filtered, and warmed to 45°C before use). For $InsP_5$, the column is first washed with 3×1 ml aliquots of 0.1 M HCl then the $InsP_5$ is eluted with 3×1 ml aliquots of 2 M HCl. Recoveries are about 85% but, unfortunately, these methods give poorer yields for other inositol phosphates. The bicarbonate or HCl is directly removed by lyophilization. Occasionally some bicarbonate may persist. This can be removed by adding 2 ml of water to the dried extract, and lyophilizing again. One major advantage of the bicarbonate technique is that it preserves cyclic inositol phosphates (S. B. Shears and N. S. Wong, unpublished data). It is also less laborious than methods based on elution from columns with LiCl (21, 54).

A more recently developed method of desalting appears to be particularly useful (L. R. Stephens, personal communication). This method involves the elution of inositol phosphates from Amprep SAX columns (Amersham International) with the relatively volatile eluant triethylamine bicarbonate. This is prepared by placing triethylamine (Fluka) on ice and diluting it to the desired concentration (see below). CO_2, from dry ice, is then bubbled into the solution for about 20 min. The SAX columns are wetted with water, washed with 10 ml of freshly prepared 1 M triethylamine bicarbonate and then rinsed with 20 ml water. The columns are best pumped under pressure (either positive or negative) and the precise flow rate is not critical, so long as it is of the order of several ml/min. The inositol phosphate in solution is diluted 20-fold with water and then applied to the column. Residual salts may be eluted with 10 ml freshly prepared 0.15 M triethylamine bicarbonate. The inositol phosphate can then be eluted with 10 ml 2M triethylamine bicarbonate, which can be lyophilized directly.

6.5 Commercially available inositol phosphates

Both New England Nuclear (DuPont) and Amersham International supply a variety of [3H]-labelled inositol phosphates. At the time of writing, these include $Ins1P$, $Ins4P$, $Ins(1,4)P_2$, $Ins(1,3,4)P_3$, $Ins(1,4,5)P_3$, $Ins(1,3,4,5)P_4$, and $InsP_6$. [14C]-$Ins3P$ is also available. However, the only [32P]-labelled inositol phosphate offered by either supplier is [32P-4,5]-$Ins(1,4,5)P_3$. The relatively short half-life of 32P means that [32P]-labelled inositol phosphates must be prepared frequently. The much longer half-life of 14C means that inositol phosphates labelled with this isotope are better suited as markers to identify cell-derived [3H]-labelled inositol phosphates. Unfortunately, the current high cost of [14C]-Ins is prohibitive to the large-scale preparation of [14C]-labelled inositol phosphates.

The list of non-radioactive inositol phosphates which are commercially available continues to grow. Currently (December 1989), $Ins(1,3,4,5,6)P_5$, $Ins(1,3,4,5)P_4$, $Ins(3,4,5,6)P_4$, $Ins(1,4,5)P_3$, $Ins(1,3,4)P_3$, $Ins(1,4)P_2$, $Ins1P$, and $Ins2P$ may be purchased from Calbiochem or Sigma; $Ins(1:2$ cyclic $4,5)P_3$ is expected to be available soon.

7. Assay of inositol phosphate phosphatases and kinases

It is now possible to purchase or synthesize a number of radioactive inositol phosphates (Section 6) for use as substrates to assay several of the kinases and phosphatases present in tissue preparations. The enzymes can be assayed under zero-order conditions, but the necessary chemical quantities of substrates are often either not commercially available or too expensive to purchase. In these cases, the enzymes may be assayed under first-order conditions, by incubating the tissue with trace quantities of radioactive substrates.

The incubation buffer is a matter of choice for the experimenter, but the authors have frequently used a medium designed to imitate physiological intracellular ionic strength and pH: 120 mM KCl, 10 mM NaCl, 1 mM EGTA, 4.18 mM $MgSO_4$, 0.33 mM $CaCl_2$, and 10 mM Hepes (pH 7.2). Saponin (0.2 mg/ml) is also added to permeabilize the added tissue preparations. At 37°C, this buffer has 4 mM free Mg^{2+} and 0.1 μM free Ca^{2+} (28). When ATP is added, this is usually at 5 mM, and total $MgSO_4$ is then raised to 9.75 mM to maintain the free Mg^{2+} and Ca^{2+} concentrations at the required level. The inositol phosphate substrate is also added, and the incubations commenced with an appropriate volume of tissue. The incubations are quenched, neutralized, and chromatographed as described in Sections 3 and 4.

Sometimes it is necessary to identify the products of phosphorylation or dephosphorylation. Several techniques are described in Section 8. One useful additional method is to use [32P]- and [3H]-labelled substrates: the assay of Ins(1,4,5)P_3 5-phosphatase is a good illustration of the benefits of such procedures. In this example, [32P-4,5]-Ins(1,4,5)P_3 can be produced from erythrocytes with an asymmetric distribution of 32P between the two phosphates. A 70:30 split between the 5 and 4 phosphates works well, and the preparation of such a compound is described in Section 6.1.1. The [32P-4,5]-Ins(1,4,5)P_3 is mixed with some [3H]-Ins(1,4,5)P_3, and these are incubated with the test tissue. In separate incubations erythrocyte ghosts (Section 6.3) are added as a control, since they dephosphorylate Ins(1,4,5)P_3 to Ins(1,4)P_2 in the presence of 4 mM

Table 1. Hypothetical example of the 32P:3H ratios of the potential InsP_2 products formed by dephosphorylation of a mixture of [32P-4,5]-Ins(1,4,5)P_3 and [3H]-Ins(1,4,5)P_3

The ratio of 32P in the 4 and 5 phosphates of [32P-4,5]-Ins(1,4,5)P_3 is assumed to be 30:70. The 32P:3H ratio of the Ins(1,4,5)P_3 substrate is assumed to be 1.

InsP_2 product	32P:3H ratio
Ins(1,4)P_2	0.3
Ins(1,5)P_2	0.7
Ins(4,5)P_2	1.0

Mg^{2+} (31). By trial and error it is possible to determine the quantity of tissue that will hydrolyse about 50% of the substrate at 37°C. The $InsP_2$ product and the $InsP_3$ substrate are separated by ion-exchange chromatography (see Section 4), and the $^{32}P:^{3}H$ ratios of each are measured. Every potential $InsP_2$ product, $Ins(1,4)P_2$, $Ins(1,5)P_2$ and $Ins(4,5)P_2$, will have a characteristic $^{32}P:^{3}H$ ratio (see *Table 1*). Hence, if the $^{32}P:^{3}H$ ratio of the genuine $Ins(1,4)P_2$ is equal to the $^{32}P:^{3}H$ ratio of the $InsP_2$ produced by the test tissue, then the tissue is shown to attack the 5-phosphate of $Ins(1,4,5)P_3$.

8. Methods for determining the structure of inositol polyphosphates

It is now clear that the original characterizations of inositol phosphate structures in stimulated cells (14) seriously underestimated the range of inositol phosphates present. Much of the evidence for this complexity has come from detailed examinations of extracts from [^{3}H]-Ins-labelled cell preparations by anion-exchange HPLC (see Section 4.2). Some other techniques used to determine the structures of inositol phosphates require these compounds in greater quantities and of greater purity than may easily be prepared from the sorts of cell and tissue preparations used to study receptor signalling. Examples of such techniques are nuclear magnetic resonance spectroscopy and mass spectroscopy (59–62) and also the determination of enantiomers by their rotation of plane polarized light (63). However, the chemically-based techniques described below are suitable for the analysis of radiochemial quantities of inositol phosphates (see reference 64, for a recent review of this work).

8.1 Periodate oxidation, reduction, dephosphorylation, and the identification of the polyols thus produced

Oxidation with periodate has proved an effective means of determining the isomeric configuration of inositol phosphates (30, 54, 63, 65). Under appropriate conditions, periodiate oxidation splits carbon–carbon bonds between adjacent carbon atoms bearing unsubstituted hydroxyl groups. Substitution of either of the hydroxyl groups, in this case by phosphorylation, protects the carbon–carbon bond from attack. For example, D-$Ins(1,4,5)P_3$ is a periodate-sensitive compound which can subsequently be reduced and dephosphorylated, yielding D-iditol (*Figure 4*). Thus, identification of the resultant polyol following periodate oxidation, reduction, and dephosphorylation can provide important information about the structure of an inositol polyphosphate.

Some understanding of the stereochemistry of *myo*inositol phosphates is necessary to appreciate the strengths and limitations of the periodate oxidation technique. In particular, it should be realized that although *myo*inositol is a *meso* compound (i.e. whilst it is not optically active, it is composed of two chiral halves), the molecule does possess a plane of symmetry: the axis between the

173

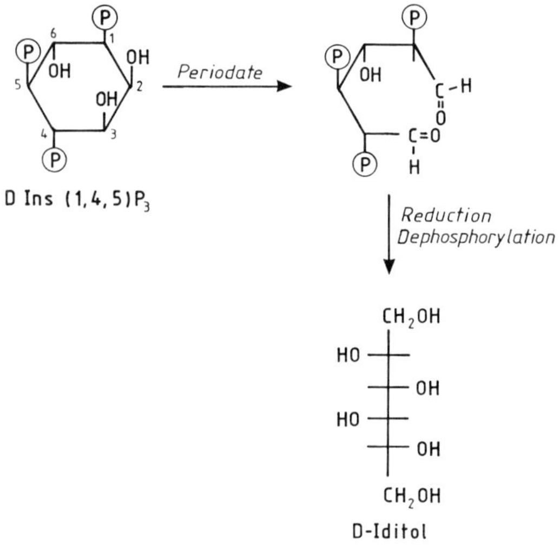

D Ins (1,4,5)P₃

Figure 4. Pathway of degradation of D-Ins(1,4,5)P_3 by periodate oxidation, reduction, and dephosphorylation.

2- and 5-OH groups. Asymmetric substitution of hydroxyl groups on either side of this plane produces optically active molecules (see references 66, 67 for reviews; and *Figure 3* for further explanation). This chirality is preserved in the polyol products of periodate oxidation, reduction, and dephosphorylation, For example, D-Ins(3,4,5,6)P_4 will ultimately yield L-iditol, whereas D-Ins(1,4,5,6)P_4 will ultimately yield D-iditol (*Figures 5*; see reference 30). Therefore the structure of these two InsP_4 molecules can only be determined provided the enantiomer of the polyol is also characterized.

Table 2 lists the products of periodate oxidation, reduction, and dephosphory-lation of all the possible isomers of *myo*inositol bis-, tris-, and tetrakisphosphate. It should be noted that all InsP_5 and InsP_6 molecules are resistant to attack. Ins$P$ isomers, and symmetrically substituted InsP_2 isomers, will be totally destroyed by this procedure. A further complication is that commercially available [³H]-*myo*inositol has the [³H] attached to the 2-carbon atom. Periodate oxidation of inositol phosphates in which the bonds between the 1- and 2- and the 2- and 3-carbon atoms are not protected by phosphorylation results in loss of this atom, and does not yield labelled polyols (*Table 2*).

The methods used in our laboratory for the periodate oxidation of inositol phosphates are essentially those of Tomlinson and Ballou (63) as adapted by Irvine *et al.* (65) and improved by Stephens *et al.* (30). Oxidation of cyclic polyols such as Ins and its phosphorylated derivitives proceeds much more slowly than that of non-cyclic polyols (compare the conditions described in Section 6.2.3 for oxidation of the glycerol moiety of Gro$P$Ins$P$s). An aqueous solution of 0.1 M Na⁺ periodate at its natural pH of around 5.0 is sufficient for oxidation of inositol

C. J. Kirk, A. J. Morris, and S. B. Shears

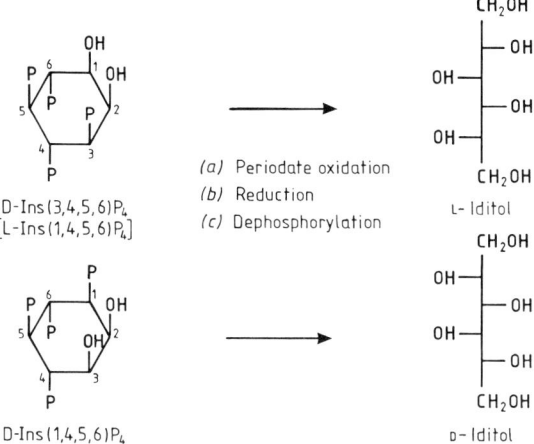

Figure 5. Conversion of D-Ins(3,4,5,6)P_4 and D-Ins(1,4,5,6)P_4 to L-iditol and D-iditol respectively by periodate oxidation, reduction and dephosphorylation.

bis- and trisphosphates. However, inositol tetrakisphosphates with vicinal hydroxyl groups are more resistant to oxidation (59, 63) so 0.1 M periodic acid (pH 3.0 with NaOH) is used. The purified and desalted inositol phosphates (Section 6) are incubated at 25°C for 36 h in the dark in 0.5 ml of the appropriate periodate solution. Then the excess periodate and the aldehydes resulting from the periodate oxidation are reduced with 0.5 ml of freshly dissolved and filtered 1 M NaBH$_4$, The samples are allowed to stand for a further 12 h, in closed tubes. Inositol, xylitol, altritol, and iditol carriers are best added at this stage to a final concentration of 20 μg/ml. The resultant solution is applied to a 3-ml column of BioRad AG 50 X8 cation exchange resin in the H$^+$ form. The solution should be added slowly to reduce the amount of effervescence. It is also vital that the cation exchange resin is thoroughly washed before use, as described in Section 6.5. The eluant is saved, and the column is allowed to stand for 30 mins before washing with 3 × 3 ml aliquots of deionized, distilled water. In order to maximize recoveries through this step, we elute the columns to dryness after the final wash by placing them in the neck of a suitable vessel and then spinning this assembly gently in a bench centrifuge. The combined eluants are evaporated to dryness and the residue resuspended in 10 ml of dry methanol. This converts unreacted borohydride to volatile trimethylborane which may be removed under vacuum. A further 10 ml of methanol is added, the sample dried under vacuum and then resuspended in 2 ml of 10 mM ethanolamine (pH 9.5 with HCl) plus 1 mM MgSO$_4$ and 20 units of bovine intestinal mucosa alkaline phosphatase. Commercially available suspensions of alkaline phosphatase should be centrifuged before use and the pellet redissolved in incubation medium. Incubations are at 25°C for 12 h, but trial runs are advantageous to ensure complete dephosphorylation. The sample is next applied to a 3-ml column of Amberlite

175

Table 2. Polyol products of periodate oxidation, reduction, and dephosphorylation of *myo*inositol bis-, tris-, and tetrakisphosphates. Inositol phosphates denoted with an asterisk lose the 2-^{3}H label derived from [2-^{3}H]-inositol upon periodate oxidation

Polyols formed		Inositol phosphates		
		D-InsP_4	D-InsP_3	D-InsP_2
HEXITOLS:	Inositol	1,3,4,5	1,3,5	
	(not oxidized)	1,2,4,5	2,4,6	
		1,2,4,6		
		1,3,5,6		
		2,3,5,6		
		2,3,4,6		
		1,3,4,6		
		1,2,3,5		
		2,4,5,6		
	D-Iditol	1,4,5,6	1,4,5	
			1,4,6	
	L-Iditol	3,4,5,6	3,5,6	
			3,4,6	
	D-Glucitol	1,2,5,6	2,5,6	
			1,2,5	
	L-Glucitol	2,3,4,5	2,4,5	
			2,3,5	
	D-Altritol	1,2,3,6	1,3,6	
			2,3,6	
	L-Altritol	1,2,3,4	1,3,4	
			1,2,4	
PENTITOLS:	Xylitol		3,4,5	3,5
			4,5,6*	4,6*
			1,5,6	1,5
	Adonitol (=ribitol)		1,2,3	1,3
	D-Arabitol		1,2,6	2,6
	L-Arabitol		2,3,4	2,4
TETRITOLS:	Eythritol			1,2
				2,3
	D-Threitol			4,5*
				1,6
	L-Threitol			5,6*
				3,4
POLYOLS DESTROYED:				1,4
				2,5
				3,6

MB-3 mixed bed deionizing resin (Sigma), left to stand for 30 min and then washed with 6 ml of deionized distilled water by centrifugation, as described above. The combined eluants from this step are finally evaporated to dryness under vacuum. Recoveries through the entire procedure should exceed 60% (30).

8.2 Separation of polyols

Paper chromatography has been used to identify the polyols produced by periodate oxidation, reduction and dephosphorylation of inositol phosphates (54, 65). This method suffers from several disadvantages, most notably it is time-consuming and, after development, the paper must be cut up and eluted before [^{3}H]-labelled polyols are detected by liquid scintillation spectrometry. Weak anion-exchange chromatography is a standard procedure for the separation of carbohydrate molecules and Stephens *et al.* (30) have employed an HPLC version of this technique for the identification of [^{3}H]-labelled polyols. Since the HPLC columns used for this procedure are extremely sensitive to impurities, it is essential to use HPLC-grade water as the eluant and advisable to employ inert HPLC hardware, such as titanium or polymers (Teflon).

Figure 6 describes the behaviour of the possible tetritol, pentitol, and hexitol products of periodate oxidation, reduction, and dephosphorylation of *myo*-inositol bis-, tris-, and tetrakisphosphates during chromatography on a Polypore

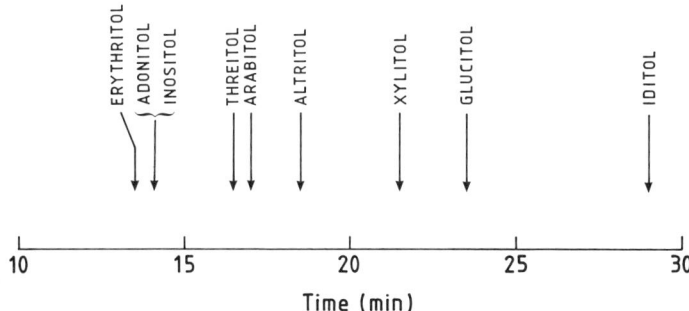

Figure 6. Elution of various polyols on a Polypore Pb^{2+} column.

Pb^{2+} column (4.6 × 220 mm main column with a 4.6 × 30 mm guard column, Pierce Chemical Co., USA). Samples should be loaded in a volume no greater than 10 μl. The column is eluted with HPLC-grade water (degassed by sparging with helium) at a flow rate of 0.2 ml/min at 80°C. The elution positions of non-radioactive polyol standards may be determined with greatest sensitivity by differential refractometry (30). However, an acceptable alternative is to monitor the absorbance of the column eluant at 200 nm. In order to maintain the back pressure necessary to keep a continuous flow of eluant through the spectrophoto-meter (and hence a stable base-line), the cell outlet should be connected to an appropriate length of narrow bore tubing. The elution characteristics can also be determined using radioactive internal standards. [^{14}C]-D-glucitol, [^{14}C]-xylitol and [^{14}C]-inositol are available from Amersham International. [^{14}C]-adonitol ('ribitol') can be made by borohydride treatment (see Section 8.1) of [^{14}C]-ribose (Amersham International).

8.3 Use of polyol dehydrogenase to distinguish between enantiomeric pairs of polyols

Where it is necessary to determine the chirality of a polyol (see above) this can sometimes be achieved by using a stereoselective NAD^+-dependent polyol dehydrogenase (L-iditol: NAD^+ oxidoreductase, E.C. 1.1.1.14; see reference 27). The enzyme is sold by Sigma, and the preparations from either sheep liver or *Candida utilis* are both suitable. This enzyme catalyses the oxidation of several acyclic fully hydroxylated polyols including pentitols, hexitols, and heptitols. The structural requirements for substrates of the enzyme are a primary hydroxyl group at C-1 with secondary hydroxyl groups at C-2 and C-4 in the L-configuration relative to C-1. For example (*Figure 7*), the enzyme will preferentially oxidize L-iditol or L-altritol (to L-sorbose or L-tagatose, which elute close to inositol on the polypore column), whereas there is little oxidation of D-iditol and D-altritol.

Polyol oxidation is measured by the methods developed by Stephens *et al.* (30, 71) using 1-ml incubations of 100 mM Tris (pH 8.3 with HCl) and 20 mM NAD, plus 1–10 units/ml of the polyol dehydrogenase, the precise amount depending upon the eventual rate of polyol oxidation. Particulate material in the reconstituted enzyme is removed by centrifugation prior to use. Once a stable base-line is attained (which may take some minutes due to impurities in the enzyme preparation), the $[^3H]$-polyol of unknown chirality is added together with a final concentration of 100 μM of the readily oxidizable enantiomer of the polyol. For example, L-iditol (Sigma) would be included when the $[^3H]$-polyol was either L- or D-iditol, and 100 μM L-altritol would be added to the assay when the $[^3H]$-polyol was either L- or D-altritol (see also *Figure 8*). Under these

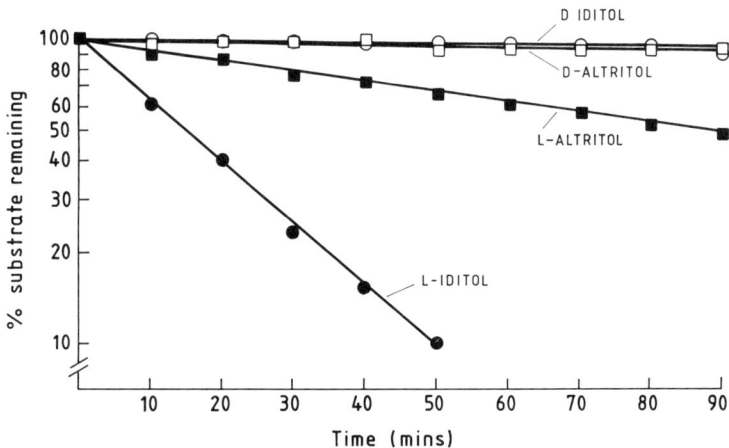

Figure 7. Stereo-selectivity of L-polyol dehydrogenase from *C. utilis*. For further details see text.

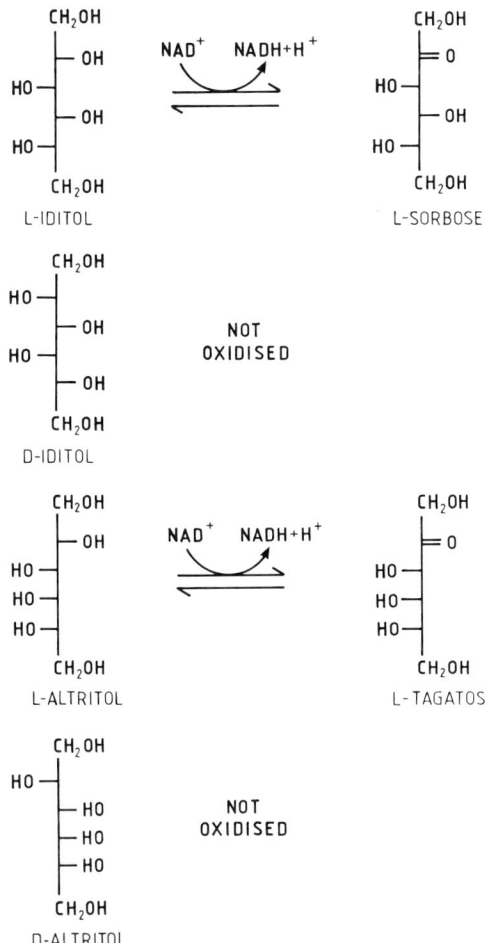

Figure 8. Structures of D- and L-enantiomers of iditol and altritol, showing their oxidation products where appropriate, following incubation with polyol dehydrogenase.

conditions, polyol oxidation obeys first-order kinetics. Oxidation of L-iditol can be measured by the absorbance change at 340 nm, assuming a 1:1 stoichiometry between this and the amount of NADH produced (the extinction coefficient for NADH is 6.22 mM cm^{-1}).

The most quantitative analysis of the progress of the reaction is obtained by fitting the absorbance changes to a first-order rate equation (30) and then extrapolating the exponential to determine the 100% value. A computer programme for this purpose ('Enzfitter') may be obtained from Elsevier Software, Cambridge, UK. Whilst these methods work well for iditol, they are not as suitable for altritol, since commercially available L-polyol dehydrogenase oxidizes L-altritol at a relatively slow rate such that substantial amounts of NADH are destroyed during the incubations (*Figure 7*; reference 71). The most accurate quantitation of L-altritol oxidation is obtained by including in the assays an internal standard of $[^{14}C]$-L-altritol, although this is not easy to prepare (71). An alternative approach is to follow the oxidation of authentic $[^{3}H]$-L-altritol in parallel incubations. It is also worth checking the specificity of the dehydrogenase by comparing the initial rates of oxidation of both D- and L-altritol (*Figure 7*). For these experiments, the $[^{3}H]$-L-altritol can be made by periodate oxidation of $Ins(1,3,4)P_3$ (see above), whereas D- and L-altritol are made by borohydride reduction of L-talose and D-altrose respectively (see Section 8.1).

Polyol oxidations are terminated, by boiling for 3 min, once the reactions are about 80–90% complete. The denatured enzyme is removed by centrifugation, and the supernatant is added to 2 ml of Amberlite MB-3 resin. After 30 min the resin is washed twice with 5 ml of water, and the eluants are combined and dried. The reduced and oxidized forms of the polyols are then separated on the Pb^{2+} polypore column as described above. The proportion of $[^{3}H]$polyol oxidized is then determined relative to non-incubated controls, and hence the enantiomeric composition of the original sample is revealed.

8.4 Specific dephosphorylation of periodate-resistant inositol phosphates to produce periodate-sensitive compounds

Examination of *Table 2* reveals that, except in the case of the periodate-sensitive inositol tetrakisphosphates, identification of the polyols produced following periodate oxidation, reduction, and dephosphorylation may not be sufficient to unequivocally determine the structure of an inositol phosphate. A complementary experimental strategy involves the selective removal of one or more of the phosphate groups yielding a periodate-sensitive compound. Alkaline phosphatase has proved useful in this respect.

Although it shows no absolute positional selectivity for phosphate groups around the inositol ring of inositol phosphates, bovine intestinal alkaline phosphatase has greater activity against solitary phosphate groups than against those adjacent to others (68). An investigation of the susceptibility of an inositol phosphates to alkaline phosphatase-catalysed dephosphorylation can therefore provide information about the relative disposition of its phosphate groups. For example, Batty *et al.*, (26) prepared an $InsP_4$ from rat cortical slices stimulated with carbamylcholine. Using alkaline phosphatase, the $InsP_4$ was partially dephosphorylated to an $InsP_3$, which was periodate oxidized, dephosphorylated, and reduced. The polyol product was identified as xylitol, so the $InsP_3$ was $Ins(3,4,5)P_3$ (*Table 2*), produced in turn by dephosphorylation of the lone 1-phosphate of $Ins(1,3,4,5)P_4$. Inositol phosphates can be partially dephosphorylated by incubation with alkaline phosphatase under the conditions described for the dephosphorylation of the phosphorylated polyols (Section 8.1), provided the incubation time is carefully controlled to optimize the yield of the products.

C. J. Kirk, A. J. Morris, and S. B. Shears

8.5 Non-selective dephosphorylation of inositol phosphates and identification of the inositol monophosphates formed

Heating with NH_4^+ results in elimination of phosphate groups from inositol phosphates (32, 68, 69). This process occurs without rearrangement of phosphate groups and would therefore be expected to produce a mixture of inositol monophosphate isomers which retain the configuration of their parent inositol polyphosphates. For example, $Ins(1,3,4,5)P_4$ would yield a mixture of Ins1P, Ins3P, Ins4P, and Ins5P. However, it must be emphasized that the alkaline dephosphorylation is not a random process. Hence the proportions of individual inositol phosphates arising from this procedure do not necessarily reflect their quantitative distribution in the precursor inositol phosphate (72), therefore these methods cannot be used to quantify the proportions of different isomers in a mixture of inositol phosphates. Nevertheless, non-selective dephosphorylation, followed by identification of the inositol monophosphate isomers formed, is a useful technique for determining the structure of pure inositol phosphates.

Radiolabelled inositol phosphates are dephosphorylated in 1 ml of freshly made 10 M NH_4OH containing 1 µmol of the NH_4^+ salt of $InsP_6$ as carrier (prepared by passing a solution of commercially available Na^+ phytate through a suitably-sized column of BioRad AG 50 X8 cation exchange resin in the NH_4^+ form). The resultant solution is placed in a 4-mm thick-walled Pyrex tube of approximately 10 ml total volume. The tube is sealed and then heated to 110°C in a thermostatically-controlled block. After the reaction is complete, the tube is broken open and its contents evaporated to dryness under vacuum or with N_2. The contents are resuspended in 1 ml of water and centrifuged to remove a precipitate which forms during the incubation. Stephens et al. (32) report that the rate of dephosphorylation is such that the optimal production of inositol phosphates takes 30, 50, and 100 h respectively for $InsP_3$, $InsP_4$ and $InsP_5$. However, trial experiments are advisable since, in prolonged incubations, the samples will be ultimately be converted to inositol. The inositol monophosphates thus produced may be separated by paper chromatography (54, 68) or anion-exchange HPLC (32) (see Section 4.2). As discussed in Section 8.1, the 1/3 and 4/6 pairs of hydroxyls of *myo*inositol are optical enantiomers. Neither of these separation systems is capable of resolving these enantiomeric pairs of inositol monophosphates. The only published method for separation of enantiomeric pairs of inositol monophosphates is that of Sherman et al. (70) which uses a chiral GLC technique. Unfortunately, this method is not suitable for use with small quantities of radiolabelled inositol phosphates.

Acknowledgements

We are grateful to Chris Barker, Len Stephens and Nai Sum Wong for helpful discussions and information on various methods which they have developed.

References

1. Michell, R. H. (1975). *Biochim. Biophys. Acta.*, **415**, 81.
2. Downes, C. P. and Michell, R. H. (1985). In *Molecular Mechanisms of Transmembrane Signalling* (ed. P. Cohen and M. D. Houslay), p. 3. Elsevier, Amsterdam.
3. Berridge, M. J. (1987). *Ann. Rev. Biochem.*, **56**, 159.
4. Whitman, N., Downes, C. P., Keeler, M., Keller, T., and Cantley, L. (1988). *Nature, Lond.*, **332**, 644.
5. Traynor-Kaplan, A. E., Harris, A. L., Thompson, B. L., Taylor, P., and Sklar, L. A. (1988). *Nature, Lond.*, **334**, 353.
6. Tarver, A. P., King, W. B., and Rittenhouse, S. E. (1987). *J. Biol. Chem.*, **262**, 17268.
7. Palmer, S., Hughes, K. T., Lee, D. Y., and Wakelam, M. J. O. (1988). *Cellular Signalling*, **1**, 147–56.
8. Daniel, J. L., Dangelmaier, C. A., and Smith, J. B. (1987). *Biochem. J.*, **246**, 109.
9. Creba, J. A., Downes, C. P., Hawkins, P. T., Brewster, G., Michell, R. H., and Kirk, C. J. (1983). *Biochem. J.*, **212**, 733.
10. Palmer, S., Hawkins, P. T., Michell, R. H., and Kirk, C. J. (1986). *Biochem. J.*, **238**, 491.
11. Koreh, K. and Monaco, M. E. (1986). *J. Biol. Chem.*, **261**, 88.
12. Michell, R. H., King, C. E., Piper, C. J., Stephens, L. R., Bunce, C. M., Guy, G. R., and Brown, G. (1988). In *Cell Physiology of Blood* (ed. R. B. Gunn and J. C. Parker), pp. 345–55. Rockefeller University Press.
13. Prpic, V., Blackmore, P. F., and Exton, J. H. (1982). *J. Biol. Chem.*, **257**, 11315.
14. Berridge, M. J., Dawson, R. M. C., Downes, C. P., Heslop, J. P., and Irvine, R. F. (1983). *Biochem. J.*, **212**, 473.
15. Sharpes, E. D. and McCarl, R. L. (1982). *Anal. Biochem.*, **124**, 421.
16. Akhtar, R. A. and Abdel-Latif, A. A. (1980). *Biochem. J.*, **192**, 783.
17. Bone, E. A., Fretton, P., Palmer, S., Kirk, C. J., and Michell, R. H. (1984). *Biochem. J.*, **221**, 803.
18. Wreggett, K. A., Howe, L. R., Moore, J. P., and Irvine, R. F. (1987). *Biochem. J.*, **245**, 933.
19. Sekar, M. C., Dixon, J. F., and Hokin, L. E. (1987). *J. Biol. Chem.*, **262**, 340.
20. Wong, N. S., Barker, C., Shears, S. B., Kirk, C. J., and Michell, R. H. (1988). *Biochem. J.*, **252**, 1.
21. Hawkins, P. T., Berrie, C. P., Morris, A. J., and Downes, C. P. (1987). *Biochem. J.*, **243**, 211.
22. Lips, D. L., Bross, T. E., and Majerus, P. W. (1987). *Proc. Natl. Acad. Sci. USA*, **85**, 88.
23. Andrews, W. V. and Conn, P. M. (1987). In *Methods in Enzymology* (ed. P. M. Conn and A. R. Means), Vol. 141, p. 156. Academic Press, London and New York.
24. Allan, D. and Michell, R. H. (1978). *Biochim. Biophys. Acta*, **508**, 277.
25. Shears, S. B., Storey, D. J., Morris, A. J., Cubitt, A. B., Parry, J. B., Michell, R. H., and Kirk, C. J. (1987). *Biochem. J.*, **242**, 393.
26. Batty, I. R., Nahorski, S. R., and Irvine, R. F. (1985). *Biochem. J.*, **232**, 211.
27. Irvine, R. F., Letcher, A. J., Heslop, J. P., and Berridge, M. J. (1984). *Biochem. J.*, **223**, 237.
28. Shears, S. B., Parry, J. B., Tang, E. K. Y., Irvine, R. F., Michell, R. H., and Kirk, C. J. (1987). *Biochem. J.*, **246**, 139.

29. Johnson, L. F. and Tate, M. E. (1969). *Can. J. Biochem.*, **47**, 63.

30. Stephens, L., Hawkins, P. T., Carter, N., Chahwala, S. B., Morris, A. J., Whetton, A. D., and Downes, C. P. (1988). *Biochem. J.*, **249**, 271.

31. Downes, C. P., Mussat, M. C., and Michell, R. H. (1982). *Biochem. J.*, **203**, 169.

32. Stephens, L., Hawkins, P. T., Barker, C. J., and Downes, C. P. (1988). *Biochem. J.*, **253**, 721.

33. Morgan, R. A., Chang, J. P., and Catt, K. J. (1987). *J. Biol. Chem.*, **262**, 1166.

34. Dean, N. M. and Moyer, J. D. (1987). *Biochem. J.*, **242**, 361.

35. Low, M. G., Carroll, R. C., and Cox, A. C. (1986). *Biochem. J.*, **237**, 139.

36. Schacht, J. (1978). *J. Lipid Res.*, **19**, 1063.

37. Bligh, E. G. and Dyer, W. J. (1959). *Can. J. Chem.*, **37**, 473.

38. Schacht, J. and Agranoff, B. W. (1972). *J. Biol. Chem.*, **247**, 771.

39. Jolles, J., Zwiers, H., Dekker, A., Wirtz, K. W. A., and Gispen, W. H. (1981). *Biochem. J.*, **194**, 283.

40. King, C. E., Stephens, L. R., Hawkins, P. T. L., Guy, G. R., and Michell, R. H. (1987). *Biochem. J.*, **244**, 209.

41. Hubscher, G., Hawthorne, J. N., and Kemp, P. (1960). *J. Lipid. Res.*, **1**, 433.

42. Clarke, N. G. and Dawson, R. M. C. (1981). *Biochem. J.*, **195**, 301.

43. Hawkins, P. T., Stephens, L. R., and Downes, C. P. (1986). *Biochem. J.*, **238**, 507.

44. Underwood, R. H., Greeley, R., Glennon, G. T., Menachery, A. I., Braley, L. M., and Williams, O. H. (1988). *Endocrinology*, **123**, 211.

45. Stephenson, G. R., Holmes, A. B., Burgess, K., and Prince, R. H. (1987). *Chem. Ind.*, **12**, 421.

46. Hokin, L. E. and Hokin, M. R. (1963). *Biochim. Biophys. Acta*, **67**, 470.

47. Allan, D. and Michell, R. H. (1978). *Biochim. Biophys. Acta*, **508**, 277.

48. Spat, A., Bradford, P. G., McKinney, J. S., Rubin, R. P., and Putney, J. W. Jr. (1986). *Nature, Lond.*, **319**, 514.

49. Morris, A. J., Storey, D. J., Downes, C. P., and Michell, R. H. (1988). *Biochem. J.*, **254**, 655.

50. Dawson, R. M. C., Frienkel, N., Jungalwala, F. B., and Clarke, N. (1971). *Biochem. J.*, **122**, 605.

51. Wilson, D. B., Bross, T. E., Sherman, W. R., Berger, R. A., and Majerus, P. W. (1985). *Proc. Natl. Acad. Sci. USA*, **82**, 4013.

52. Lapetina, E. G. and Michell, R. H. (1973). *Biochem. J.*, **131**, 433.

53. Dawson, R. M. C. and Clarke, N. M. (1972). *Biochem. J.*, **127**, 113.

54. Grado, C. and Ballou, C. (1961). *J. Biol. Chem.*, **236**, 64.

55. Brockerhoff, H. and Ballou, C. E. (1962). *J. Biol. Chem.*, **237**, 49.

56. Brown, D. M., Hall, G. E., and Letters, R. (1959). *J. Chem. Soc.*, 3547.

57. Brown, D. M. and Stewart, J. C. (1966). *Biochim. Biophys. Acta*, **125**, 413.

58. Irvine, R. F., Letcher, A. J., Lander, D. J., and Berridge, M. J. (1986). *Biochem. J.*, **240**, 301.

59. Johnson, L. F. and Tate, M. E. (1969). *Can. J. Chem.*, **47**, 63.

60. Lindon, J. C., Baker, D. J., Williams, J. M., and Irvine, R. F. (1987). *Biochem. J.*, **244**, 591.

61. Cerdan, S., Hansen, C. A., Johanson, R., Inubushi, T., and Williamson, J. R. (1986). *J. Biol. Chem.*, **261**, 14676.

62. Sherman, W. R., Ackerman, K. E., Berger, R. A., Gish, B. G., and Zinbo, M. (1986). *Mass. Spec.*, **13**, 333.

63. Tomlinson, R. V. and Ballou, C. E. (1962). *Biochemistry*, **1**, 166.
64. Irvine, R. F. (1986). In *Receptor Biochemistry and Methodology* (ed. J. W. Putney, Jr.), Vol. 7, p. 89. Alan Liss, New York.
65. Irvine, R. F., Letcher, A. J., Lander, D. J., and Downes, C. P. (1984). *Biochem. J.*, **206**, 587.
66. Parasarathy, R. and Eisenberg, F. (1986). *Biochem. J.*, **235**, 313.
67. Agranoff, B. W. (1978). *Trends Biochem. Sci.*, **3**, N283.
68. Tomlinson, R. V. and Ballou, C. E. (1961). *J. Biol. Chem.*, **236**, 1902.
69. Pizer, F. L. and Ballou, C. E. (1959). *J. Am. Chem. Soc.*, **81**, 915.
70. Sherman, W. R., Leavitt, A. L., Honchar, M. P., Hallcher, M. B., and Phillips, B. E. (1981). *J. Neurochem.*, **36**, 1947.
71. Stephens, L. R., Hawkins, P. T., and Downes, C. P. (1989). *Biochem. J.*, **262**, 727.
72. Shears, S. B. (1989). *J. Biol. Chem.*, **264**, 19879.

6

Lipid-related second messengers

ROBERT V. FARESE and DENISE R. COOPER

1. Introduction

The realization that many, if not all, peptide hormones rapidly stimulate changes in phospholipid metabolism has led to exponential advances in our understanding of the molecular mechanisms whereby these hormones regulate metabolic processes in their target tissues. Although most interest has focused upon hormones and other agonists which mobilize intracellular Ca^{++} by activating a specific phospholipase C that hydrolyses phosphatidylinositol-4', 5'-bisphosphate (PIP_2) (1, 2), it is clear that this hydrolysis is only one of several mechanisms which results in the generation of diacylglycerol, and activation of protein kinase C. The subject of PIP_2 hydrolysis, and release of inositol–phosphate headgroups and their subsequent effects on Ca^{++} mobilization, will be reviewed in other chapters of this book. The present review will focus upon intracellular signalling substances which are contained in the diacylglycerol domain of phospholipids, including both the entire diacylglycerol molecule itself, and constituent fatty acids (in particular, arachidonic acid), and their subsequent activation of certain enzyme systems. To the greatest extent, this review will emphasize practical issues that investigators must face in exploring relevant areas experimentally, and the reader is referred to other recent reviews for more background information (3, 4).

2. Diacylglycerol

2.1 Chemistry and molecular forms

Diacylglycerols, by definition, contain two fatty acids esterified to glycerol (*Figure 1*). The diacylglycerols, which contain fatty acids in the 1 and 2 positions of glycerol, appear to be more important, as only these have been reported to activate protein kinase C (5). Analogues of diacylglycerol which contain an ether-linked alkyl or alkenyl, instead of an acyl, group at the 1 position of glycerol, while similar to diacylglycerol in some respects, reportedly do not activate protein kinase C (6), a major enzyme which is activated by diacylglycerol. Nevertheless, these 1-alkyl or 1-alkenyl ether analogues, are phosphorylated by diacylglycerol kinase (7), and possibly other enzymes which metabolize

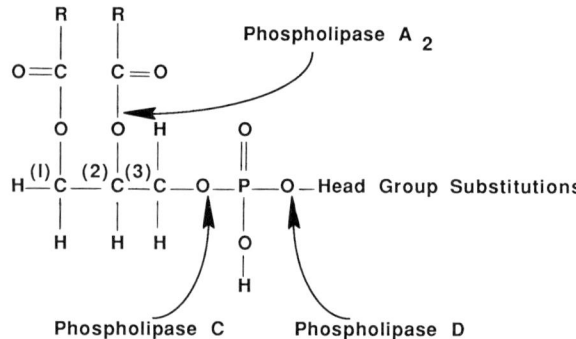

Figure 1. Structure and hydrolysis of glycerophospholipids. Glycero-phospholipids contain two fatty acids esterified at the 1 and 2 positions of glycerol (diacylglycerol), and a polar head group, containing a phosphate ester (phosphatidic acid) or phosphate diester linked to a variety of substances including: inositol, inositol–PO_4, inositol–$(PO_4)_2$, glucosamine with other glycosides (i.e. a glycan), choline, ethanolamine, serine, etc.

DAG, and they may thereby alter diacylglycerol metabolism in intact tissues, and, provoke increases in diacylglycerol content. In addition, there are preliminary indications that the ether analogues may stimulate protein kinase C at higher intracellular Ca^{++} concentrations, or may be inhibitory. Clearly, we need to learn more about the ether analogues, which seem to be increased particularly during hydrolysis of phosphatidylcholine (see below).

The ability of diacylglycerol to activate protein kinase C appears to be influenced only moderately by variations in the chain length or degree of unsaturation of the fatty acids in the 1 and 2 positions of glycerol (5, 6). Although diacylglycerols with unsaturated fatty acids at the 2 position may be more potent activators of protein kinase C (8), this is not an absolute requirement for enzyme activation. For example, any two fatty acids having chain length of 5 carbon atoms or greater at the 1 and 2 positions of glycerol, or having sufficient bulk at the 1 position, e.g., 1,2-dipentanoyl-glycerol (DiC_5), 1,2-dioctanoyl-glycerol (DiC_8), and 1-oleoyl, 2-acetyl-glycerol (OAG), appear to be active (5, 6). DiC_8 and OAG have been commonly used to activate protein kinase C in intact tissues, and this usefulness probably reflects the fact that these short-chain diacylglycerol analogues have reasonably good solubility or dispersibility in aqueous media, and yet they are sufficiently lipophilic to penetrate and localize in cellular membranes, and thereby activate intracellular protein kinase C. Longer chain fatty acids such as dioleoyl-diacylglycerol (diolein) are able to activate protein kinase C directly in enzyme assays, but, presumably because of poor solubility or dispersibility in aqueous media, they may be less useful for activating protein kinase C in intact cells. However, there are exceptions to this rule, and diolein has been successfully used to activate glucose transport in rat adipocytes (9), perhaps because of the large amount of lipids in these cell suspensions.

2.2 Sources and mechanisms for induced increases

In order of relative importance for the present review, diacylglycerol may be produced by four mechanisms: (a) phospholipase C-mediated hydrolysis of any phospholipid; (b) phosphatase action on phosphatidic acid; (c) lipase action on triacylglycerol; and (d) acyltransferase action on monacylglycerol. Only the first two mechanisms (see *Figure 2*) appear to have relevance for intracellular signaling, and the latter two mechanisms will not be discussed in the present review.

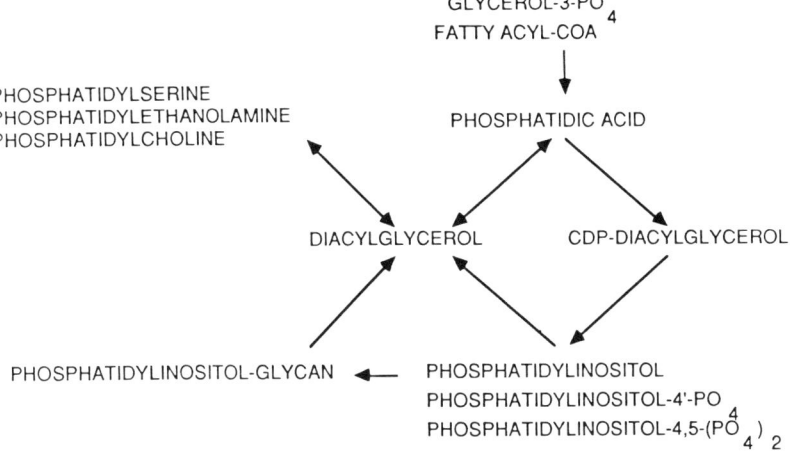

Figure 2. Pathways of phospholipid metabolism.

As alluded to above, most interest has focused upon generation of diacylglycerol from phospholipase C-mediated hydrolysis of PIP_2. However, it has been recognized that, even in the action of agonists which provoke this PIP_2 hydrolysis response, increases in diacylglycerol (or ether analogues) appear to be considerably greater than that which can reasonably be expected to be derived from the PIP_2 hydrolysis pathway (10). In addition, in the action of some hormones such as insulin (11) and adrenocorticotropin (12), this dissociation was found to be particularly noteworthy, as these hormones provoke large increases in diacylglycerol, but do not increase inositol-phosphates appreciably. Clearly, this hormone-induced diacylglycerol must be derived from sources other than PIP_2 hydrolysis, and, in the case of insulin, it was subsequently found that these sources include: (a) *de novo* synthesis of phosphatidic acid and its immediate conversion to diacylglycerol (13); (b) hydrolysis of a phosphatidylinositol-glycan complex (PI-glycan) (14); and (c) hydrolysis of phosphatidylcholine (87). In addition to these potential sources and mechanisms for generation of diacylglycerol, there is experimental evidence that in some systems phosphatidylinositol may be hydrolysed separately from PIP_2 by a different phospholipase C or D

(15–17). Note: diacylglycerol may be produced from phosphatidic acid, which is released by phospholipase D action). Whether or not other mechanisms for producing diacylglycerol are also operative and important in the action of peptide hormones is presently unknown.

The activation of specific types of phospholipase C which hydrolyse PIP_2 and the PI-glycan occurs in the plasma membrane, and probably requires a GTP-binding protein (G-protein), which apparently serves as a transducer and couples the activated receptor to phospholipase C (18, 19). Presumably, phospholipase C acts on the inner side of the plasma membrane to release headgroups into the cytosol, but the possibility also exists that phospholipase C may act upon inositol-lipids which are on the external surface of the plasma membrane (20). The diacylglycerol which is produced on the internal or external surface of the membrane may be relatively free to move within the membrane both laterally and by flip-flop action after loss of the polar headgroup.

The subcellular site for hydrolysis of phosphatidylcholine is less certain. Hormone-induced hydrolysis of phosphatidylcholine from plasma membrane preparations has been reported in the actions of platelet-derived growth factor (21), and vasopressin, angiotensin II, and epidermal growth factor (22), and while this appears to be a GTP-requiring phenomenon, presumably through a 'G-protein', there is some uncertainty as to whether this hydrolysis is due to activation of phospholipase C or phospholipase D. In either case, however, it would be expected that increases in diacylglycerol (or ether analogues) would result, since phosphatidic acid is readily converted to diacylglycerol, and vice versa. In addition to the plasma membrane, phosphatidylcholine may also be hydrolysed in other subcellular locations, since phosphatidylcholine is ubiquitously found in all intracellular membranes. The activation of a phosphatidylcholine-specific phospholipase seems to be stimulated by protein kinase C, as suggested by studies with phorbol esters (21, 22), which act as diacylglycerol analogues and directly activate protein kinase C (see below). Thus, it would appear that a local increase in diacylglycerol through hydrolysis of PIP_2 or the PI-glycan in the plasma membrane, may activate protein kinase C and thereby activate a phosphatidylcholine-specific phospholipase C or D. The latter activation would amplify the local production of diacylglycerol and contribute to further activation of protein kinase C. Moreover, the diacylglycerol may also contribute additional substrate to resynthesize PI, PIP_2, and/or the PI-glycan. Similarly, production of diacylglycerol through *de novo* synthesis of phosphatidic acid and subsequent phosphatase action, may activate protein kinase C in the endoplasmic reticulum (the major site for *de novo* phosphatidic acid synthesis), and this in turn may activate a local phosphatidylcholine-specific phospholipase C or D, and generate more diacylglycerol, etc. These amplification mechanisms for diacylglycerol production and protein kinase C activation must obviously operate only during hormone or other agonist action, and are probably limited by regulatory factors, or a self-perpetuating mechanism would occur. At the present time, there is no insight concerning the factors which limit or inhibit the amplifying mechanism, but perhaps G-proteins are involved, as appears to be the case for protein-kinase-C-mediated inhibition of the phospholipase C which hydrolyses PIP_2.

As stated above, *de novo* synthesis of phosphatidic acid is thought to occur in the endoplasmic reticulum, and, in most tissues, this newly synthesized phosphatidic acid is largely directly converted to diacylglycerol, which is an obligatory intermediate for the *de novo* synthesis of phosphatidylcholine and phosphatidylethanolamine in mammalian tissues (23). Activation of *de novo* phosphatidic acid synthesis has been observed in the actions of a number of polypeptide hormones, including insulin (13, 24), and adrenocorticotropin (88), angiotensin II (89), vasopressin (90), epidermal growth factor (91), insulin-like growth factor I (91), and endothelin (92). In the case of insulin, we have determined the specific radioactivity of glycerol-3-PO_4 in [^{3}H]glycerol-labelling experiments (24), and have calculated that the amount of diacylglycerol which is produced through the *de novo* phosphatidate synthesis pathway, appears to account for only 15–30% of the increases of diacylglycerol content, as measured by the DAG-kinase method (see below). The remainder of the diacylglycerol is probably derived from hydrolysis of the PI-glycan (14) and phosphatidylcholine (13, 24, 87). Evidence for insulin-induced phosphatidylcholine hydrolysis derives from the findings that (a) phosphocholine (?release from phosphatidylcholine) is increased by insulin (87); (b) there are initial decreases in phosphatidylcholine mass (unpublished observations) followed by subsequent increases; and, (c) increases in arachidonate-labelling of diacylglycerol cannot be accounted for by hydrolysis of the PI-glycan, since the latter lipid does not contain appreciable amounts of arachidonate (25, 26). Insulin-induced increases in arachidonate-labelled diacylglycerol must therefore arise through *de novo* phosphatidic acid synthesis or phosphatidylcholine hydrolysis, or both. Along these lines, although it had been previously thought that newly synthesized phosphatidic acid contains little arachidonate in quiescent cells (27), it is possible that hormonal-induced activation of the *de novo* phosphatidate synthesis pathway may be attended either by increased arachidonate incorporation into the newly synthesized phosphatidic acid, and/or rapid remodelling (transacylation, etc.) of the derived diacylglycerol and/or phosphatidylcholine. Similarly, although it has been suggested that myristate-labelled diacylglycerol is derived from the PI-glycan during insulin action (25), there is also evidence (24) that phosphatidylcholine may be a source of this diacylglycerol, and we have even more recently obtained evidence that myristoyl-labelled diacylglycerol may be derived from *de novo* synthesis of phosphatidic acid (submitted for publication).

As alluded to above, increases in phosphatidylcholine synthesis *de novo* seem to be followed very closely by, or associated with, increases in its hydrolysis, as evidenced by release of phosphocholine (87). Thus, *de novo* phosphatidate synthesis may be closely coupled to activation of a phosphatidylcholine-specific phospholipase C, possibly because the increase in diacylglycerol due to *de novo* phosphatidate synthesis may also serve to activate protein kinase C; or, because *de novo* phosphatidate synthesis and phosphatidylcholine hydrolysis are parallel, but, causally unrelated responses to a prior event. In keeping with the first possibility, we have found that glucose-induced activation of *de novo* diacylglycerol synthesis and protein kinase C activation in rat adipocytes (see reference 93) is associated with activation of phosphatidylcholine hydrolysis (unpublished). In keeping with the second possibility, we have found that insulin-induced increases in phosphatidylcholine hydrolysis are not blocked by inhibitors of protein kinase C. Thus effects of insulin on *de novo* phosphatidic acid synthesis and phosphatidylcholine hydrolysis may be parallel responses.

The mechanism whereby *de novo* phosphatidate synthesis is activated is not entirely clear, but, in the case of insulin action, it has been reported that 'headgroup mediators', which appear to be derived from phospholipase C-mediated hydrolysis of the PI-glycan (14, 25, 26), may activate glycerol-3-phosphate acyltransferase (28). Indeed, we have recently found that insulin effects on this enzyme are mimicked by PI-glycan-specific phospholipase C, and are blocked by antibodies which inhibit this phospholipase C (submitted for publication). Whether this mechanism is also operative in the action of other polypeptide hormones which increase *de novo* phosphatidate synthesis is presently unknown. It should also be noted that simple provision of substrate for glycerol-3-phosphate acyltransferase can of itself increase *de novo* phosphatidic acid synthesis, diacylglycerol production, and activation of protein kinase C: this has been observed in tissues which readily take up glucose (93–95).

The hydrolysis of phosphatidylinositol by phospholipase C may occur either in the plasma membrane or in the endoplasmic reticulum, which contains most of the cellular phosphatidylinositol. The hydrolysis of phosphatidylinositol, or PIP_2, for that matter, in the plasma membrane would require either rapid resynthesis of phosphatidylinositol in the plasma membrane (29), or transfer of substantial quantities of phosphatidylinositol from the endoplasmic reticulum to the plasma membrane (see reference 30). In the case of the PI-glycan, this glycolipid appears to be hydrolysed in the plasma membrane (14, 25, 26), and resynthesized in the endoplasmic reticulum (31).

2.3 Measurement by labelling or direct assay

Although isotopes have been widely used to study changes in diacylglycerol production, there are many difficulties which are inherent in the interpretation of such isotopic studies, and these studies may not provide definitive evidence for increases in diacylglycerol content. The latter problem is particularly relevant in studies in which there is relatively short-term labelling of precursor pools. Fortunately, increases in diacylglycerol content may frequently be documented by direct measurement of its mass, or by labelling with glycerol or fatty acids for sufficiently long periods to allow precursor pools and relevant lipids attain constant or sufficiently high specific activity. Indeed, in the BC3H-I myocytes, we have found comparable increases in insulin-induced increases in diacylglycerol mass and increases in [^{3}H]glycerol-labelled diacylglycerol, when the myocytes have been labelled over 3–10 days. In short-term [^{3}H]glycerol labelling experiments, on the other hand (i.e. 2 h or less of pre-labelling prior to hormone addition), the labelling of diacylglycerol may not reflect changes in DAG mass, and probably better reflects the acceleration of the incorporation of label through activation of *de novo* phosphatidic acid synthesis (reference 24, and submitted for publication).

Diacylglycerol mass may be measured by several methods. These include separation of diacylglycerol from other lipids by thin layer chromatography or high pressure liquid chromatography (HPLC), followed by quantitation of diacylglycerol by elution and charring (32), charring and densitometry directly on thin layer plates (33), optical density of HPLC effluent (34), or acetylation of the free 3-hydroxyl group of diacylglycerol with labelled acetic anhydride and purification and radioactive counting of the resulting labelled derivative (35). All of these methods have been employed successfully to measure hormone-induced increases in diacylglycerol, but in the densitometric and acetic anhydride methods, relatively high blanks are encountered, and there is poor specificity, sensitivity, and precision of these assays, particularly in biological samples which may have hydroxylated substances that are not fully separated from diacylglycerol by chromatographic methods. Better purification of diacylglycerol undoubtedly results after high pressure liquid chromatography or two-dimensional, thin layer chromatography, but these are obviously more expensive, laborious, and time-consuming methods. Hydrolysis of diacylglycerol to yield fatty acids is another method for measuring diacylglycerol mass, but the fatty acids must be further quantitated after derivitization and separation by gas liquid chromatography (10). This, too, is a cumbersome procedure, but offers further insight into the fatty acid composition, and potential sources, of diacylglycerol. More recently, molecular species of diacylglycerol (in which the composition of both fatty acids or ethers is ascertained) has been determined by conversion to dimethyl phosphoric acid esters (96) or t-butyl-dimethylsilyl esters, followed by chromatographic resolution and quantitation by ^{32}P content or flame ionization detection, respectively.

From a practical point of view, the diacylglycerol-kinase method for measuring diacylglycerol mass (7, 36, 37) is well suited to precisely and accurately measure diacylglycerol in a large number of samples. (The latter is important particularly in time-course and dose-response studies.) In this method, purified diacylglycerol kinase from bacterial sources (which may be obtained commercially from Lipidex, Wisconsin, USA) is incubated with [γ-^{32}P]ATP, detergent, and extracted lipids from samples of interest (see reference 37 and outline below for more details). The incorporation of ^{32}P into diacylglycerol results in the formation of ^{32}P-phosphatidic acid, which can be extracted into organic solvents and readily purified by one-dimensional thin layer chromatography. The level of radioactivity in phosphatidic acid serves as an accurate index of the diacylglycerol content of the sample, and, by comparison to known standards (e.g. diolein), it is possible to calculate the diacylglycerol content of the sample. Indeed, we have obtained virtually the same values for diacylglycerol content of tissues by this method, as compared to the acetic anhydride-derivative and the charring methods. The only apparent drawback for the kinase method is the fact that the diacylglycerol kinase will phosphorylate ether-linked analogues (7), as well as diacylglycerol *per se*. Although the latter may result in overestimation of diacylglycerol, the DAG-kinase method is the simplest and easiest to use, and, at the present time, it is uncertain whether other described methods are also measuring ether-linked derivatives, since these substances may not be readily separated from diacylglycerol by commonly-employed chromatographic techniques. Another advantage of the diacylglycerol kinase method is that it specifically measures sn-1, 2-diacylglycerol, rather than sn-1, 3-diacylglycerols, and only the former are capable of activating protein kinase C (5, 6).

Protocol 1. Assay of diacylglycerol mass by the kinase method

Reagents

● Cardiolipin (5 mg/ml in organic solvent)
● Octyl glucopyranoside (solid)
● Diethylenetriamine pentaacetic acid (DETAPAC) (1 mM in H_2O)
● Dithiothreitol (DTT) (100 mM in H_2O)
● Imidazole (100 mM in 1 mM DETAPAC)
● Imidazole (100 mM in H_2O)
● ATP (5.072 mg/ml of Imidazole–DETAPAC solution)
● 1,2-dioleoyl-sn-glycerol (diacylglycerol standard) [10 mg/ml in $CHCl_3$: MeOH, 2/1 (v/v)]
● Diacylglycerol (DAG) kinase (supplied by Lipidex).

Solutions (for an assay of 60 tubes)

Solution I:
   690 µl cardiolipin (dry under N_2 or air)
   0.090 g octylglucopyranoside
   1.2 ml 1 mM DETAPAC solution

Solution II:
   100 mM imidazole (NOT in DETAPAC)
   100 mM NaCl
   25 mM $MgCl_2$
   2 mM EGTA
   Adjust pH to 6.6

Solution III:
   120 µl DTT

Protocol 1. *Continued*

    360 μl DAG kinase

    780 μl H_2O

Solution IV: $[\gamma\text{-}^{32}P]$-ATP

    5.072 mg 'cold' ATP/ml in 100 mM imidazole/1 mM DETAPAC

    Mix 600 μl 'cold' ATP and 18 μl $[\gamma\text{-}^{32}P]$-ATP (keep on ice).

Diacylglycerol standards

● Prepare 4 serial 1:10 dilutions of DAG stock solution (10 mg/ml).

● Dry down 15 μl of each DAG standard in a glass tube (in duplicate) including solvent alone for a blank.

● Diacylglycerol content for each standard decreases with dilution from 25 000 to 2500, 250, and 25 pmol.

Samples

Organic lipid extracts ($CHCl_3$ phases) from tissues. These must contain sufficient diacylglycerol (ideally between 250–25 000 pmol) to be easily determined and calculated from the standard curve. Tissue contents of diacylglylcerol vary considerably and must be evaluated by assaying increasing amounts of lipid extracts.

Assay

1. Dry samples and standards under air.

2. Add 20 μl Solution I and sonicate for 1 min in bath sonicator.

3. Let tubes sit 15 min at room temperature.

4. Add 20 μl Solution III + 50 μl Solution II.

5. Add 10 μl ^{32}P-ATP Solution IV and incubate 30 min at room temperature in shaking water-bath.

6. Add 6.0 ml $CHCl_3$:MeOH (2:1, v/v) and 2.0 ml H_2O, mix with a vortexer, obtain lipid extract (lower $CHCl_3$ phase) and wash three times with 2.0 ml H_2O.

7. Dry down lipid extracts and apply aliquot to thin-layer chromatography Silica gel H plates containing 1% potassium oxalate. Develop with $CHCl_3$: CH_3OH : CH_3COCH_3 : CH_3COOH : H_2O (10:4:2:2:1, v/v) using phosphatidic acid standard. Scrape and count the ^{32}P-labelled phosphatidic acid.

8. Plot the linear regression of radioactivity in diacylglycerol standards versus the $\log_{10}$ of the diacylglycerol content of the standards.

9. Determine diacylglycerol content of samples from their radioactivity versus that of diacylglycerol standards.

As stated above, isotopes have also been commonly used for measuring changes in diacylglycerol turnover and content. To reiterate, it is important to keep in mind that in short-term labelling studies, changes in isotopic labelling may reflect simple turnover (increased hydrolysis or metabolism, and replacement of unlabelled substance with labelled substance), rather than changes in diacylglycerol content. Such changes in turnover, on the other hand, may be more readily apparent in short-term, than in long-term, labelling experiments, which primarily reflect changes in diacylglycerol content. If the diacylglycerol content of the entire tissue changes sufficiently, clearly the long-term labelling methods are preferable. However, in some cases, it is possible that only a small pool of diacylglycerol will change in response to hormone treatment, and it may not be possible to observe changes in total cellular diacylglycerol content. In these cases, short-term labelling experiments may be the only alternative to reflect changes in the smaller, metabolically active pool; nevertheless, interpretation of these experiments must be limited accordingly.

The isotopes which may be used for labelling diacylglycerol include labelled glycerol and fatty acids. Labelling of diacylglycerol with glycerol is obviously least specific, and glycerol-labelled diacylglycerol may be derived from virtually any source, including phosphatidic acid, PIP_2, phosphatidylinositol, the PI-glycan and phosphatidylcholine. On the other hand, if the [^{3}H]glycerol is labelled in 2-position, it will specifically reflect diacylglycerols rather than ether analogues, which are synthesized exclusively from dihydroxyacetone-phosphate. In short-term labelling experiments, in which the labelled glycerol is added, either directly with the hormone, or after a relatively short pre-labelling period, (the duration of which may vary from tissue to tissue, depending on its glycerokinase activity for labelling the glycerol-3-phosphate precursor pool), subsequent hormone-induced increases in glycerol-labelled diacylglycerol may primarily reflect *de novo* phosphatidate synthesis. However, if there is significant incorporation into phosphatidylcholine, phosphatidylinositol, or other lipids, these too may be sources of labelled diacylglycerol. This is particularly true for phosphatidylcholine, since this lipid may be heavily labelled in relatively short periods of time. Thus, glycerol-labelling of diacylglycerol cannot necessarily be equated with an increase in *de novo* phosphatidate synthesis. Evidence that the diacylglycerol is derived through *de novo* phosphatidate synthesis pathway can only be obtained if (a) the isotope is found to accumulate in phosphatidic acid and diacylglycerol prior to significant labelling in other phospholipids, and (b) total glycerolipid labelling is increased (including DAG, other neutral lipids, and total phospholipids). Assuming that degradation is not decreased, increases in total glycerolipid labelling can only be explained by increases in *de novo* phosphatidate synthesis, since all glycerolipids are obligatorily synthesized by this pathway. However, this experimental approach does not necessarily provide insight into the relative proportion of labelled diacylglycerol, which is derived directly from the newly synthesized phosphatidic acid, as opposed to that which may be re-entering the diacylglycerol pool from hydrolysis of recently labelled phospholipids, such as phosphatidylcholine.

Protocol 2 is an example of an acute [³H]glycerol labelling procedure which was used in BC3H-1 myocytes (13, 24).

Protocol 2. Acute[³H]glycerol labelling

1. BC3H-1 myocytes are cultured to confluence in 35- or 100-mm plates (approximately 0.5–1.5 mg protein).

2. Medium is changed to 1–3 ml of phosphate-buffered saline containing 1 mg/ml albumin and 1 mg/ml glucose.

3. 10 μCi of [2-³H]glycerol is added to each plate and incubation is continued for 10–120 min to label the glycerol-3-phosphate pool sufficiently to facilitate observance of rapid increases in lipid synthesis. Note the length of this pre-labelling period will vary in different tissues and must be evaluated experimentally.

4. Agonist in vehicle or vehicle alone is added and incubation is continued for varying times.

5. Reactions are stopped by adding an equal volume of ice-cold methanol.

6. Plates are scraped with a plastic policeman and contents are quantitatively transferred to glass tubes with capillary pipettes.

7. $CHCl_3$ is added to yield a $CHCl_3 : CH_3OH : H_2O$ ratio of 2:1:1. Samples are mixed and centrifuged.

8. The upper aqueous phase is removed (and saved if it is to be used for glycerol-3-phosphate analysis; see below) and the lower $CHCl_3$ layer is washed three times with 1 vol. H_2O.

9. The $CHCl_3$ extracts are dried and aliquots are spotted on silica gel thin layer plates, which are developed first with ethyl ether:benzene:ethanol:acetic acid (40:50:2:2, v/v) and then with ethyl ether:hexane (6:94).

10. Diacylglycerol, monoacylglycerol, triacylglycerol, and total phospholipid (origin) areas are scraped into vials and counted for radioactivity.

11. Diacylglycerol mass can be determined by assaying (see above) an aliquot of the original lipid ($CHCl_3$) extract, or an aliquot of an extract of the chromatographically purified diacylglycerol. The specific activity of diacylglycerol can be calculated from the radioactivity found in purified diacylglycerol and the diacylglycerol content of the sample. For this calculation, it is important to correct for the reduced counting efficiency of the samples from the chromatograms as these samples contain silica gel, which generally causes quenching.

12. The aqueous phase (from step 8 above) may be examined for labelling of the glycerol-3-phosphate pool. An aliquot is lyophilized (to remove methanol), redissolved in 5 ml H_2O and applied to a short Dowex 1-4X (formate form) (Biorad) column. The column is washed with 10–20 ml H_2O to remove free [³H]glycerol. [³H]Glycerol-3-phosphate is then eluted with 10 ml 0.2 M

Protocol 2. *Continued*

ammonium formate in 0.1 M formic acid. Aliquots are examined for radioactivity and glycerol-3-phosphate content (see reference 24), and specific activity is determined.

13. From the specific activity of glycerol-3-phosphate (G-3-PO$_4$) and the total radioactivity found in DAG, the amount of DAG derived from glycerol-3-phosphate (i.e. the *de novo* pathway) can be calculated by the following equation:

> c.p.m. of diacylglycerol sample $\div$ specific activity of G-3-PO$_4$
> = Diacylglycerol derived from *de novo* pathway.

As alluded to above, to gain some insight into the question of how much diacylglycerol is produced from the *de novo* phosphatide synthesis pathway, it is possible to measure the specific activity of glycerol-3 phosphate after labelling with [2-^{3}H]glycerol (24), and from this and the observed increase in labelled diacylglycerol, calculate the quantity of concomitantly measured diacylglycerol mass, which is derived from newly synthesized phosphatidic acid and other rapidly labelled phospholipids. Because of the latter, this method may over-estimate the amount of diacylglycerol derived directly from phosphatidic acid. In addition, it should be realized that labelling of diacylglycerol with [2-^{3}H]glycerol will not reflect synthesis of ether-linked analogues of diacylglycerol, which are measured by the diacylglycerol-kinase method, and which are synthesized *de novo* from dihydroxyacetone-PO$_4$ (the latter will not be labelled by [2-^{3}H]glycerol). Another problem with the calculation of the amount of diacylglycerol derived from the *de novo* pathway is the fact that the glycerol-3-PO$_4$ pool which is used for *de novo* phosphatide synthesis may have a specific activity quite different from that of the total glycerol-3-PO$_4$ pool, which is measured after its purification by ion-exchange column chromatography (24).

As stated above, long-term pre-labelling with [^{3}H]glycerol may be used to reflect changes in diacylglycerol content during hormone stimulation. In these long-term labelling protocols, many investigators have removed the isotope from the medium, and have assumed that hormone-induced increases in labelled diacylglycerol could be simply attributed to increases in hydrolysis of pre-formed phospholipids. However, it cannot tacitly be assumed that [^{3}H]glycerol is effectively removed from the glycerol-3 phosphate pool, which labels phosphatidic acid synthesized *de novo*. In fact, in BC3H-I myocytes (24) and several other tissues, we have not been able to effectively diminish radioactivity in the glycerol-3-PO$_4$ pool with any degree of consistency, even in prolonged 'chase' experiments (24), and it is clear that hormone-induced increases in [^{3}H]glycerol-labelling of diacylglycerol continues to reflect increased *de novo* phosphatide synthesis, as well as hydrolysis of pre-labelled phospholipids such as phosphatidylcholine.

Labelling of diacylglycerol with fatty acids has also been used in studies of diacylglycerol generation. In particular, labelled arachidonate acid has been used, since this fatty acid is very rapidly incorporated into phospholipids, such as phosphatidylcholine and phosphatidylinositol, by a transacylation process, rather than by *de novo* phosphatidate synthesis, and since hydrolysis of these phospholipids may be responsible for increases in arachidonate-rich diacylglycerol and phosphatidic acid. While increases in arachidonate-labelled diacylglycerol may in fact be due to increases in the hydrolysis of these phospholipids (but not the PI-glycan, since it is not labelled by arachidonate), for the reasons stated above, the increase in labelled diacylglycerol may also reflect *de novo* phosphatidate synthesis. Labelling with myristic acid has also been employed, and this isotope is rapidly incorporated into phosphatidic acid, diacylglycerol, phosphatidylcholine, and, after a lag period, into phosphatidylinositol and the PI-glycan (24). Increases in myristate-labelled diacylglycerol can in fact be shown to occur rapidly after insulin treatment (13, 14, 24, 25), but, contrary to initial suggestions that this diacylglycerol reflects PI-glycan hydrolysis (25), this myristate-labelled diacylglycerol may also reflect hydrolysis of phosphatidylcholine and *de novo* phosphatidic acid synthesis. Labelling of diacylglycerol with other fatty acids has been less commonly employed, but is nevertheless possible. In our experience, studies with labelled fatty acids, such as palmitate and oleate, have revealed only small hormone-induced effects on incorporation into diacylglycerol and other lipids. This may reflect the fact that hormones may also increase endogenous fatty acid synthesis, and exogenous fatty acids must be coupled to coenzyme A prior to utilization for phospholipid synthesis.

2.4 Biological effects

The major known biological effect of diacylglycerol is the activation of protein kinase C. The 30 kd regulatory domain of protein kinase C (which can be cleaved from the 50 kd catalytic domain by proteolytic enzymes) contains binding sites for diacylglycerol, Ca^{++}, and acidic phospholipids (3, 4). As a result of binding diacylglycerol or its analogues, the activity of the enzyme is changed in two ways, first, an increase in sensitivity for activation by Ca^{++} (i.e., a decrease in the K_a), and, second, an increase in intrinsic enzyme activity, i.e., an increase in the V_{max} (for example, see reference 38), or a decrease in the K_m (for example, see reference 39).

The fact that most diacylglycerol in intact cells resides within cellular membranes causes cytosolic protein kinase C to bind to membrane sites, where it is also degraded proteolytically to smaller fragments, some of which retain full or partial enzymatic activity to phosphorylate substrates of C-kinase, at least transiently during their limited life-span (3, 4). The movement of protein kinase C from cytosol to membrane has been called 'translocation' and it has been used (see below) as experimental evidence for activation of protein kinase C.

Diacylglycerol has been reported to activate not only protein kinase C, but also a proteolytic-activated enzyme referred to as 'PAK II' (39). The latter enzyme seems to be very similar to protein kinase C in its activation by diacylglycerol, chromatographic characteristics, and molecular size, i.e. 80 kd, and it is possible that it is a form of protein kinase C, perhaps related to epsilon-protein kinase C (98) which has a lesser or no dependence upon Ca^{++}. Regardless of whether the latter is true, this enzyme(s) is activated by diacylglycerol (which decreases the K_m), and increases in diacylglycerol could be responsible for hormone-induced increases (39) in the activity of 'PAK II' or epsilon-protein kinase C.

The question of whether diacylglycerol regulates other enzymes is presently uncertain. With loss of polar headgroups by phospholipase C action, resultant increases in diacylglycerol could change the physical characteristics of membranes, and provoke changes in the activity of resident enzymes, transporters, receptors, and other proteins. Large increases in diacylglycerol due to Ca^{++} ionophore treatment, for example, have been shown to increase vesicle formation in red blood cells (40), but this may not be of physiological importance, and it is uncertain whether it occurs in other cells.

2.5 Use of phorbol esters, diacylglycerols, and other analogues

The role of hormone-induced increases in diacylglycerol in activating biological responses has been evaluated in several ways: (a) correlations between changes in diacylglycerol content or labelling and the biological response under study in varying experimental conditions; (b) addition of diacylglycerol or its analogues to intact cells to provoke biological responses that are similar to those of the hormone; (c) use of diacylglycerol kinase inhibitors to increase tissue diacylglycerol levels and induce biological responses; (d) addition of phospholipase C to intact cells to increase endogenous diacylglycerol content; and (e) use of inhibitors of diacylglycerol-stimulated enzymes, i.e. protein kinase C, to determine the importance of this enzyme in hormone-induced biological responses. The last experimental approach will be discussed in greater detail in the section on protein kinase C (see below).

Of the above experimental approaches, the addition of diacylglycerol, its analogues, or phospholipase C to mimic hormone action, has been used to the greatest extent, and has probably provided the most meaningful results. In particular, phorbol esters have been particularly helpful in providing *clues* on the importance of diacylglycerol–protein kinase C signalling during hormone action. However, there are many caveats in the interpretation of experimental findings obtained with this approach. These caveats largely derive from the fact that there are at least four (41) and possibly seven (42) or more, protein kinase C isozymes (usually 80 kd or slightly smaller), which, when activated, are hydrolysed proteolytically to smaller molecular forms that are short-lived, but, nevertheless, enzymatically active and unrestrained by the C-kinase regulatory domain (see below). Unfortunately, there is evidence (43) that these isozymes vary in their ability to be activated (particularly as related to phosphorylation of specific proteins and biological functions) by phorbol esters and diacylglycerol (43), and the relative abundance of these protein kinase C isozymes varies from tissue to tissue. Thus, phorbol esters would be expected to provoke hormone-like effects on biological processes in tissues which have an abundance of protein kinase C isozymes which are more responsive (with respect to a particular biological function) to phorbol esters, but the phorbol esters would be less potent in this regard if the isozymes are less responsive to the phorbol esters. An example of this problem has become apparent in studies of glucose transport. In the BC3H-1 myocyte, for example, we have found an abundance of phorbol ester-sensitive isozymes of protein kinase C, and phorbol esters are as effective as insulin, exogenous diacylglycerol, or phospholipase C for increasing glucose transport. On the other hand, phorbol ester-induced increases in glucose transport in the rat adipocyte and diaphragm are quantitatively much less (10–40%) than those provoked by insulin, and this has been interpreted to suggest that insulin largely operates through a C-kinase-independent mechanism. The latter conclusion, however, may be incorrect, since the major forms of protein kinase C in the rat adipocyte and diaphragm may be poorly responsive to phorbol esters, but more responsive to diacylglycerol. Indeed, addition of exogenous diolein, OAG or phospholipase C to rat adipocytes or diaphragm provokes increases in glucose transport which may approximate to those of insulin. The latter finding suggests that insulin-induced increases in diacylglycerol may be fully responsible for increases in glucose transport, and, if this is true, it may be surmised that the results with phorbol esters in the adipocytes and diaphragm have been misleading. In brief, the protein kinase C isozyme pattern may determine whether phorbol esters will be as effective as hormone-induced increases in endogenously produced, or exogenously added, diacylglycerol in provoking a biological response. This would not be too surprising since there are countless examples of biological responses that are differentially activated by agonists which activate protein kinase C, such as phorbol esters, teleocidin, bryostatin, mezerein, and exogenously-added phospholipase C or diacylglycerols. These findings suggest that the activating ligand may alter substrate recognition by various forms of protein kinase C, and this, in fact, seems likely from certain studies (see, e.g., references 44, 45).

The above considerations suggest that diacylglycerol may be preferable to phorbol esters when these substances are added to intact cells to activate protein kinase C and observe biological effects. While this is true in a general sense, it is much easier to solubilize or disperse phorbol esters in aqueous media, and there is always some uncertainty whether exogenously added diacylglycerols effectively increase intracellular diacylglycerol levels. This has led to the use of synthetic diacylglycerols which have shorter fatty acid chains and, thus, greater aqueous solubility. As stated above, DiC_8 and OAG have been used extensively in this regard, but the former has also, on occasion, been found to provoke detrimental effects on cellular metabolism, which may cloud the interpretation of experimental results. Moreover, even these synthetic diacylglycerols may not be able to penetrate the membranes of some tissues. In the latter case, addition of phospholipase C may be the only alternative for increasing diacylglycerol, but this experimental approach also has its limitations, since enzyme preparations may be impure and a variety of headgroups are released.

The demonstration of biological effects of phorbol esters or diacylglycerol, which are comparable to those of a hormone (or other agonists), does not necessarily mean that the hormone provokes these effects by activating protein kinase C. Phorbol esters and diacylglycerol (probably through changes in protein kinase C), are known to alter many factors and general processes which have widespread consequences, including changes in G-protein activity, membrane potential, membrane ion transport, phospholipase C activity, adenylate cyclase activity, phospholipase A_2 activity, protein synthesis, gene expression (mRNA synthesis), and DNA synthesis, in addition to more specific changes in the activities of many enzymes, receptors, and cytoskeletal proteins, which are substrates for protein kinase C. Thus, effects of hormones and phorbol esters (or diacylglycerol) may be identical or similar, either because they both activate protein kinase C, or because they activate separate, but terminally convergent, pathways. In short, similarity of biological effects of hormones and phorbol esters (or diacylglycerol) can only be construed as suggestive evidence, at best, for a role of protein kinase C during hormone action.

The difficulties that are encountered in the experimental use of phorbol esters, also pertain to the use of other diacylglycerol analogues, such as mezerein and teleocidin. Like active phorbol esters, these substances possess structural moieties which resemble 1,2-diacyglycerol, and bind to and activate protein kinase C. In using all of these protein kinase C activators, it is important to employ doses which are not in excess of those known to activate protein kinase C, and, in the case of phorbol esters, it is important to show that the biological effects are provoked only by those chemical forms which are known to activate protein kinase C, e.g. 4α-phorbol 12, 13-didecanoate does not activate protein kinase C and serves as an appropriate control.

Phorbol esters, and occasionally exogenously-added diacylglycerol, have also been used experimentally to 'deplete' protein kinase C and thereafter determine whether hormone effects on biological processes are compromised. The rationale for this experimental paradigm is that long-term, high-dosage phorbol ester treatment would be expected to increase the binding of protein kinase C to membranes, and thereby increase proteolytic degradation and depletion of protein kinase C. This experimental approach is valid if: (a) phorbol ester-sensitive forms of protein kinase C predominate in the tissue under study; and (b) the rate of degradation of protein kinase C greatly exceeds its rate of synthesis. If these conditions are met, protein kinase C should be greatly decreased, and, if the hormone operates through protein kinase C, its biological effects should be diminished. On the other hand, if hormonal effects persist when protein kinase C is depleted, it may be surmized that the hormone effects are independent of protein kinase C.

Unfortunately, the evidence for protein kinase C 'depletion' in the above experimental paradigm may be misleading, and, in some tissues, it may be necessary to deplete protein kinase C below a poorly-defined threshold value, before there is limitation in a particular biological response. First, the fact that the phorbol ester itself no longer has biological effectiveness only means that the phorbol ester-sensitive forms of protein kinase C have been depleted, and forms of the enzyme that are less responsive to phorbol esters may still be present (or, in fact, enriched if there is compensatory synthesis) for the hormone to activate through increases in diacylglycerol. Second, the loss of histone phosphorylation after phorbol ester treatment may be more apparent than real, as phorbol esters apparently decrease the ability of protein kinase C to recognize or phosphorylate histone (44, 45). More definitive evidence for protein kinase C depletion can be derived from immunological studies, providing that the antibody used does not have particular sensitivity for recognizing phorbol ester-responsive forms of the enzyme. (Each isozyme has different antigenic properties and some forms may not be recognized by certain antibodies.) Obviously, this may be more of a problem with monoclonal antibodies than polyclonal antibodies or antisera, as the former have more narrow specificity. These problems may account for the fact that in BC3H-1 myocytes, we have not been able to demonstrate true depletion of protein kinase C with our polyclonal antiserum (99), whereas others (46) have reported such depletion when their antiserum is used. The latter group, however, has also reported loss of immunoreactive protein kinase C in other cells (47), but inspection of their immunoblots shows that, while the 80 kd immunoreactive protein kinase C was diminished after long-term phorbol ester 'desensitization', there was no decrease (an increase was in fact apparent) in slightly lower molecular weight immunoreactive proteins. The latter may be precursors to (100), derivatives of, or other forms of (e.g., see reference 98) protein kinase C, and, may in fact be activated during hormonal treatments (e.g., see reference 100).

Although the above considerations may temper the investigator's enthusiasm for using phorbol esters to experimentally test the importance of protein kinase C in hormone action, it should not be inferred that the use of phorbols is without merit. Indeed, if the above-mentioned caveats are kept in mind, and if the investigator is aware of the protein kinase C isozyme profile of the tissue which is being studied, and the responsiveness of these isozymes to phorbol ester treatment, meaningful information may be derived from studies in which phorbol esters are used to activate protein kinase C, and thereby mimic hormone effects on biological processes. Again, it may be even more meaningful if exogenous or endogenous diacylglycerol is used as the activator of protein kinase C, particularly if it has been established that the hormone or agonist under study provokes increases in diacylglycerol, and activates protein kinase C, *in vivo*.

3. Protein kinase C

3.1 General aspects

The Ca^{++} and phospholipid-dependent protein kinase originally purified by Nishizuka and co-workers (3, 4, 48), has become an intensely studied enzyme system, due largely to the fact that agonist-induced diacylglycerol and tumour promoters, such as phorbol esters, bind to the regulatory site and activate this enzyme, which in turn phosphorylates many important enzymes and regulatory proteins. The activation of the enzyme is complex in that multiple cofactors are required for full activity, namely, Ca^{++}, Mg^{++}, phospholipid, diacylglycerol, ATP, and a suitable substrate such as lysine-rich histone. Protein kinase C activity is often difficult to assess in crude extracts from certain tissues, due to the presence of endogenous inhibitors or other confounding kinases (50, 51). For example, phospholipid- and Ca^{++}- dependent activity cannot be detected in crude adrenal glomerulosa cell extracts (52), but after DEAE-cellulose chromatography, the Ca^{++}- and phosopholipid-dependent protein kinase activity is unmasked. On the other hand, in many cell types, there is no problem in detecting protein kinase C activity in crude extracts, particularly if the specific activity of the enzyme is relatively high and low concentrations of protein are assayed. However, even in the latter cell types, it is desirable to show that hormonal effects are demonstrable in column purified enzyme preparations as well, to be certain that these effects are on protein kinase C *per se*.

Several methods have been used to evaluate agonist-induced changes in protein kinase C activity and subcellular distribution: (a) enzymic transfer of ^{32}P from $[\gamma-^{32}P]$ ATP to lysine-rich histone in the presence of phosphatidylserine and Ca^{++}, with or without added diolein; (b) binding of $[^3H]$phorbol-dibutyrate (PDBu) to the regulatory site of protein kinase C, (c) histone- phosphorylating activity of the catalytic fragment following treatment of intact protein kinase C with trypsin, and (d) immunoblots of the enzyme, using polyclonal antisera or monoclonal antibodies to various forms of protein kinase C.

The same principles apply for examining activity and subcellular distribution of protein kinase C in all of the above-mentioned methods. The most stable preparations of protein kinase C are obtained from brain. In other tissues and cells, enzyme activity is much less stable due to difficult-to-control proteolysis, and experiments should be designed to extract and assay the enzyme as quickly as posible in the presence of protease inhibitors, preferably on the same day that samples are obtained.

3.2 Hormonal effects on enzyme activity and subcellular distribution

Following hormone-induced increases in intracellular Ca^{++}, diacylglycerol, or after phorbol ester treatment, protein kinase C translocates from the cytosol to cellular membranes (53). This 'translocation' phenomenon or redistribution of the enzyme between cytosol and membrane is a reflection of enzyme activation by diacylglycerol or phorbol esters, which are both primarily located in membranes in intact cells. This translocation is followed rapidly by protease action, leading to decreases in the absolute concentration of protein kinase C, especially in the cytosol fraction. Many hormones provoke this redistribution of the enzyme from the cytosol to the membrane, but the ability to observe this translocation response, or loss of cytosolic protein kinase C, appears to be better in the actions of hormones which mobilize Ca^{++} from intracellular stores (54). Some hormones or agonists may not provoke a typical translocation pattern if the enzyme is measured by histone phosphorylation, but true translocation is evident if the enzyme is assayed by immunoblot (for example, as in the case of insulin— unpublished observations). The apparent distribution pattern of protein kinase C activation in response to a hormone may therefore be different from that observed with phorbol esters, which invariably provoke a typical translocation response. Along these lines, it should be realized that phorbol esters exaggerate the translocation pattern by virtue of the fact that cytosolic enzyme may have diminished ability to phosphorylate histone after phorbol ester treatment (44, 45). Nevertheless, studies with phorbol esters and a variety of agonists (to which the tissue is sensitive) should be performed, to serve as positive controls, and provide a frame of reference for comparison with effects of the hormone under study.

Two different assay procedures for protein kinase C have been used to assess an agonist's effect on the activity and subcellular distribution of the enzyme. To prepare cellular enzyme preparations for the first method (i.e. the 'standard assay'—see below), the treatment (usually 1–30 min) is terminated by rapidly decanting the medium and rinsing the cells with ice-cold phosphate buffered saline. Cells are harvested by rapid centrifugation (2000 r.p.m. for 3 min) and suspended in homogenization buffer containing 0.25 M sucrose, 20 mM Tris-HCl, pH 7.4, 1.2 mM EGTA (the latter may be omitted in initial studies to evaluate chelator effects), 50 μM mercaptoethanol, 25 μg/ml leupeptin, and 1.0 mM phenylmethylsulfonylfluoride. Cell disruption methods (which should be compared for their effects on subcellular distribution) include sonication, Dounce-type homogenization, or hypotonic shock with ice-cold double-distilled water (55). The addition of chelators to the cell lysate buffer ensures that only tightly bound, detergent-extractable enzyme remains associated with the membrane fraction. In the absence of chelators, endogenous Ca^{++} will increase the amount of enzyme which (?non-specifically) remains bound to the membrane. The cell lysate is centrifuged at 100 000g for 30–60 min to obtain cytosol and membrane fractions. The membrane fraction is then rehomogenized with the above buffer containing high chelator concentrations (2 mM EGTA, 5 mM EDTA) plus 1% Triton X-100. After standing for 30 min at 4°C, the membrane fraction is centrifuged at 100 000g for 60 min. Protein kinase C activity is then measured (see standard assay below) in crude cytosol and membrane fractions, or following purification by chromatography on DEAE-cellulose or Mono Q columns. The latter may be eluted with a continuous salt gradient, or stepwise, with buffers containing increasing concentrations of salt (56).

3.3 Standard assay

Protein kinase C is most commonly assayed by measuring the transfer of ^{32}P from [γ-^{32}P]-ATP to histone, in the presence of Ca^{++} and phosphatidylserine, with or without diolein. (Note—addition of diolein may mask hormonal effects that may be due to increases in diacylglycerol.) Basal activity (due to other kinases) is measured by omission of Ca^{++} and phosphatidylserine, and addition of an excess of EGTA. The following assay procedure (57) is widely used for measuring types I (γ), II (β_1 and β_2) and III (α) protein kinase C.

Protocol 3. Protein kinase assays

Reagents

- Tris–HCl, 0.5 M, pH 7.5
- Magnesium acetate, 100 mM
- [γ-^{32}P] ATP, 0.25 mM, specific activity 15–25 $\times$ 10^4 c.p.m./nmol
- Histone III-S (Sigma), 5 mg/ml, in H$_2$O
- Phosphatidylserine (Sigma), 10 μg/ml, in chloroform
- 1,2-diolein (Sigma), 0.2 μg/ml, in chloroform
- CaCl$_2$, 2.5 mM
- EGTA, 2.5 mM
- Trichloroacetic acid, 25% (w/v)
- Trichloroacetic acid, 10% (w/v)

Procedure

1. The final volume of the assay mixture is 0.25 ml. To each tube add:
   - 10 μl Tris–HCl
   - 50 μl magnesium acetate
   - 10 μl [γ-32P] ATP
   - 10 μl histone
   - 50 μl CaCl$_2$ or 50 μl EGTA for measuring basal activity
   - 20 μl of lipid mixture containing 10 μg of phosphatidylserine with 0.2 μg of diolein. Or add 20 μl of 20 mM Tris–HCl for basal activity.

   (Immediately prior to assay, lipids are dried under nitrogen and resuspended in 20 mM Tris–HCl and sonicated for 1 min at 4°C.)
   - 100 μl of the enzyme fraction to be assayed.

2. After 3–5 min at 30°C, the reaction is stopped by adding 1 ml of 25% trichloroacetic acid. Acid-precipitable substances are collected on nitrocellulose membranes, 0.45 μm pore size, using a suction apparatus. The filters are washed four times with 1 ml of 10% trichloroacetic acid. Radioactivity of the acid-precipitable substances is counted in a liquid scintillation counter, and the activity of the enzyme is calculated on the basis of nmol or pmol of ^{32}P transferred per min per mg of protein in enzyme extract.

Comments

It is important to conduct assays in the absence and presence of increasing amounts of diolein in both control and hormone-treated preparations. Hormonal effects on enzyme activity due to increases in diacylglycerol may only be apparent in the absence of added diolein, and addition of exogenous diolein to enzyme preparations frequently provokes a complex biphasic response, with inhibition of activity following peak stimulation. The effects of exogenous diolein are more evident following chromatography, preferably using Mono Q (Pharmacia) type columns. Some less characterized forms of protein kinase C (i.e., δ and ε) are not activated to the same extent by Ca^{++}. Translocation of protein kinase C may be masked by increases in cytosolic enzyme activity that exceed decreases in protein kinase C content; Mono Q column chromatography obviates this problem by removing enzyme activators, such as diacylglycerol.

3.4 Translocation assay

This is an alternative method for assaying protein kinase C which avoids some of the potential artifacts encountered with cell homogenization, diminishes the problem of proteolytic degradation of protein kinase C, and involves permeabilization of cells with digitonin (58). An additional advantage of this method is that small numbers of cells are needed, as the assay volume is only 25 μl, and cells can be grown on 24-well culture dishes.

Protocol 4. Protein kinase C translocation assay

- Buffer A: 0.5 mg of digitonin/ml, 20 mM Mops, pH 7.2, 10 mM EGTA, 5 mM EDTA
- Buffer A plus 0.5% Triton X-100
- Trypsin, 0.1 mg/ml
- Trypsin inhibitor, 3 mg/ml
- Protein kinase inhibitor (Sigma), 1 U/5 μl
- 250 μM [γ-^{32}P]-ATP (2–3 × 10^5 c.p.m./5 μl)
- 75 mM MgCl$_2$

Protocol 4. *Continued*

- Histone III-S (5 mg/ml)
- 25% trichloroacetic acid (TCA).

Procedure

A. After treating cells with agonists, the cells are permeabilized with 0.2 ml digitonin in Mops [3-(N-morpholino)-propanesulphonic acid] buffer with EGTA and EDTA for 5 min at 4°C. The cytosol is removed and cells are then treated for another 5 min with 0.2 ml of the digitonin-containing buffer plus Triton X-100 to obtain membrane-associated protein kinase C.

Principle

Protein kinase C activity is assessed by activating the enzyme with trypsin to remove the regulatory (inhibitory) portion of protein kinase C, stopping the reaction with soybean trypsin inhibitor, and then assaying the catalytic portion of C-kinase for histone phosphorylation in the presence of an inhibitor of cAMP-dependent protein kinase.

B. Trypsin activation (two sets of sample tubes are needed for 'activated' and 'basal' measurements)

1. Add: 90 μl sample
      5 μl trypsin solution (to another set of sample tubes, add 5 μl buffer for
      basal values)

2. Incubate: 10 min at 30°C

3. Stop reaction with 10 μl trypsin inhibitor solution

C. Protein kinase Assay

1. Add: 5 μl [γ-^{32}P]-ATP
      5 μl protein kinase inhibitor solution
      5 μl histone solution
      10 μl trypsin-activated or unactivated sample

2. Incubate: 10–20 min at 30°C.

3. Stop the reaction by adding 0.5 ml 25% TCA.

4. Process samples as described in the Protein kinase C section (above). Protein kinase C activity is calculated as the difference between trypsin activated and basal activity.

Comments

This method provides data which are generally comparable to those obtained with the standard assay method discussed above. However, this 'translocation' assay primarily measures the activity of the 'M-kinase', the 50 kd protein released by trypsin treatment of the 80 kd holoenzyme. This assay theoretically should not be useful for measuring hormone-induced changes in enzymatic activity, if the change (e.g. an increase in diacylglycerol) activates the 30 kd regulatory domain of protein kinase C which is separated from the 50 kd catalytic fragment

Protocol 4. *Continued*

by trypsin. Thus, this assay may be better suited to measure translocation, rather than enzymic activation of protein kinase C.

Depending on the agonist and treatment period used, protein kinase C activity found in membranes and cytosol can vary because of the nature and turnover of the generated diacylglycerol, and subsequent changes in enzymic activity and subcellular distribution. Following addition of phorbol esters, recoveries of apparent enzyme activity can vary not only because of translocation, but also because of masking of the enzyme in assays which use histone as the substrate. The latter masking of enzyme may be less apparent if other substrates, or binding or immunoassays are used (see below).

3.5 Phorbol ester binding assay

The observation that phorbol esters and diacylglycerol bind to the same site on the regulatory domain of protein kinase C suggested that this enzyme was, in fact, 'the phorbol ester receptor' (59). The fact that treatment of cells with phorbol esters causes a decrease in cytosolic protein kinase C and, in many cases, an increase in membrane-bound protein kinase C, further suggested that phorbol esters stabilize the association of protein kinase C with membrane phospholipids (53). Several methods for phorbol ester binding have been described (60, 61). Cell extracts can be obtained as described above following hormone treatment, although membranes need not be solubilized with detergent. The same buffers used for the assay of enzyme activity can be used in the binding assay, with the exception that histone and ATP are omitted. Phosphatidylserine is necessary for demonstrating cytosolic binding activity, but is not necessary for binding of phorbol 12,13-dibutyrate (PDBu) in membrane extracts. Non-specific binding is determined with excess unlabelled PDBu, and is subtracted from the total binding to yield specific binding. Conditions should be optimized with respect to protein concentrations, as this function may not be linear. Generally, 10–20 μg of cytosolic and 50–100 μg of membrane protein are used in each sample. It should be realized that hormonally-induced increases in diacylglycerol will compete with PDBu for binding sites on protein kinase C, and a decrease in PDBu receptor affinity with no change in receptor density may also occur with hormone treatment. In addition, it should be realized that the 30 kd regulatory protein, which is released from protein kinase C during activation, also binds labelled-PDBu. Thus, this method may not truly reflect the status of protein kinase C, as such.

Protocol 5. Protein kinase C assay by phorbol ester binding

Reagents

- Buffer: 50 mM Tris–HCl, pH 7.4, 22.5 mM magnesium acetate, and bovine serum albumin (1.5 mg/ml).

Protocol 5. *Continued*

- [³H]Phorbol 12,13-dibutyrate (PDBu) (dry-down under nitrogen), 300 nM solution in buffer. Serial dilutions of two-fold are made. Unlabelled PDBu (60 μM) is added to an aliquot of each solution to be used for measurement of non-specific binding.
- Phosphatidylserine (600 μg/ml in 50 mM Tris–HCl).
- Bovine immunoglobulin (12 mg/ml in 50 mM Tris–HCl).
- Polyethylene glycol 6000 (33.3% in 50 mM Tris–HCl).

Procedure
A. Add solutions to microfuge tubes in the following order:

1. 50 μl sample.
2. 50 μl phosphatidylserine.
3. 50 μl[³H]PDBu solution ± unlabelled PDBu.
4. 10 μl CaCl₂ (the concentration should be calculated to be in excess of chelators to achieve 1 mM).
5. Incubate for 15 min at 37°C, return to ice for 5 min.

B. Precipitation with polyethylene glycol:

1. Add 50 μl immunoglobulin solution and 150 μl polyethylene glycol solution. Mix thoroughly, allow samples to sit for 15 min on ice.
2. Centrifuge 15 min full speed in microfuge. Return to ice. Aspirate supernate.
3. Rinse pellet with phosphate-buffered saline (300 μl).
4. Warm tubes to facilitate slicing off bottom with pellet.
5. Transfer pellet to scintillation vials. (Allow samples to equilibrate 3 h before counting.)

Comments
Data are analysed using a Scatchard analysis. PDBu binding measurements can be adjusted to measure binding in intact cells, as well as subcellular fractions.

3.6 Antisera and immunoassays

Polyclonal antisera to rat brain protein kinase C are easily raised in rabbits. The enzyme can be purified from rat brain using DE-52 chromatography followed by phenyl-Sepharose. At this point, the enzyme can be separated as an 80 kd band by sodium dodecylsulphate-polyacrylamide gel electrophoresis (SDS-PAGE) and identified as protein kinase C on the basis of Ca^{++}/phospholipid-dependent autophosphorylation (62). This preparation can be used to immunize rabbits after electroelution or electroblotting onto nitrocellulose, or by injecting the enzyme immobilized in acrylamide. The antisera raised by this procedure usually cross-react with the described isoforms of rat brain protein kinase C, viz., types I, II, and III (63–65) which are separated by chromatography on hydroxylapatite columns. Antisera to specific isozymes can be raised using specific peptide regions. Parker *et al.* (101) have successfully developed antisera to isozyme types I, II, and III. The antisera are useful for identifying and quantitating protein kinase C in Western blots, and this can be used to assay protein kinase C in cytosol and membrane fractions after SDS-PAGE and Western blots (66). This assay is particularly useful for examining changes in amounts and distribution (e.g. translocation) of protein kinase C. Further quantitation of protein kinase C may be accomplished by subjecting immunoblots to (a) densitometric scanning after horse-radish peroxidase or alkaline phosphatase staining, or (b) counting of radioactive ^{125}I-protein A or (c) avidin-biotin, which binds to the antigen-antibody complex. Alkaline phosphatase staining is more permanent.

The general procedure for SDS-polyacrylamide gel electrophoresis and transfer of proteins to nitrocellulose for immunodetection is given in *Protocol 6*.

Protocol 6. Western blotting analysis of protein kinase C

Reagents

- Acrylamide: BIS (30:0.8) solution containing 30 g acrylamide, 0.8 g *N'N'*-BIS methylene acrylamide in 100 ml with water. Filter and store at 4° in dark.
- 1.5 M Tris–HCl, pH 8.8.
- 0.5 M Tris–HCl, pH 6.8.
- Separating gel preparation: 9% gel, 0.375 M Tris, pH 8.8.

Distilled water	13.1 ml
1.5 M Tris–HCl, pH 8.8	7.5 ml
10% (w/v) SDS	0.3 ml
Acrylamide: BIS	9.0 ml
10% ammonium persulphate	0.1 ml
TEMED	75 μl

- Stacking gel preparation: 3% gel, 0.125 M Tris, pH 6.8.

Distilled water	6.3 ml
0.5 M Tris–HCl, pH 6.8	2.5 ml
10% (w/v) SDS	0.1 ml
Acrylamide: BIS	1.0 ml
10% ammonium persulphate	0.1 ml
TEMED	75 μl

- Sample buffer—Dilute the sample 1:5 with sample buffer and heat in boiling water bath for 2 min.

Distilled water	4.7 ml
0.5 M Tris–HCl, pH 6.8	1.0 ml
Glycerol	1.0 ml

Protocol 6. *Continued*

10% (w/v) SDS	1.0 ml
2-mercaptoethanol	0.1 ml
0.05% bromphenol blue	0.2 ml

● Electrode buffer, pH 8.3

Tris-base	6.0 g
Glycine	28.8 g
10% (w/v) SDS	10 ml
q.s. to 1 litre with distilled water	

● Blotting buffer

Tris	3.03 g
Glycine	14.4 g

q.s. to 1 litre with distilled water.
Use reagent grade methanol to make the buffer 20% v/v methanol.

Procedure

1. Prepare gel solutions on ice. Add all reagents except for TEMED, de-aerate under vacuum for 5 min. To initiate polymerization, add TEMED and swirl. Pour the solution into the assembled gel apparatus. Allow gel to polymerize overnight. Polymerize the stacking gel 30–45 min.

2. Electrophorese at 2.5 mA per gel to stack proteins. After the dye front enters the separating gel, raise the current to 30 mA per gel.

3. Transfer the acrylamide gel immediately to the blotting apparatus to avoid diffusion of proteins. Depending on the instructions that accompany the apparatus, transfer proteins to nitrocellulose.

4. Immunoblot procedure: The membrane with bound antigen is incubated with the first antibody specific for protein kinase C. The membrane is washed to remove unbound antibody, incubated with goat anti-rabbit alkaline phosphatase conjugate (second antibody), and washed again. Colour development will show the antigen as a purple band with a molecular weight of 74–80 kd.

Solutions

● Tris buffered saline, 1 × TBS (20 mM Tris, 500 mM NaCl, pH 7.5).

● Tween-20 wash solution, 1 × TTBS (20 mM Tris, 500 mM NaCl, 0.05% Tween-20, pH 7.5).

● Blocking solution (3% gelatin–TBS).

● Antibody buffer (1% gelatin–TTBS).

● First antibody solution: dilute the antisera 1:100–500 in 100 ml of the antibody buffer.

● Second antibody: Dilute second antibody, alkaline phosphatase conjugate 1:3000 in 100 ml of antibody buffer.

Protocol 6. *Continued*

● Carbonate buffer: $5 \times (0.3$ M $NaHCO_3$, 1.2 mg/ml $MgSO_4$, pH 9.7).

● Colour reagent: 25 mg each dianisidine (tetrazotized) and β-naphthyl acid phosphate in 20 ml of 5 × carbonate buffer. Bring total volume to 100 ml with distilled water.

● Stop solution: 23 ml methanol, 4 ml AcOH, 23 ml distilled water.

Immunoblot assay

1. Immerse the membrane in blocking solution. Agitate the solution for 30 min. to 1 h.

2. Transfer the membrane to TTBS. Wash 2 × for 5 min.

3. Transfer to first antibody solution and incubate 1 to 2 h with gentle agitation. Overnight incubation may be preferred to increase sensitivity.

4. Remove unbound first antibody by washing membrane in TTBS for 5 min with gentle agitation. Transfer the nitrocellulose to second antibody solution. Incubate for 1 h with agitation.

5. Wash membrane in TTBS for 2 × 5 min. Do a final wash in TBS to remove residual Tween-20 from the membrane surface.

6. Develop colour with reagent for 15–30 min. Stop reaction after 30 min. by transferring membrane to stop solution, then to distilled water and soak.

3.7 Phosphorylation of substrates in intact cells

Activation of protein kinase C *in vivo* should result in the site-specific phosphorylation of putative substrates. In many, but not all cases, some functional property of the substrate may change in response to this phosphorylation, and dephosphorylation by phosphatases should reverse this functional change. The demonstration that *in vivo* phosphorylation sites are identical to *in vitro* phosphorylation sites by protein kinase C is also desirable.

There are several techniques for demonstrating protein-kinase-C-mediated phosphorylation in intact cells. In addition to hormone treatment, as positive controls, protein kinase C may be activated by treating cells with biologically active phorbol esters, such as phorbol 12-myristate 13-acetate (PMA), also known as 12-O-tetradecanoyl phorbol 13-acetate (TPA), phorbol 12,13-dibutyrate (PDBu), or phorbol 12,13-didecanoate (PDD). These compounds are usually dissolved in a concentrated form (20 mM) in dimethylsulphoxide (DMSO) and stored at $-20°C$. It is extremely important to use the same concentration of DMSO (diluted in an identical manner) as a control, and a negative control compound (such as 4α-phorbol 12,13-didecanoate) with no tumour-promoting, or C-kinase-activating, activity should also be tested. Other compounds which are commonly used *in vitro* to directly activate protein kinase C are synthetic diacylglycerols, such as 1-oleoyl-2-acetylglycerol (OAG) and dioctanoyl-glycerol (DiC8). However, it should be kept in mind that these compounds can be metabolized to phosphatidic acid by diacylglycerol kinase, as well as to monoacylglycerol by diacylglycerol lipase in intact cells. The phorbol esters are also metabolized to varying, but lesser, degrees in intact cells, depending on the cell type.

While phorbol-esters induce phosphorylation of C-kinase-specific proteins in many cells, it should be noted that these compounds can activate other protein kinases, presumably through the action of protein kinase C and alterations in G-protein activity. Also, there are no specific inhibitors of protein kinase C which are effective in all cells, although several compounds have been employed with moderate levels of success in some cases (see *Table 1*). Inhibitors are usually defined by their effect on the assay of protein kinase C *in vitro*, but this does not necessarily mean that the inhibitors will be equally effective *in vivo*.

Table 1. Some inhibitors of protein kinase C

H-7	gossypol
C-1	sphingosine
amiloride	sphinganine
retinal	sangivamycin
mellitin	staurosporin
polymixin B	tetracaine
	trifluoroperizine

The effects of hormones and phorbol esters on protein phosphorylation are followed in ^{32}P-labelled cells. This is accomplished by prelabelling cells in a phosphate-free buffer with 50–100 μCi of ^{32}PO$_4$ per ml of medium for 2–16 h. After hormone treatment, cytosol and membrane fractions are dissolved in $5 \times$ Sample Buffer. Phosphorylated proteins are then separated using one- (see procedure above for acrylamide gels) or two-dimensional gel electrophoresis (67, 68). It is becoming increasingly evident that phorbol esters and certain other non-diacylglycerol C-kinase activators are less effective for activating certain isozymes of protein kinase C (43). Thus, some proteins may be better substrates for TPA-induced phosphorylation, and other proteins may be better phosphorylated by the activation of protein kinase C by endogenous or exogenous diacylglycerol. Likewise, some cells express forms of protein kinase C which are relatively insensitive to TPA stimulation, or which change their substrate specificity when treated with TPA. For example, as stated above, histone phosphorylation is lost or masked when cells have been treated with TPA (44, 45). TPA treatment may also yield an inactive form of the enzyme which can be recovered in some instances by treatment with DP-40 followed by DEAE-cellulose chromatography (55). These caveats must be kept in mind if hormonal effects on protein phosphorylation are compared to those of TPA.

3.8 Phosphoprotein markers

Only a few specific phosphoproteins have been identified as specific markers for activation of protein kinase C *in vivo*. The best described marker is a 40 kd phosphoprotein, which has been suggested to be the 5′-monophosphoesterase which hydrolyses inositol-trisphosphate (69). The *in vivo* phosphorylation of the 40 kd protein has been followed in platelet responses to thrombin, collagen, platelet-activating factor, phorbol esters, and diacylglycerol analogues, and this phosphorylation parallels the release of serotonin. Trypsin-generated ^{32}P-phosphopeptide fragments of the 40 kd protein after *in vivo* phosphorylation correlate very well to those obtained from direct *in vitro* phosphorylation of the 40 kd protein by protein kinase C. An 80 kd (or 87 kd in some tissues) protein has also been used as a specific phosphoprotein marker for activation of protein kinase C *in vivo*. ^{32}P-phosphoprotein maps of proteolytic fragments of the 80 kd protein from both *in vivo* studies with phorbol ester treatment, and *in vitro* treatment with protein kinase C, are also identical. Although the function of this 80 kd protein is unknown, it is found in virtually all cell types examined, although the brain is the richest source. This protein can be purified relatively easily (70) and can be readily identified in two-dimensional SDS-PAGE gels (71).

3.9 Use of alternative substrates in *in vitro* assays

As mentioned above, the 80–87 kd protein appears to be a specific substrate for protein kinase C. This protein, which is phosphorylated on a serine residue, can be purified 500-fold from bovine brain (70) and can be used *in vitro* (instead of histone) to assay protein kinase C (72). Vinculin, a cytoskeletal protein which is also phosphorylated *in vivo* by protein kinase C (73), is also easily purified (74) and can be used to assay protein kinase C activity *in vitro*. Additionally, vinculin has been shown to be an excellent *in vitro* substrate, when assaying extracts or purified protein kinase C preparations derived from phorbol ester-treated (or 'down-regulated' or 'desensitized') cells: phorbol esters mask histone, but not vinculin phosphorylation (44, 45).

3.10 Inhibitors

Specific inhibitors of protein kinase C *in vivo* have not been described, although most of the compounds listed in *Table 1* have been shown to inhibit the enzyme *in vitro*. Unfortunately, there are always questions regarding effectiveness and specificity of the inhibitor in *in vivo* studies.

3.11 'Desensitization' or 'down-regulation' by phorbol esters

Certain groups have observed that by treating cells *in vivo* with high doses of TPA (1–16 μM) (or other phorbol esters) for prolonged periods (24–48 h), some cell types are rendered protein kinase C-deficient (71, 75). As discussed above, such deficiency or depletion, however, can only be surmized if there are losses of (a) biological responses to TPA (i.e. 'desensitization'), (b) $Ca^{++}/$ phospholipid-dependent histone phosphorylation, (c) PDBu binding, and (d), most importantly, all immunoreactive protein kinase C, including the 80 kd holoenzyme and its proteolytic derivatives. It is emphasized that at least several of these criteria (including the last) be validated with each cell type used, as one criterion alone may be misleading (see below), and there are many cell types which do not become protein-kinase-C-deficient with TPA treatment. The reasons for the latter failure may be that there are certain forms of protein kinase C that are only poorly activated and depleted by TPA (see above). Thus, TPA treatment may only deplete those forms of protein kinase C that are most sensitive, and this may explain why TPA is no longer effective in provoking biological responses, whereas diacylglycerol remains effective in activating remaining enzymes.

Comments

In assessing evidence for protein kinase C depletion, the following should be kept in mind. First, phorbol esters may mask histone-phosphorylating activity of protein kinase C. Second, labelled-PDBu binding, if used as an indicator of protein kinase C, may be inhibited by high levels of endogenous diacylglycerol (76). Third, in evaluating immunoassay data, it should be realized that some antisera recognize only certain forms of protein kinase C: thus, unless the specificity of the antiserum is known, it is not possible to equate lack of immunoreactivity to full loss of protein kinase C. Fourth, it is also possible that TPA may cause changes in antigen recognition. To summarize, 'TPA-desensitization' or 'down-regulation' experiments are useful only if the tissue protein kinase C isozyme profile, and its immunoreactivity and responsiveness to TPA are known. For example, in bovine adrenal glomerulosa cells, as reported by Lang *et al.* (52), the only significant protein kinase C activity isozyme found was type III, a TPA-sensitive form (43). Tissues which have a preponderance of type III protein kinase C may be easier to 'down-regulate' with prolonged TPA treatment, but other criteria mentioned above should be fulfilled before any conclusions are drawn.

3.12 Purification of isoforms by hydroxylapatite chromatography

In general, procedures which stress simplicity and rapidity of analyses will prove to be the most satisfactory for obtaining purified preparations of intact 80 kd protein kinase C. The eventual use of the enzyme will often dictate the number of purification steps required. If the goal is to run hydroxylapatite column chromatography to identify isoforms, only partial preliminary purification of the enzyme by DEAE–cellulose chromatography is usually necessary. The most common initial step for partially purifying protein kinase C is DEAE–cellulose (DE-52) chromatography (57) or Mono Q chromatography. The enzyme can be eluted from either of these columns by using either a linear gradient or batchwise elution with 90 to 100 mM NaCl or KCl. There are often two or more peaks of PKC activity which elute from these or other ion exchange columns. The second peak usually has a higher basal (non-protein kinase C) kinase-activity than the first, possibly reflecting co-elution of this PKC and M-kinase. Following DE-52 or Mono Q chromatography (see *Figure 3*), several approaches can be taken. If the enzyme is relatively free of other contaminating kinases, the eluate can be concentrated and subjected to gel filtration, followed by phenyl–Sepharose chromatography. In some cases, the DE-52 eluate [or eluates from Mono Q (Pharmacia) columns] can be concentrated and applied directly to a hydroxyla-patite column (see *Figure 3*) (65). The hydroxylapatite column itself, and conditions used to elute the enzyme are crucial for the successful separation of protein kinase C isoforms. Most procedures published are adequate for identifying types I, II, and III. Type I is the γ form, type II contains β_I and β_{II}, and III is the α form. The quality of hydroxylapatite used for the column packing influences the separation of isoforms. The most preferable is spherical hydroxylapatite (Koken, Japan). These columns are supplied by Regis in the United States. An automatic programmable liquid chromatography (FPLC) system to program the gradients needed to separate the three isoforms described is also helpful for obtaining reproducible separations. The hydroxylapatite columns should be run at a slow flow rate (0.4 ml/min or less) to achieve the best separation (43, 65). A shallow, curvilinear gradient also improves separation of types I and II. Chromatographic separation of the four known isoforms of the 80 kd enzyme from rat brain has been demonstrated by Pelosin *et al.* (43). Other protein kinase C gene products (presumably other isozymes) have also been reported (77), but their elutions on hydroxylapatite chromatography have not been described.

4. Arachidonate, prostaglandins, and leukotrienes

4.1 General aspects

Studies of the cellular metabolism of labelled arachidonate yield data that reflect precursor pools, as well as production of prostanoid and eicosanoid substances (potential intracellular signalling substances or mediators derived from arachi-donate). Arachidonate, derived either directly from agonist-induced hydrolysis of phospholipids by phospholipase A_2 (see *Figure 1*), or indirectly by sequential hydrolysis of phospholipids by phospholipase C, followed by diacylglycerol lipase, is rapidly converted to prostanoids and eicosanoids by the cyclooxygenase and lipoxygenase pathways. The biologically active 'end-products' and their 'inactive' metabolites appear to be specific for each cell type examined. The agonists which stimulate phospholipase A_2 activity are frequently the same as those which activate PIP_2-specific phospholipase C, and thus stimulate inositol-trisphosphate production. The latter is attended by increases in calcium flux,

Mono Q (Pharmacia) chromatography

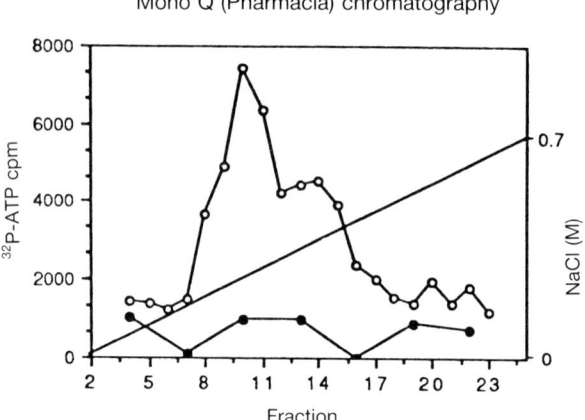

Hydroxyapatite chromatography

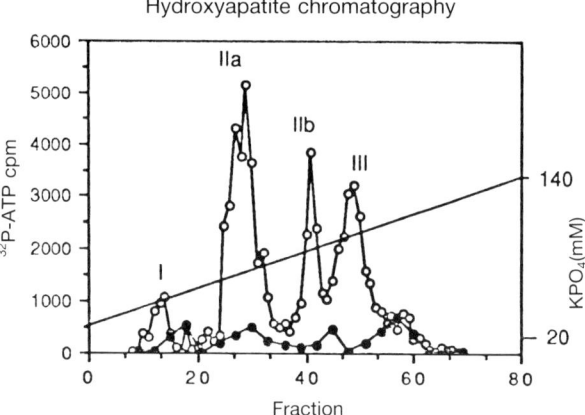

Figure 3. Characteristic elution profiles of rat brain protein kinase C following Mono Q and hydroxyapatite chromatography. Mono Q (Pharmacia HR 5/5) chromatography was performed on 15 mg of rat brain cytosolic protein. Breakthrough proteins were eluted prior to initiating the NaCl gradient which was generated by an FPLC system. Protein kinase activity was determined in the presence of phosphatidylserine, diolein, and Ca^{++} (open circles) or EGTA (closed circles) as described. The separation of protein kinase C isozymes was achieved using an hydroxyapatite (Koken S-type, 10 cm) column. Chromatography was performed on protein kinase C which had previously been purified by Mono Q chromatography (0.7 mg protein). Breakthrough proteins were eluted prior to initiating the KPO_4 gradient using a programmable FPLC system. The four peaks of activity represent protein kinase C types I, II, and III in order of elution. Types IIa and IIb were identified immunologically using an antisera specific for type II protein kinase C.

which activate phospholipase A_2. In addition, the diacylglycerol produced by phospholipase C is relatively rich in arachidonate, which may be released by diacylglycerol lipase action. PKC activation may also result in phospholipase A_2 activation.

Prostanoids and eicosanoids are released into the incubation medium shortly after they are synthesized from non-esterified arachidonate. The prostaglandins and leukotrienes formed can be analysed by a variety of methods, including gas chromatography with mass spectrometry (78), high-pressure liquid chromatography (HPLC) with UV or fluorescence detection (79, 82) or radioimmunoassay (RIA) (80, 81).

4.2 Methods for examining products of arachidonate metabolism

4.2.1 Tissue incubation

Reagents: Hank's balanced salt solution, pH 7.4

    ([^{3}H] arachidonate is light-sensitive and care should be taken to protect solutions and samples, as autooxidation can change HPLC, TLC, and RIA results.)

i. Procedure

Cellular products of arachidonate metabolism can be examined by pre-labelling cells with tracer (1 μCi/ml) doses of [^{3}H]arachidonate, stimulating its release from phospholipids by addition of agonists, and following subsequent conversion of labelled-arachidonate to cyclooxygenase and lipoxygenase products. Cells in culture may be rinsed and incubated in Hank's balanced salt solution, pH 7.4 (HBSS). Rapid labelling of phospholipids by arachidonic acid, largely by transacylation, occurs in the presence or absence of fatty acid-free bovine serum albumin, but this labelling is faster in the absence of albumin. Initially, uptake and labelling time should be monitored by sampling the medium, and examining incorporation into tissue phospholipids. After labelling has reached equilibrium (1–3 h or more) in relevant phospholipid precursor pools, cells may be rinsed with HBSS and then challenged with agonists. The time-course of arachidonate release should be examined, as metabolically active, arachidonate-labelled phospholipid pools may be rapidly depleted. A23187 (1 μM) elicits the release of arachidonate in most cells, and treatment with this calcium ionophore can be used as a positive control. After incubation, the medium can be extracted and analysed by HPLC for labelled products of arachidonate metabolism.

ii. Comments

This procedure for using [^{3}H]arachidonate is successful only if relevant phospholipid precursor pools are well-labelled and quite often, this is not the case. Usually phosphatidylcholine is the most rapidly labelled precursor, but cells often release more arachidonate from phosphatidylethanolamine than phosphatidylcholine for prostanoid or eicosanoid synthesis, and the true source of the latter may not be accurately represented by losses of radioactivity from phosphatidylcholine (longer periods of pre-labelling may be needed to label phosphatidylethanolamine). Another factor that should be kept in mind is that the amount of arachidonate released or converted to prostanoids or other eicosanoids is not apparent from isotopic studies, as the specific activity of the phospholipid pool which is hydrolysed is usually unknown.

4.2.2 Extraction and analysis of prostaglandins

(a) Samples are acidified to pH 3 with HCl and extracted twice with 2–4 vol of ethyl acetate. Inclusion of a trace amount of an unrelated [^{3}H]prostaglandin such as $PGF_{1\alpha}$ is useful for calculating recovery.

(b) The extracts are dried under N_2. Other extraction procedures using C_{18} Sep-Pak (Waters) columns are also available.

(c) Isocratic, reverse phase HPLC analysis can be run on extracts using CH_3CN in H_2O with formic acid (82). Labelled standards are used to establish elution times. The HPLC analysis of prostaglandins is generally only applied to samples in which [^{3}H]arachidonate is used and converted to the labelled prostanoids being studied. Principles for measuring prostaglandins by radioimmunoassay are described below.

4.2.3 Measurement of prostaglandins by radioimmunoassay

The measurement of prostaglandins by radioimmunoassay (RIA) requires specific antisera with low cross-reactivity to other prostanoid species, unless these compounds are extracted and purified by column chromatography on silica gel (83). Antisera characterized as to cross-reactivity are available commercially (suppliers include Dupont NEN and Amersham).

Comments

In some tissues and fluids, interfering substances may preclude the direct assay of prostaglandins, and extraction then becomes necessary. Generally, balanced salt solutions and culture media without sera can be assayed without extraction. Samples for analysis should be stored at $-70°C$ unless assayed immediately. In the case of serum, which contains prostaglandins, samples must be extracted and serum 'blanks' included for reference. Antisera to many prostaglandin species are commercially available and radioimmunoassay procedures have been reviewed elsewhere (84). Samples should be stored in silanized glass to avoid losses, especially in samples containing low levels of prostaglandins. Plasticware should be avoided as extracted plasticizers can interfere with assays.

4.2.4 Lipoxygenase product analysis

Lipoxygenase products, in particular, the leukotriene (LT) LTB_4, and derivatives, 20-OH-LTB_4 and 5-HETE, can be extracted into organic solvents, along with prostaglandins, or by using solid-phase columns (available from J. T. Baker or Waters). Solid-phase extractions are preferable for lipoxygenase products, as phospholipids and other polar contaminants can be removed (85). An internal standard such as PGB_2 should be used to correct for recovery. Lipoxygenase products can be analysed by reverse-phase HPLC. Since lipoxygenase products possess conjugated diene and triene moieties, they can be detected in chromatographic effluents by UV absorption, as well as by following the conversion of [^{3}H]arachidonate to the purified, labelled leukotrienes. Aspects of HPLC analysis and resolution of metabolites have been discussed elsewhere (86). Antisera to leukotrienes and radioimmunoassay kits are also available commercially (Dupont NEN) for the analysis of lipoxygenase products.

References

1. Downes, P. and Michell, R. H. (1982). *Cell Calcium*, **3**, 467.
2. Berridge, M. J. (1987). *Ann. Rev. Biochem.*, **56**, 159.
3. Nishizuka, Y. (1986). *Science*, **233**, 305.
4. Kikkawa, U. and Nishizuka, Y. (1986). *Ann. Rev. Cell Biol.*, **2**, 149.
5. Hannun, Y. A., Loomis, C. R., and Bell, R. M. (1986). *J. Biol. Chem.*, **261**, 7184.
6. Ganong, B. R., Loomis, C. R., Hannun, Y. A., and Bell, R. M. (1986). *Proc. Natl. Acad. Sci. USA*, **83**, 1184.
7. Kennerly, D. A. (1987). *J. Biol. Chem.*, **262**, 16305.
8. Nishizuka, Y. and Takai, Y. (1981). In *Protein Phosphorylation* (ed. O. M. Rosen and E. G. Krebs), Vol. 8, p. 237. Cold Spring Harbor Conferences on Cell Proliferation, Cold Spring Harbor Laboratory Press, New York.
9. Christensen, R. L., Shade, D. L., Graves, C. B., and McDonald, J. M. (197). *Int. J. Biochem.*, **19**, 259.
10. Hughes, B. P., Rye, K. A., Pickford, L. P., Barritt, G. J., and Chalmers, A. H. (1984). *Biochem. J.*, **222**, 535.
11. Farese, R. V., Davis, J. S., Barnes, D. E., Standaert, M. L., Babishkin, J. S., Hock, A., Rosic, N. K., and Pollet, R. J. (1985). *Biochem. J.*, **231**, 269.
12. Farese, R. V., Fanjul, L. F., de Ruiz Galarreta, C. M., Davis, J. S., and Cooper, D. R. (1987). *Life Sci.*, **41**, 2631.
13. Farese, R. V., Konda, T. S., Davis, J. S., Standaert, M. L., Pollet, R. J., and Cooper, D. R. (1987). *Science*, **236**, 586.
14. Saltiel, A. R., Fox, J. A., Sherline, P., and Cuatrecasas, P. (1986). *Science*, **233**, 967.
15. Wilson, D. B., Neufeld, E. J., and Majerus, P. W. (1985). *J. Biol. Chem.*, **260**, 1046.
16. Farese, R. V., Orchard, J. L., Larson, R. E., Sabir, M. A., and Davis, J. S. (1985). *Biochim. Biophys. Acta*, **846**, 296.
17. Cockcroft, S. (1984). *Biochim. Biophys. Acta*, **795**, 37.
18. Cockcroft, S. and Gomperts, B. D. (1985). *Nature, Lond.* **314**, 534.
19. Luttrell, L. M., Hewlett, E. L., Romero, G., and Rogol, A. D. (1986). *J. Biol. Chem.*, **263**, 6134.
20. Romero, G., Luttrell, L., Rogol, A., Zeller, K., Hewlett, E., and Larner, J. (1988). *Science*, **240**, 509.
21. Besterman, J. M., Duronio, V., and Cuatrecasas, P. (1986). *Proc. Natl. Acad. Sci. USA*, **83**, 6785.

22. Bocckino, S. B., Blackmore, P. F., Wilson, P. B., and Exton, J. H. (1987). *J. Biol. Chem.*, **262**, 15309.
23. Kennedy, E. (1986). In *Lipids and Membranes* (ed. J. A. Op Den Kamp, B. Roelofsen, and K. W. Wirtz), Chap. 8, p. 171. Elsevier Science Publishers, Amsterdam and New York.
24. Farese, R. V., Cooper, D. R., Konda, T. S., Nair, G., Standaert, M. L., Davis, J. S., and Pollet, R. J. (1988). *Biochem. J.*, **256**, 175.
25. Saltiel, A. R., Sherline, P., and Fox, J. A. (1987). *J. Biol. Chem.*, **262**, 1116.
26. Mato, J. M., Kelly, K. L., Abler, A., and Jarrett, L. (1987). *J. Biol. Chem.*, **262**, 2131.
27. Prescott, S. M. and Majerus, P. W. (1981). *J. Biol. Chem.*, **256**, 579.
28. Stevens, E. J. and Husbands, D. R. (1987). *Arch. Biochem. Biophys.*, **258**, 361.
29. Imai, A. and Gershengorn, M. C. (1987). *J. Biol. Chem.*, **262**, 6457.
30. Gerber, D., Davis, M., and Hokin, L. E. (1973). *J. Cell Biol.*, **56**, 736.
31. Low, M. G. and Saltiel, A. R. (239). *Science*, **239**, 268.
32. Kabara, J. J. and Chen, J. S. (1976). *Anal. Chem.*, **48**, 814.
33. Takuwa, Y., Takuwa, N., and Rasmussen, H. (1986). *J. Biol. Chem.*, **261**, 14670.
34. Bocokino, S. B., Blackmore, P. F., and Exton, J. H. (1985). *J. Biol. Chem.*, **260**, 14201.
35. Banschbach, M. W., Geison, R. L., and O'Brien, J. F. (1974). *Anal. Chem.*, **59**, 617.
36. Kennerly, D. A., Parker, C. W., and Sullivan, T. J. (1979). *Anal. Biochem.*, **98**, 123.
37. Preiss, J., Loomis, C. R., Bishop, W. R., Stein, R., Neidel, J. E., and Bell, R. M. (1986). *J. Biol. Chem.*, **261**, 8597.
38. Cooper, D. R., Konda, T. S., Standaert, M. L., Davis, J. S., Pollet, R. J., and Farese, R. V. (1987). *J. Biol. Chem.*, **262**, 3633.
39. Gonzatti-Haces, M. I. and Traugh, J. A. (1986). *J. Biol. Chem.*, **261**, 15266.
40. Harary, I., Renaud, J. F., Sato, E., and Wallace, G. A. (1976). *Nature, Lond.*, **261**, 58.
41. Ono, Y., Kikkawa, U., Ogita, K., Fujii, T., Kurokawa, T., Asaoka, Y., Sekiguichi, K., Ase, K., Igarashi, K., and Nishizuka, Y. (1987). *Science*, **236**, 1116.
42. Ono, Y., Fujii, T., Ogita, K., Kikkawa, U., Igarashi, K., and Nishizuka, Y. (1987). *FEBS Lett.*, **226**, 125.
43. Pelosin, J. M., Vilgrain, I., and Chambaz, E. M. (1987). *Biochem. Biophys. Res. Commun.*, **147**, 382.
44. Cochet, C., Souvignet, C., Kermidas, M., and Chambaz, E. M. (1986). *Biochem. Biophys. Res. Commun.*, **134**, 1031.
45. Cooper, D. R., de Ruiz Galaretta, Fanjul, L. F., Mojsilovic, L., Standaert, M. L., Pollet, R. J., and Farese, R. V. (1987). *FEBS Lett.*, **214**, 122.
46. Spach, D. H., Nemenoff, R. A., and Blackshear, P. J. (1986). *J. Biol. Chem.*, **261**, 12750.
47. Blackshear, P. J., Nemenoff, R. A., Hovis, J. G., Halsey, D. L., Stumpo, D. J., and Huang, J. K. (1987). *Molec. Endocrinol.*, **1**, 44.
48. Nishizuka, Y. (1984). *Nature, Lond.*, **308**, 693.
49. Nishizuka, Y. (1984). *Science*, **225**, 1365.
50. McDonald, J. R. and Walsh, M. P. (1985). *Biochem. Biophys. Res. Commun.*, **2**, 603.
51. McDonald, J. R. and Walsh, M. P. (1985). *Biochem. J.*, **232**, 559.
52. Lang, U. and Vallotton, M. B. (1987). *J. Biol. Chem.*, **262**, 8047.
53. Kraft, A. S. and Anderson, W. B. (1983). *Nature, Lond.*, **301**, 621.
54. Hirota, K., Hirota, T., Aguilera, G., and Catt, K. J. (1985). *J. Biol. Chem.*, **218**, 531.

55. Thomas, T. P., Gopalakrishna, R., and Anderson, W. B. (1987). *Methods Enzymol.*, **141**, 399.
56. Skoglund, G., Hansson, A., and Ingelman-Sundberg, M. (1985). *Eur. J. Biochem.*, **148**, 412.
57. Kikkawa, U., Minakuchi, R., Takai, Y., and Nishizuka, Y. (1983). *Methods Enzymol.* **99**, 288.
58. Pelech, S. L., Meier, K., and Krebs, E. G. (1986). *Biochemistry*, **25**, 8348.
59. Castagna, M., Takai, Y., Kaibuchi, K., Sano, K., Kikkawa, U., and Nishizuka, Y. (1982). *J. Biol. Chem.*, **257**, 7847.
60. Jaken, S. (1987). In *Methods in Enzymology* (ed. P. M. Conn and A. R. Means), Vol. 141, p. 275. Academic Press, London and New York.
61. Kikkawa, U., Takai, U., Tanaka, Y., Miyake, R., and Nishizuka, Y. (1983). *J. Biol. Chem.*, **258**, 11442.
62. Ballester, R. and Rosen, O. M. (1985). *J. Biol. Chem.*, **260**, 15194.
63. Huang, K., Nakabayashi, P., and Huang, F. L. (1986). *Proc. Natl. Acad. Sci. USA*, **83**, 8535.
64. Jaken, S. and Kiley, S. C. (1987). *Proc. Natl. Acad. Sci. USA*, **84**, 4418.
65. Ono, Y., Kikkawa, U., Ogita, K., Fujii, T., Kurokawa, T., Asaoka, Y., Sekiguchi, K., Ase, K., Igarashi, K., and Nishizuka, Y. (1987). *Science*, **236**, 1116.
66. Girard, R. R., Mazzei, G. J., and Kuo, J. F. (1986). *J. Biol. Chem.*, **261**, 370.
67. Laemmli, U. K. (1970). *Nature, Lond.*, **227**, 364.
68. Blackshear, P. J., Nemonoff, R. A., and Avruch, J. (1983). *Biochem. J.*, **214**, 11.
69. Connolly, T. M. and Majerus, P. W. (1986). *Clin. Res.*, **34**, 656A.
70. Albert, K. A., Nairn, A. C., and Greengard, P. (1987). *Proc. Natl. Acad. Sci. USA*, **84**, 7046.
71. Blackshear, P. J., Wen, L., Blynn, B. P., and Witters, L. A. (1986). *J. Biol. Chem.*, **261**, 1459.
72. Walaas, S. I., Horn, R. S., Adler, A., Albert, K. A., and Walaas, O. (1987). *FEBS Lett.*, **220**, 311.
73. Werth, D. K. and Pastan, I. (1984). *J. Biol. Chem.*, **259**, 5264.
74. Feramisco, J. R. and Burridge, K. (1980). *J. Biol. Chem.*, **255**, 1194.
75. Rodriguez, A., Rozengurt, P., and Rozengurt, E. (1984). *Biochem. Biophys. Res. Commun.*, **120**, 1053.
76. Jaken, S. (1987). In *Methods in Enzymology* (ed. P. N. Conn and A. R. Means), Vol. 141, p. 275. Academic Press, London and New York.
77. Ono, Y., Fujii, T., Ogita, K. Kikkawa, U., Igarashi, K., and Nishizuka, Y. (1987). *FEBS Lett.*, **226**, 125.
78. Fitzpatrick, F. A., Gorman, R. R., and Wynalda, M. A. (1977). *Prostaglandins*, **13**, 201.
79. Wynalda, M. A., Lincoln, F. H., and Fitzpatrick, F. A. (1979). *J. Chromatog.*, **176**, 419.
80. Maclouf, J. (1982). In *Methods in Enzymology* (ed. W. E. M. Lands and W. L. Smith), Vol. 86, p. 273. Academic Press, London and New York.
81. Fitzpatrick, F. A. (1982). In *Methods in Enzymology* (ed. W. E. M. Landes and W. L. Smith), Vol. 86, p. 286. Academic Press, London and New York.
82. Whorton, A. R., Carr, K., Smigel, M., Walker, L., Ellis, K., and Oates, J. A. (1979). *J. Chromatog.*, **163**, 64.
83. Campbell, W. B. and Ojeda, S. R. (1987). *Methods Enzymol.*, **141**, 323.

84. Lands, W. E. M. and Smith, W. L. (eds). (1982). In *Methods in Enzymology*, Vol. 86, p. 1.
85. Powell, W. S. (1982). In *Methods in Enzymology*, Vol. 86, p. 467.
86. Shak, S. (1987). In *Methods in Enzymology* (ed. P. H. Conn and A. R. Means), Vol. 141, p. 355. Academic Press, London and New York.
87. Nair, G., Standaert, M. L., Pollet, R. J., Cooper, D. R., and Farese, R. V. (1988). *Biochem. Biophys. Res. Commun.*, **154**, 1345.
88. Cozza, E. N., del Carmen Vila, M. D., Gomez-Sanchez, C. E., and Farese, R. V. (1989). *J. Steroid Biochem.* (in press).
89. Foster, R. H. and Farese, R. V. (1989). *Life Sci.*, **45**, 2015.
90. Cooper, D. R., Hernandez, H., Kuo, J., and Farese, R. V. (1989). *Arch. Biochem. Biophys.* (in press).
91. Farese, R. V., Nair, G., Sierra, C., Standaert, M. L., Pollet, R. J., and Cooper, D. R. (1989). *Biochem. J.*, **261**, 927.
92. Lee, T. S., Chao, T., Kang-Quan, Hu., and King, G. L. (1989). *Biochem. Biophys. Res. Commun.*, **162**, 381.
93. Ishizuka, T., Hoffman, J., Cooper, D. R., Watson, J. E., Pushkin, D. B., and Farese, R. V. (1989). *FEBS Lett.*, **249**, 234.
94. Peter-Riesch, B., Fathi, M., Schlegel, W., and Wollheim, C. B. (1988). *J. Clin. Invest.*, **81**, 1154.
95. Lee, T. S., Saltsman, K. A., Ohashi, H., and King, G. L. (1989). *Proc. Natl. Acad. Sci. USA*, **86**, 5141.
95a. Craven, P. and DeRubertis, F. (1989). *J. Clin. Invest.*, **83**, 1667.
96. Kennerly, D. A. (1987). *J. Biol. Chem.*, **262**, 16313.
97. Pessin, M. S. and Raben, D. M. (1989). *J. Biol. Chem.*, **264**, 8729.
98. Schaap, D., Parker, P. J., Bristol, A., Kriz, R., and Knopf, J. (1989). *FEBS Lett.*, **243**, 351.
99. Cooper, D. R., Watson, J. E., Acevedo-Duncan, M., Pollet, R. J., Standaert, M. L., and Farese, R. V. (1989). *Biochem. Biophys. Res. Commun.*, **161**, 327.
100. Borner, C., Filipuzzi, I., Wartmann, M., Eppenberger, U., and Fabbro, D. (1989). *J. Biol. Chem.*, **264**, 13902.
101. Marais, R. M. and Parker, P. J. (1989). *Eur. J. Biochem.*, **182**, 129.

7

The insulin receptor tyrosine kinase

MORRIS F. WHITE

1. Introduction

Insulin controls metabolism and growth by modulating the activity of cellular enzymes and transport systems. The molecular events are initiated by insulin binding to the α-subunit of the insulin receptor, which stimulates tyrosyl autophosphorylation and activates a tyrosyl specific phosphotransferase in the β-subunit of the receptor (1). Thus, signal transduction may involve tyrosyl phosphorylation of cellular substrates by the activated insulin receptor kinase (2). Several putative substrates have been identified in the intact cell (3–6), including a protein which we call pp185 that is immunoprecipitated from insulin-sensitive cells with antiphosphotyrosine antibodies (7–9).

Our studies on the structure and function of the insulin receptor and the search for the cellular substrates have utilized a variety of techniques, cells, and tissues. In particular, many of our original experiments were carried out with Fao hepatoma cells which are physiologically sensitive to insulin and express about 80 000 receptor molecules; we also used rat hepatocytes and adipocytes, and human fibroblasts and placenta. Since the cDNA of the human insulin receptor was obtained, we have prepared various mammalian cell lines which express human receptor molecules (2, 10). Eukaryotic expression of the insulin receptor offers a number of advantages over previous systems including rapid and easy production of high concentrations of insulin receptor molecules. Moreover, mutant receptor cDNA can be expressed in host cells to allow analysis of the biological activity of the altered molecules. Preparation of transfected cells and techniques for analysing the activity of the expressed receptor molecules are the topics of this chapter.

2. Expression of the human insulin receptor in cultured cells

2.1 Expression plasmids used for mammalian cell transfection

Expression plasmids containing the normal and mutant human insulin receptor cDNA have been prepared in the laboratories of Ullrich (11), Rutter (12), and

Whittaker (13). In our experiments, we use an expression plasmid prepared by Dull and Ullrich called pCVSVHIRc (14). This 10 kb vector contains the origin of replication and the ampicillin-resistance gene (Ap®) of the *E. coli* plasmid pBR322, and the complete coding sequence for the wild-type human placental insulin receptor (11) controlled by the SV40 early promoter. A brief diagram of the plasmid including some unique restriction sites is shown in *Figure 1*. The dihydrofolate reductase (DHFR) gene is contained on this plasmid to provide a means of gene-linked co-amplification of the insulin receptor (15).

Similar plasmids for the expression of mutant insulin receptor molecules were generated by oligonucleotide-directed mutagenesis. The details of this procedure are beyond the scope of this chapter, and the reader is directed to Kunkel *et al.* (16) for a summary of the subject. Briefly, a portion of the insulin receptor cDNA (i.e. originally a $bgl\text{II}_{2146}$ to $hind\text{III}_{6777}$ fragment, but more recently a $bam\text{HI}_{2492}$ to $spe\text{I}_{5138}$ fragment from pCVSVHIRc) encoding the catalytic domain of the insulin receptor is subcloned into double-stranded M13mp19. Competent *E. coli* cells containing dut⁻ and ung⁻ mutations (BW313) are transformed with the double-stranded virus, and single-stranded uracil-containing viral template is isolated by standard methods from the supernatant (16). The single-stranded template DNA is annealed to oligonucleotide primers which contain one or more base substitutions, and complete double-stranded circular DNA is prepared *in vitro* with T7 polymerase and T4 ligase (16). Competent *E. coli* JM103 cells are transformed with double-stranded DNA, and plaques are identified under stringent conditions using the mutagenesis primer as a probe. The mutations are confirmed by M13 dideoxy sequencing (17) and the

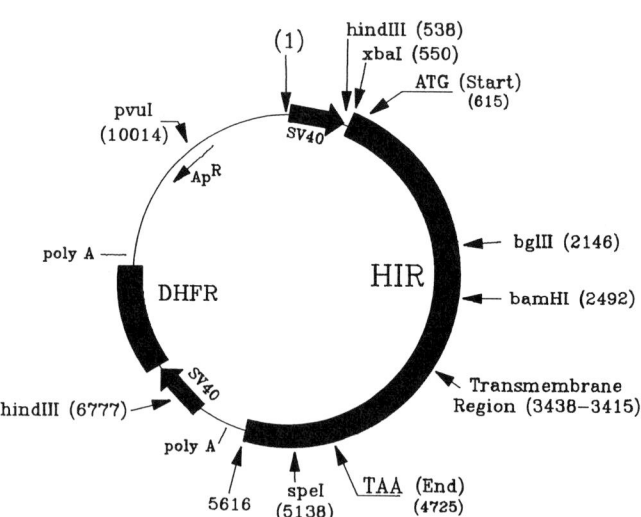

Figure 1. A schematic drawing of the pCVSVHIRc eukaryotic expression plamid used to transfect CHO cells. This contains the full-length cDNA of the human placental insulin receptor (HIR) and dihydrofolate reductase (DHFR), both under control of the SV40 early promoter.

mutant cDNA fragment is reintroduced into the pCVSVHIR vector. The yield of altered sequences is greater than 50%, because the wild-type uracil-containing strand is degraded, whereas the mutant-containing DNA strand is stable (16).

2.2 Transfection of chinese hamster ovary (CHO) cells

The introduction of foreign DNA into eukaryotic cells can be accomplished in several ways, including DEAE-dextran or calcium-phosphate-mediated trans-fection, protoplast fusion, or electroporation (15). For Chinese hamster ovary cells, we have used the calcium phosphate method. This method is simple and effective, and works well with a variety of mammalian cell lines including HeLa cells, Ltk⁻, CHO, CV1, NIH, 3T3, and C127F cells.

CHO cells (10^6 cells/10-cm dish) are co-transfected with pCVSVHIRc and pSVEneo. The *neo* gene in pSVE*neo* confers resistance to the aminoglycoside antibiotic G418 which provides a dominant selectable marker in all mammalian cells. Co-transfection of cells with pSVE*neo* at a 10- to 20-fold excess of pCVSVHIRc permits the isolation after about two weeks of colonies which constitutively express the insulin receptor. Prepare both plasmids by standard methods and purify them twice by CsCl–ethidium bromide banding (18). Transfect adherent CHO cells with these plasmids by the following procedure:

Protocol 1. Transfection of CHO cells with the human insulin receptor

1. Grow wild-type CHO cells to 50% confluence in 10-cm dishes containing F12 medium (GIBCO) supplemented with 10% fetal bovine serum (GIBCO). Maintain the cells at 37°C in a humidified atmosphere composed of 95% air and 5% CO_2.

2. Add fresh serum-containing medium to the cells 12 h before transfection; Hepes (25 mM, pH 7.4) can be added to help maintain the pH during transfection.

3. Combine sterile portions of pCVSVHIRc (10 µg in 10 µl) and pSVE*neo* (≤1 µg) with 0.55 ml of 1 mM Tris–HCl (pH 7.5) containing EDTA (0.1 mM), and $CaCl_2$ (250 mM). While gently agitating this solution with a slow stream of sterile air, add slowly 0.5 ml of 50 mM Hepes (exactly pH 7.12) containing NaCl (280 mM) and Na_2HPO_4 (1.5 mM). Leave the mixture undisturbed for 30–45 min while the DNA/$Ca_3(PO_4)_2$ precipitate forms.

4. Add all of the precipitated DNA in small drops over the entire surface of subconfluent CHO cells. Incubate the cells with the DNA/$Ca_3(PO_4)_2$ for 24 h at 37°C.

5. After 24 h, incubate the cells in 10 ml of fresh serum-containing F12 medium. Then add 800 µg/ml G418 (GIBCO) to select cells which expressed the *neo* gene. Ten to 14 days later, harvest the surviving colonies and culture them in the presence of G418 to amplify the cell line.

2.3 Selection of CHO cells expressing the human insulin receptor by fluorescence-activated cell sorting (FACS)

CHO cells which survive in G418 have been successfully transfected with pSVE*neo*. However, transfected DNA which is stably incorporated by cells is frequently expressed inefficiently, or may become methylated and transcriptionally inactive (15). Thus simple co-transfection generally results in relatively low levels of constitutive gene expression; it has been our experience that only a small fraction of the *neo*-resistant cells express the human insulin receptor at high levels. Since CHO cells contain about 3000 rodent insulin receptors, it is necessary to select cells which overexpress the human insulin receptor by at least tenfold before the cells are useful. One way to select CHO cells that express a high number of receptors is fluorescence-activated cell sorting (FACS) (*Protocol 2*).

Protocol 2. FACS sorting of CHO cells transfected with the human insulin receptor

1. The day before FACS, culture each cell line to be sorted in a 15-cm dish at 20% confluence.

2. After 16 h of culture, prepare a single-cell suspension of CHO cells by washing the monolayers with phosphate buffered saline (PBS) followed by incubation of the cells at 22°C for 2 to 5 min with PBS containing 0.05% trypsin and 0.5 mM EDTA.

3. Wash the cells in 10 ml of serum-containing F12 medium and then incubate the cells at 4°C in 0.5 ml of this medium containing a monoclonal anti-insulin receptor antibody (1 μg) directed against epitopes in the α-subunit of the human insulin receptor. We have used two different monoclonal antibodies successfully, αIR-1 provided by Dr Steven Jacobs of Burroughs Wellcome, and MA-51 provided by Dr Ira Goldfine of the University of California, San Francisco. A non-specific mixture of mouse IgG can be used as the negative control.

4. Wash the cells twice with 10 ml of serum-containing F12 medium and suspend them in 1.0 ml of this medium. Add 40 μl of fluorescein-conjugated goat anti-mouse IgG (Cappel/Worthington) and incubate at 4°C for an additional 30 min.

5. Finally, wash the cells with 15 ml of serum-containing F12 medium and suspend them in 1 ml of this medium for FACS (Coulter Epics V flow cytometer). The brightest cells are collected and amplified in tissue culture. Repeat the sorting two or three times to obtain cell lines with significant increases in insulin binding, and obtain pure clones by limiting dilution.

The results of a typical series of cell sorting experiments are shown in *Figure 2*.

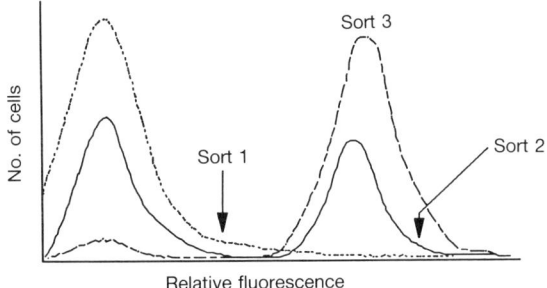

Figure 2. Fluorescence-activated cell sorting of CHO cells transfected with pCVSVHIRc. The arrows indicate the cut-off for selection of cell during the first and second sort.

After the first sort, no distinct peak of cells showing an increased amount of fluorescence emerge. However, cells with higher than average fluorescence occur in the tail of the main peak, which are collected and grown to confluence (all cells to the right of the arrow). During the second sort, a distinct peak of highly fluorescent cells occurs. The brightest fraction of this peak is collected (all cells to the right of the arrow) and grown to confluence in culture. Finally, the third sort reveals that the majority of cells express a high number of insulin receptors.

2.4 Insulin-binding to transfected CHO cells

Insulin-binding measurements provide an easy way to monitor the expression of the insulin receptor during a series of FACS experiments. In many cases specific binding of ~ 0.01 nM [^{125}I]insulin determined in the absence or presence of 10^{-6} M insulin is adequate. However, the following procedure includes the intermediate insulin concentrations and is suitable to obtain data for a Scatchard plot of insulin-binding to transfected CHO cells.

Protocol 3. Insulin-binding to transfected CHO cells

1. Grow the CHO cells to confluence in 24-well cluster trays (Costar), and incubate the cells with F12 medium (no serum) for 15 h before the experiment.

2. Prepare the following solutions:
   - Hepes binding buffer (*Table 1*).
   - Stock solutions of insulin in Hepes binding buffer (*Table 1*) with the following concentrations: 10^{-4} M (600 μg/ml), 10^{-5} M (60 μg/ml), 10^{-6} M (6 μg/ml), 10^{-7} M (0.6 μg/ml), and 10^{-8} M (0.06 μg/ml).
   - Prepare [^{125}I]insulin at a final activity of 10 000 c.p.m./ml by diluting [^{125}I]insulin (2000 Ci/mmol, Amersham) into 10 ml of Hepes binding buffer. The concentration of insulin in this solution is ≤0.01 nM.
   - Combine unlabelled and labelled insulin solutions as described in *Table 2*, and cool to 4°C.

Protocol 3. *Continued*

Table 1. Composition of Hepes binding buffer

	Concentration	
Component	mM	g/litre
Hepes (pH 7.4)	100	11.9
NaCl	118	6.9
KCl	5.0	0.37
MgSO$_4$	1.2	0.15
Dextrose	8.8	1.6
BSA		10.0

Table 2. Dilution of labelled and unlabelled insulin for Scatchard plots

	Unlabelled insulin		[^{125}I]Ins	Hepes buffer	
Tube No.	Conc. (μM)	Vol. (μl)	Vol. (μl)	Vol. (μl)	Final insulin concentration (nM)
1	—	0	250	50	~0.01
2	—	0	250	50	~0.01
3	—	0	250	50	~0.01
4	10	4	250	46	0.10
5	10	8	250	42	0.20
6	10	16	250	34	0.40
7	10	24	250	26	0.60
8	10	32	250	18	0.80
9	100	4	250	46	1.0
10	100	8	250	42	2.0
11	100	12	250	38	3.0
12	100	16	250	34	4.0
13	100	24	250	26	6.0
14	100	32	250	18	8.0
15	1000	4	250	46	10
16	1000	8	250	42	20
17	1000	12	250	38	30
18	1000	16	250	34	40
19	1000	24	250	26	60
20	1000	32	250	18	80
21	10 000	4	250	46	100
22	100 000	4	250	46	1000
23	100 000	4	250	46	1000
24	100 000	4	250	46	1000

The total volume in each tube is 300 μlitres which is sufficient to cover one well in a 24-well cluster tray. The five different stock solutions of unlabelled insulin should be made in Hepes binding buffer (*Table 1*). The [^{125}I]insulin is made in Hepes binding buffer and contains about 40 000 c.p.m./ml, approximately 0.01 nM insulin. The distribution of insulin concentration in this table works well for a Scatchard plot of insulin binding to transfected CHO cells.

Protocol 3. *Continued*

3. Remove F12 medium and wash the cell three times (2 ml each) with PBS at 4°C. Keep the cluster tray on ice.

4. Add 0.3 ml of the insulin solutions prepared according to *Table 2* to each well of the cluster tray. Incubate the cluster tray for 15 h at 4°C with gentle rocking.

5. Wash the cells once at 4°C with 2 ml of PBS containing 1% bovine serum albumin, and twice at 4°C with 2 ml of PBS.

6. Solubilize cells at 22°C with 1 ml 0.1 M NaOH containing 0.1% SDS. Measure the radioactivity in each extract in a γ-counter.

7. Determine the protein concentration in 10 μl of the extract with the Bradford Protein Assay (Bio-Rad).

8. Calculate the ratio of bound insulin to free insulin and plot this quantity against the amount of insulin bound per milligram protein of cell number. Prepare a Scatchard plot and read the number of insulin receptors from the abscissa. These data can be further analysed as described by Rodbard to resolve curvilinear plots and calculate binding affinities (19).

Transfection of CHO cells with pSVE*neo* only, and growth of these cells (CHO/neo) in G418-containing medium has no detectable effect on the number of insulin receptors (*Figure 3*). However, insulin binding increases following transfection of the CHO cells with the normal human insulin receptor cDNA. Scatchard analysis indicates that CHO/HIRC$_1$ cells, which were sorted three times by FACS, express about 40 000 insulin receptors (*Figure 3*). A similar

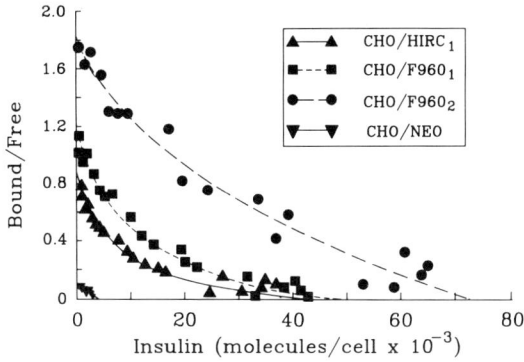

Figure 3. A Scatchard plot of insulin binding to the transfected CHO cells. The cells were grown to confluence in 24-well cluster trays and incubated at 4°C for 15 h with 10^{+11} M [^{125}I]insulin and various concentrations of unlabelled insulin. Then the cells were washed and the bound radioactivity was measured in a γ-counter. The calculated affinity constants are reported in the text.

analysis indicates that two cell lines (CHO/F960$_1$ and CHO/F960$_2$) expressing a mutant cDNA, pCVSVHIRF/Y960 (2), which encodes a substitution of phenylalanine for tyrosine in position 960 of the insulin receptor precursor, contain about 45 000 and 70 000 mutant receptors, respectively (*Figure 3*). Analysis of the binding data using a two-sites model show that the high affinity binding constants ($\pm$SD) for the wild-type and mutant receptors were 0.6 ± 0.2 nM and 0.8 ± 0.1 nM, respectively; the values for the low affinity binding constant were 79 ± 48 nM and 50 ± 25 nM, respectively. Thus, CHO cells expressing 10 to 20 times more insulin receptors are obtained by transfection and fluorescence-activated cell sorting.

3. Analysis of the phosphorylated insulin receptor in transfected CHO cells

3.1 Introduction

We study tyrosyl phosphorylation of the insulin receptor and other proteins in transfected CHO cells by labelling the cells with [^{32}P]orthophosphate, stimulating the cells with insulin, and identifying the phosphoproteins by immunoprecipitation with anti-insulin receptor (α-IR) or anti-phosphotyrosine (α-PY) antibodies, and sodium dodecyl sulphate polyacrylamide gel electrophoresis (SDS-PAGE). Detection of the insulin receptor is difficult in the [^{32}P]-labelled wild-type CHO cells since they contain less than 3000 receptors per cell (*Figure 4*, lanes a and b). However, following transfection and selection of high expressing cell lines by FACS, the phosphorylated receptor immunoprecipitated with α-PY is prominent (*Figure 4*, lanes e–h). Moreover, additional phosphotyrosine-containing protein have been identified in these cells with the α-PY; pp185 is detected in lanes f and h of *Figure 4* (2). The following *in vivo* phosphorylation protocol was initially developed for use with the Fao cell and has since proven to be useful for CHO and other cells.

3.2 Phosphate labelling

3.2.1 Safety with [^{32}P]phosphate

^{32}P in unshielded form represents a significant external radiation exposure hazard (18). Using hundreds of microcuries to tens of millicuries in the following protocol, maximum permissible occupational doses (US or international standards) are received within seconds to minutes. In working with ^{32}P, use the factors of time, distance, and shielding to minimize radiation exposure. The recommended shielding material for ^{32}P is 1.0-cm-thick Plexiglas; it is inexpensive, lightweight, and transparent. Adequate shielding protects all persons working in the laboratory, including those behind and to all side of the source. Plexiglas boxes are convenient for containing and shielding ^{32}P during incubation of tissue culture dishes.

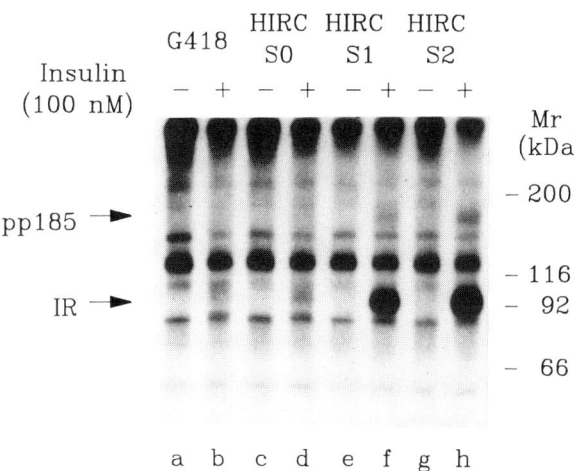

Figure 4. Immunoprecipitation of phosphotyrosine-containing proteins from [^{32}P]phosphate-labelled CHO cells. CHO cells transfected with pSVEneo (G418), or co-transfected with both pSVEneo and pCVSVHIRc (HIRC) and not sorted by FACS (S0) or sorted one (S1) or two (S2) times, were labelled for 2 h with [^{32}P]orthophosphate. The cells were incubated without (−) or with 100 nM insulin (+) for 1 min and solubilized as described in Section 3.2.2. The phosphotyrosine-containing proteins were immunoprecipitated with the aPY, reduced with DTT and separated by SDS-PAGE. The autoradiogram was obtained during a 6 h exposure.

Protocol 4. Radiolabelling cells with ^{32}P$_i$

1. Incubate confluent cells in 10-cm dishes for 2 h with 10 ml of phosphate-free and serum-free RPMI 1640 medium (GIBCO) containing 0.5 mCi/ml carrier-free [^{32}P]orthophosphate (New England Nuclear). Three millilitres of medium is sufficient to cover the monolayer if the dish is slowly rocked in the incubator. A Hoeffer 'Red Rocker' (Model No. PR50) is excellent for this purpose.

2. Add insulin (30 μl of 10^{-5} M insulin to 3 ml of medium) and continue the incubation at 37°C for the desired time-intervals.

3. Stop phosphorylation and dephosphorylation quickly by removing the incubation medium and freezing the cell monolayers by pouring liquid nitrogen into the dishes.

4. Before the monolayer of cells thaw, add 2 ml of *extraction buffer* (*Table 3*).

5. Scrape the cells from the dishes and thaw on ice. Sediment the insoluble material by centrifugation at 50000 r.p.m. in a Beckman 70.1 Ti rotor for 30 min. Separate and save the supernatant and the pellet for further analysis. In the absence of Triton X-100, the insulin receptor is in the pellet, whereas in the presence of Triton X-100 the receptor is in the supernatant.

Table 3. Phosphoprotein extraction buffer

	Concentration	
Component	mM	g/litre
Hepes (pH 7.4)	50	11.9
NaCl	118	6.9
NaF	100	4.2
Na_3VO_4	2	0.37
$Na_3P_2O_7$	10	4.4
EDTA	4	1.5
PMSF		0.34
Aprotinin		0.10

3.3 Immunopurification of the insulin receptor with anti-insulin receptor antibody and separation of the phosphoproteins by SDS-PAGE

3.3.1 Purification of anti-insulin receptor antibodies from human sera

Many sources of anti-insulin receptor antibody are available including human serum (20), monoclonal antibodies (21), and antipeptide antibodies (22). Most of our work has been carried out with human serum containing autoantibodies against the insulin receptor; our most potent sera are designated B-2 or B-9. These sera, obtained from patients with an autoimmune form of insulin resistance (20), immunoprecipitate more than 90% of labelled receptors across all species at 1:200 dilution. For phosphorylation experiments, the antibody works best if the IgG fraction is affinity purified on immobilized protein A before use (*Protocol 5*).

Protocol 5. Affinity purification of IgG on protein A agarose

1. Equilibrate 5 ml of protein A-agarose (Pierce) with phosphate buffered saline.
2. Apply 10 ml of serum to the column and wash it with 100 ml of PBS.
3. Elute the IgG fraction with 100 mM glycine, pH 2.5. Collect 1 ml fractions in 1.0 ml of 1 M Hepes (pH 7.4) to neutralize the antibody immediately.
4. Determine the elution by measuring the protein concentration in 10 μl portions of eluate using the Bradford protein assay (BioRad).
5. Combine the IgG-rich fractions and dialyse them against PBS. Adjust the protein concentration to 1 mg/ml, and distribute the purified IgG in 100 μl portions and freeze at $-70°C$.

3.3.2 Preparation of anti-insulin receptor antibodies from synthetic peptides

Since the amino-acid sequence of the human insulin receptor became available, it is possible to prepare antibodies against specific domains of the insulin receptor by immunization of rabbits with synthetic peptides. Three regions of the β-subunit have been found to yield antisera which immunoprecipitate the native receptor [numbered according to Ullrich *et al.* (11)]: residues 1314–1324, Arg Ser Tyr Glu Glu His Ile Pro Tyr Thr His; residues 1143–1152, Arg Asp Ile Tyr Glu Thr Asp Tyr Tyr Arg; residues 952–962, Leu Tyr Ala Ser Ser Asn Pro Glu Tyr Leu Ser (22, 23). We used the method shown in *Protocol 6* to prepare the polyclonal domain-specific antibodies in rabbits:

Protocol 6. Preparation of anti-insulin receptor antibodies from synthetic peptides

1. Reaction of the peptide with bromoacetyl bromide.

  (a) Dissolve the 5 mg of peptide in 1 ml of 100 mM $NaBO_4$, pH 8.5.

  (b) Add a 2 molar excess of bromoacetyl bromide (Aldrich Chemical Co.), about 20 μl, in 1 μl portions at 22°C with vigorous mixing. During this step the pH falls rapidly and must be maintained between 7.5 and 8.5 with 5 M LiOH.

  (c) Purify the derivatized peptides on a C_{18} Sep-Pak (Waters). After activating the Sep-Pak with acetonitrile and equilibrating it with water containing 0.5% trifluoroacetic acid, apply the sample and wash the Sep-Pak with 10% acetonitrile/0.5% trifluoroacetic acid in water. Elute the peptides with water containing 50% acetonitrile and 0.05% trifluoro-acetic acid. By reverse-phase HPLC nearly 100% of the peptide reacts with bromoacetyl bromide.

2. Conjugate the peptide to Keyhole limpet hemocyanin.

  (a) Incubate the derivatized peptide (5 mg) for 4 days at 22°C with 10 mg of Keyhold limpet hemocyanin (Calbiochem) in 1 ml of 10 mM $NaBO_4$. Measure the pH daily and maintain it at 9.0 by addition of 5 M LiOH.

  (b) Dialyse the reaction mixture against PBS, and concentrate the KLH–peptide conjugate to 2 mg/ml. Distribute the KLH conjugate in 0.5 ml aliquots and freeze at −20°C.

3. Inoculate rabbits.

  (a) Inoculate New Zealand white rabbits with 1 mg of the KLH–peptide conjugate suspended in 0.5 ml of complete Freund's adjuvant.

  (b) Subsequent inoculations are made with incomplete Freund's adjuvant at 3-week intervals.

  (c) Collected 25 ml of serum from the rabbits two weeks after each inoculation.

4. Affinity purify the antipeptide antibody.

  (a) Immobilize 1 mg of peptide on 1 ml of Affi-Gel (Bio-Rad) as described in the manufacturers instructions. The reaction is most efficient in dimethyl

Protocol 6. *Continued*

sulphide rather than water. After the reaction, wash, and equilibrate the column with PBS.

(b) Pass the serum (10 ml) over the 1-ml affinity column several times. Wash the column with 50 ml of PBS at 4°C.

(c) Elute the bound IgG from the washed column with 100 mM glycine, pH 2.5, and immediately neutralize the fractions by collecting 0.5-ml portions of eluate in 1 ml of 1 M Hepes, pH 7.4.

(d) Dialyse the eluate against PBS, pH 7.4. The yield of specific antibodies varies widely among the rabbits, but our usual yield is about 0.5 mg per 10 ml of serum. Store the antibody at 0.1 mg/ml in PBS at $-70°C$.

Protocol 7. Immunoprecipitation and SDS-PAGE separation of phosphorylated insulin receptor

1. Add anti-insulin receptor antibody (≤ 10 μg of purified B2; ≤ 1 μg of purified antipeptide antibody (the exact amount depends on the antibody affinity and concentration, and must be determined by trial and error) to 1 ml of whole cell extract and incubate the solution at 4°C for at least 3 h (or overnight) in a 1.5-ml microfuge tube.

2. Precipitate the antibody–receptor complex with Protein A. Purified Protein A immobilized on agarose works fine, although we routinely use Pansorbin (Calbiochem) as it is inexpensive and forms a 'tight' pellet during centrifugation. Pansorbin must be washed before use as follows:

   (a) Suspend 0.5 g of Pansorbin in 50 ml of calcium and magnesium-free PBS containing 3% SDS and 10% β-mercaptoethanol.

   (b) Boil the suspension for 20 min and sediment the Pansorbin by centrifugation at 10 000g for 15 min.

   (c) Wash the Pansorbin twice with 50 ml of calcium and magnesium-free PBS containing 3% SDS without β-mercaptoethanol.

   (d) Wash the Pansorbin once with 50 ml of 25 mM Hepes, pH 5.5, and three times with 25 mM Hepes, pH 7.4. Finally, suspend the Pansorbin in 5 ml of 25 mM Hepes, pH 7.4, containing 10 mM sodium azide.

3. Incubate the cell extract with 100 μl of 10% Pansorbin for 1 h at 4°C.

4. Sediment the Pansorbin by centrifugation for 1 to 2 min in a microfuge.

5. Wash the pellet three times by resuspension and centrifugation with Pansorbin wash buffer (*Table 4*). Suspend the Pansorbin with micro-tip sonicator or vigorous vortexing.

6. Elute the oligomeric forms of the insulin receptor ($M_r \geq 350$ kd) from the

Protocol 7. *Continued*

washed precipitates with Laemmli sample buffer (24) without dithiothreitol (DTT), or include 100 mM DTT to dissociate the α- and β-subunits.

7. Separate the non-reduced oligomers on 5% polyacrylamide gels, or separate the reduced α- and β-subunits on 7.5% gels (25).

8. Stain and dry the gels, and identify phosphoproteins by autoradiography at $-70°C$ using Kodak X-Omat film and an intensifying screen. Quantify the radioactivity in the gel fragments by Cerenkov counting.

Table 4. Pansorbin wash buffer

Component	Concentration	
	mM	g/litre
Hepes (pH 7.4)	50	11.9
NaCl	118	6.9
NaF	100	4.2
Na_3VO_4	2	0.37
SDS 0.1%		1.0
Triton X-100 1%		10.0

3.4 Immunopurification of the insulin receptor and other phosphotyrosyl-containing proteins with antiphosphotyrosine antibody

The anti-phosphotyrosine antibody (α-PY) which we have used in our experiments was originally described by Pang *et al.* (26, 27). It is obtained from the serum of rabbits that were immunized with phosphotyramine coupled covalently to Keyhole limpet hemocyanin (KLH):

Protocol 8. Preparation of polyclonal anti-phosphotyrosine antibody

1. Synthesize phosphotyramine (Tym(P)).

   (a) Dissolve 1 g of tyramine (Aldrich) and 4.2 g of phosphorus pentoxide in 4.1 ml of 85% phosphoric acid.

   (b) Heat the mixture to 100°C for 72 h in a sealed tube.

   (c) Dilute the reaction to 50 ml with water and apply it on to a Dowex 50 column (20 ml of resin) that was equilibrated with water. During elution of the resin with water, two peaks are observed at 268 nm, a relatively small initial peak and a very broad second peak containing phosphotyramine.

Protocol 8. *Continued*

(d) Concentrate the second fraction by lyophilization to about 100 ml and crystallize the phosphotyramine from 50% ethanol during 24 h at 4°C. Collect the crystals by filtration and dry them *in vacuo*. The product migrates as a single peak on reverse-phase HPLC (Bio-Rad, RP-318), and separates from tyramine which is retarded on the column during isocratic elution with 0.5% trifluoroacetic acid in water.

2. Prepare *N*-bromoacetyl-O-phosphotyramine

(a) Dissolve phosphotyramine (1 g) in 20 ml of 100 mM sodium borate and adjust the pH to 8.5 with 5 M LiOH.

(b) Slowly add 8.8 g of bromoacetyl bromide (Aldrich) over a 15-min time-interval. Stir the reaction vigorously. The pH falls rapidly and must be maintained between 7.5 and 8.5 by addition of 5 mM LiOH.

(c) Dry the reaction solution by lyophilization and extract the residue three times with ethyl ether, and then three times with ethanol. The product migrates as a single peak on C18 reverse-phase HPLC, and clearly separates from phosphotyramine which elutes from the column more quickly.

3. Couple *N*-bromoacetyl-O-phosphotyramine to Keyhole limpet hemocyanin.

(a) Dissolve 30 mg of KLH in 5 ml of 100 mM sodium borate, pH 9.25.

(b) Add 350 mg of *N*-bromoacetyl-O-phosphotyramine and incubate at 22°C for 96 h.

(c) Check the pH daily and maintain it above 9.0 by adding small portions of 5 M LiOH. After 96 h, the solution is dialysed against PBS, pH 7.4, and the protein concentrated to 2 mg/ml.

4. Immunize New Zealand white rabbits as described in Section 3.3.1.

5. Purify α-PY from the rabbit serum by affinity chromatography on immobilized phosphotyramine.

(a) Couple 1 g Tym(P) to 5 ml of Affi-Gel (Bio-Rad) in 10 ml of 50 mM Hepes as described in the manufacturers instructions. After the reaction, equilibrate the Affi-Gel with PBS.

(b) Apply up to 50 ml of serum over the 5-ml column several times and wash the resin successively with 50 ml of PBS, 50 ml of 200 mM phosphate buffer (pH 7.0), and 50 mM Hepes containing 150 mM NaCl.

(c) Elute the antibody specifically with 100 mM *p*-nitrophenyl phosphate (Sigma) in 50 mM Hepes, pH 7.4, containing 150 mM NaCl. Identify the IgG-containing fractions by measuring the protein concentration in each fraction. Allow the column to sit overnight in the elution buffer and collect a second fraction of α-PY. Combine the antibody-containing fractions.

Protocol 8. *Continued*

(d) Dialyse the eluate exhaustively against **PBS** at 4°C to remove the
p-nitrophenyl phosphate. Ordinarily, 5 changes of 1 litre each are
sufficient to remove all of the *p*-nitrophenyl phosphate. Fifty millilitres of
serum usually yields about 10 mg of purified IgG.

Protocol 9. Immunoprecipitation of phosphotyrosine-containing
proteins with the *a*PY

1. Incubate the phosphorylated cell extracts (1–2 ml) with 1 to 2 μg of affinity-
purified α-PY for at least 2 h at 4°C.

2. Precipitate the antibody complex during a 1 h incubation with 50 μl of 10%
Pansorbin prepared as described in Section 3.3.3.

3. Wash the Pansorbin three times by centrifugation and resuspension in
Pansorbin wash solution (*Table 4*).

4. Elute the proteins from the washed precipitates with Laemmli sample buffer.
Reduce the protein with 100 mM dithiothreitol and separate them by SDS-
PAGE on 7.5% gels.

5. Alternatively, elute the phosphotyrosine-containing proteins with
p-nitrophenyl phosphate (PNPP) by incubating the washed Pansorbin at 4°C
with 50 μl of 20 mM PNPP in 50 mM Hepes, pH 7.4. Centrifuge and remove
the supernatant and add 50 μl of 2X Laemmli buffer for SDS-PAGE. Use this
method when problems are encountered with nonspecific phosphoproteins.

6. Identify the phosphoproteins by autoradiography at −70°C of the stained
and dried gels using Kodak X-Omat film and an intensifying screen. Quantify
the radioactivity in the gel fragments by Cerenkov counting.

Western blot analysis identifies specific proteins resolved by SDS-PAGE by
binding with specific antisera. This technique is particularly useful with the α-PY
to identify putative substrates of the insulin receptor. The procedure described
below requires ^{125}I-labelled protein A, but it is also possible to employ
enzyme-labelled reagents in order to avoid the use of radioactivity for the
identification of phosphotyrosine-containing proteins.

Protocol 10. Immunoblotting with the *a*PY

1. Prepare a cell extract as described in Section 3.2.2. Separate the proteins in the
extract by SDS-PAGE. Specific proteins can be purified from the extract
before the electrophoresis by affinity chromatography or immunoprecipitation, but this step is not required.

Protocol 10. *Continued*

2. Transfer the proteins to nitrocellulose by electroblotting using published techniques (28); use standard transfer buffer composed of Tris (1.12 g) and glycine (14.4 g) in 800 ml of water plus 200 ml of methanol. To transfer large protein (≥ 100 kd) such as the insulin receptor subunits, carry out electroblotting for 10 h at 75 V. Add SDS (0.1%) to the transfer buffer if necessary to obtain a more complete transfer.

3. After the transfer, wash the nitrocellulose briefly with deionized water. Stain proteins on the nitrocellulose with 0.5% Ponceau S prepared by dissolving 1.5 g of Ponceau S (Sigma P-3504) in 300 ml of water containing 3 ml acetic acid. Destain the nitrocellulose with water.

4. Incubate nitrocellulose sheet in PBS containing bovine serum albumin (10 g/l) (PBS/BSA) for 2 h at 37°C to block non-specific binding.

5. Incubate the nitrocellulose in PBS/BSA containing αPY (1–3 µg/ml) for 2 h at room temperature on a Hoeffer 'Red Rocker'. This step requires about 7–10 ml of antibody solution, so use a small container or sealed plastic bags. After incubation remove and store the antibody solution with 0.02% azide as it can be reused.

6. Wash nitrocellulose three times, 5 min each, and incubate for 30 min, in PBS/BSA.

7. Incubate nitrocellulose with [^{125}I]-protein A (10 uCi) in 10 ml of PBS/BSA for 1 h at room temperature on a Hoeffer 'Red Rocker'.

8. Wash nitrocellulose four times, 5 min each, with PBS/BSA and then air-dry the nitrocellulose. Identify the labelled proteins by autoradiography.

---

4. Purification of the insulin receptor on immobilized wheat germ agglutinin and *in vitro* autophosphorylation

4.1 Purification of the insulin receptor on immobilized wheat germ agglutinin

Several methods have been employed to purify the insulin receptor including gel filtration, ion-exchange of DEAE, lectin affinity chromatography on immobilized concanavalin A or wheat germ agglutinin, immunoaffinity chromatography using specific antibodies, and immobilized insulin. For much of our *in vitro* work, the partially purified insulin receptor eluted from a wheat germ agglutinin affinity column has been adequate. The *N*-acetylglucosamine used for elution does not interfere with any assays carried out with the receptor, so dialysis is not indicated. The receptor is stable for about 3 months at $-70°C$, but after 3 months, the preparation loses insulin sensitivity. The yield of insulin receptor from this procedure depends on the level of expression of the receptor in the cell.

Protocol 11. Purification of insulin receptor on wheat germ agglutinin

1. Grow 10 confluent dishes (15-cm) of CHO cells. Incubated the cells with serum-free F12 medium 15 h before the isolation.

2. Remove the medium and add 3 ml of Hepes (50 mM, Sigma), containing Triton X-100 (1.0%, New England Nuclear), aprotinin (0.1 mg/ml, Sigma) and PMSF (2 mM, Sigma) to each dish at room temperature. The cells are immediately scraped from the dish, and the extracts are combined in a 50-ml plastic conical test-tube at 4°C.

3. Vortex the extract for 1 min and sediment the insoluble material by centrifugation at $100\,000g$ in a 70.1 Ti rotor using a Beckman ultracentrifuge.

4. Apply the supernatant (~ 30 ml) on to a 2-ml wheat germ agglutinin agarose (Vector) column at 4°C. Pass the extract over the column three times and then wash the agarose with 100 ml of 50 mM Hepes containing 0.1% Triton X-100.

5. Elute the insulin receptor and other membrane proteins from the column with 2 ml of 100 mM N-acetylglucosamine (Sigma) in 50 mM Hepes containing 0.1% Triton X-100.

4.2 Autophosphorylation of the WGA-purified insulin receptor

Autophosphorylation of the WGA-purified insulin receptor has become a standard method used in most studies addressing the function of the insulin receptor and the mechanism of signal transmission. The following procedure describes the phosphorylation of the WGA-purified insulin receptor obtained from transfected CHO cells.

Protocol 12. Autophosphorylation of WGA-purified insulin receptor

1. Incubate 4 μg of WGA-purified protein in 45 μl of 50 mM Hepes containing 5 mM $MnCl_2$ in the absence or presence of 100 nM insulin at 22°C for 10 min.

2. Initiate phosphorylation by adding 5 μl of 250 μM ATP containing 20 μCi of [γ-^{32}P]ATP (New England Nuclear). Continue the incubation for the desired time-interval: ≤ 1 approximates initial rate, whereas steady state is reached after 10 min.

3. Stop the autophosphorylation reaction by adding 50 μl of 2X Laemmli buffer to the phosphorylation reaction, followed immediately by SDS-PAGE. Alternatively, stop the reaction by dilution with 0.5 ml of 50 mM Hepes containing 0.1% Triton X-100 and 2 mM Na_3VO_4, and immunoprecipitate the receptor with 3 μg of α-PY as described in Section 3.4; typical results are shown in *Figure 5*.

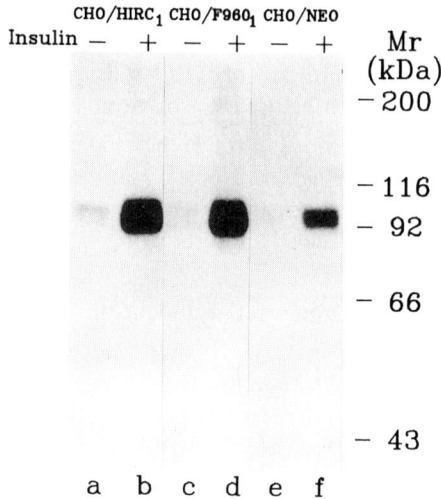

Figure 5. Phosphorylation of the β-subunit of the insulin receptor purified from transfected CHO cells on WGA-agarose. The insulin receptor was purified partially from CHO cells transfected with pSVE*neo* (CHO/NEO), or co-transfected with both pSVE*neo* and pCVSVHIRc and sorted three times (CHO/HIRC$_1$) or co-transfected with both pSVE*neo* and pCVSVHIRF/Y960 (2) and sorted five times (CHO/F960$_1$). Purified protein from each cell line (2 μg) was incubated in the absence ($-$) or presence ($+$) of 100 nM insulin. Phosphorylation was initiated by adding 25 μM ATP containing 20 μCi of [γ-^{32}P]ATP and the incubation was continued for 30 min. The reaction was terminated and the phosphorylated receptor was immunoprecipitated with the αPY, reduced with DTT and separated by SDS-PAGE. An autoradiogram obtained after 5 h exposure is shown.

4.3 WGA purification of the insulin receptor from phorbol ester-treated cells

We have described various modifications of the standard WGA purification. One involves the purification of the insulin receptor from phorbol ester-treated cells (29). The main concern here is that the serine/threonine phosphorylation state obtained during incubation of the intact cells with 12-O-tetradecanoylphorbol-13-acetate (TPA) must be preserved during isolation. We address this problem by adding phosphatase inhibitors to the extraction buffer. Insulin receptors prepared from TPA-treated cells according to this protocol retain an elevated level of serine phosphorylation (29), and a decreased level of insulin-stimulated tyrosine phosphorylation (*Figure 6*, A and B). The Ser(P) can be removed from the β-subunit by reaction with immobilized alkaline phosphatase, which also removes the inhibition of insulin-stimulated autophosphorylation (*Figure 6*, C).

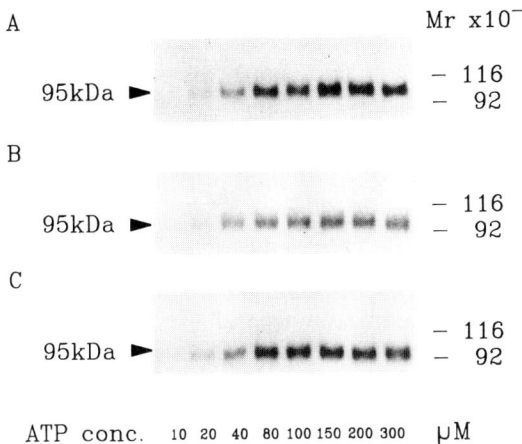

A

95kDa ▶

Mr x10⁻³

$-$ 116
$-$ 92

B

95kDa ▶

$-$ 116
$-$ 92

C

95kDa ▶

$-$ 116
$-$ 92

ATP conc.   10 20 40 80 100 150 200 300   µM

Figure 6. Autophosphorylation of the insulin receptor purified from control and TPA-treated Fao cells. The WGA-purified insulin receptor from control cells (Panel A), from TPA-treated cells incubated with heat-inactivated alkaline phosphatase-Sepharose (Panel B), and from the TPA-treated cells incubated with active alkaline phosphatase-Sepharose (Panel C) was incubated with 100 nM insulin at 4°C for 10 min. The initial velocity of autophosphorylation was measured during a 30 sec time-interval at the indicated concentrations of $[\gamma$-^{32}P]ATP. In each case, the reaction was terminated by adding 50 mM Hepes containing 2 mM Na_3VO_4 at 4°C. The insulin receptor was immunoprecipitated with the anti-phosphotyrosine antibody, reduced with DTT, separated by SDS-PAGE, and identified by autoradiography.

Protocol 13. WGA purification of insulin receptor from phorbol-ester-treated cells

1. Incubate cells with 1 µg/ml TPA for 30 min at 37°C.

2. Follow the standard WGA purification, but add phosphatase inhibitors to the extraction buffer of Section 4.1: NaF (100 mM), EDTA (4 mM) and $Na_2P_2O_7$ (10 mM).

3. Prepare and apply the cell extract to the WGA column as described in Section 4.1, but wash the column with 50 mM Hepes containing 0.1% Triton X-100 and the phosphatase inhibitors.

4. Finally, wash the WGA agarose with 10 ml of Hepes (50 mM) containing Triton X-100 (0.1%) and EDTA (4 mM), and elute the column with this solution containing 100 mM N-acetylglucosamine.

5. The β-subunit of the insulin receptor in the eluate contains an elevated level of Ser(P). This receptor can be used immediately for autophosphorylation experiments. Use 10 mM Mn^{2+} during autophosphorylation as the preparation contains EDTA.

6. The Ser(P)-containing receptor can be dephosphorylated on immobilized alkaline phosphatase as follows:

Protocol 13. *Continued*

(a) Prepare immobilized alkaline phosphatase by reacting 0.5 ml of Affi-gel 10 (BioRad) with 5 mg (1000 U/mg) of alkaline phosphatase from bovine intestinal mucosa (Sigma, P-2276) in 100 mM Hepes as described in the instructions.

(b) Wash the immobilized alkaline phosphatase with Hepes (50 mM) containing Triton X-100 (0.1%).

(c) Incubate 4 μg of WGA-purified insulin receptor with immobilized alkaline phosphatase (500 U of alkaline phosphatase) for 1 h at 4°C.

(d) Separate the supernatant from the Affi-Gel by centrifugation and carry out the autophosphorylation reaction.

4.4 WGA purification of the insulin receptor from trypsin-treated cells

Trypsin activates the tyrosine kinase in the β-subunit by removing the regulatory domain of the α-subunit (30). The action of trypsin on the insulin receptor *in situ* is very predictable, cleaving only the α-subunit. Thus this procedure can be used to obtain constitutively activated tyrosine kinase.

Protocol 14. WGA purification of the insulin receptor from trypsin-treated cells

1. Wash confluent 15-cm dishes of cells with 10-ml portions of phosphate-buffered saline three times at 4°C.

2. Incubate the cells with 2.0 ml of PBS containing 0.5 mg/ml trypsin at 4°C for 30 min.

3. Remove the cell suspension, and wash the cell three times at 4°C with PBS containing 0.5% soybean trypsin inhibitor (Sigma).

4. Solubilize the trypsinized cells at 4°C in 50 mM Hepes (pH 7.4) containing 1% Triton X-100, 0.1 mg/ml aprotinin and 2 mM PMSF.

5. Remove the insoluble material by centrifugation, and purify the insulin receptor on immobilized wheat germ agglutinin as described in Section 4.1.

5. Analysis of the phosphorylated insulin receptor

5.1 Tryptic peptide mapping of the β-subunit of the insulin receptor

Multiple tyrosyl residues undergo autophosphorylation in the β-subunit of the insulin receptor during insulin stimulation. The role of each auto-phosphorylation event has not been determined; apparently phosphorylation of tyrosyl residues 1146, 1150, and 1151 activate the phosphotransferase (1). We use the method in *Protocol 15* to resolve some of the tryptic phosphopeptides that contain these residues:

Protocol 15. Tryptic mapping of the insulin receptor β-subunit

1. Fixed and dried gel fragments containing the β-subunit are placed in 1.5 ml microfuge tubes and rehydrated with 1 ml of 50 mM NH_4HCO_3 (pH 8.2).

2. Add TPCK-treated trypsin (100 μg, Worthington) to this solution and incubated for 6 h at 37°C.

3. Add an additional 100 μg of trypsin and incubate for 12 h at 37°C.

4. Collect the supernatant, centrifuge to remove residual gel fragments, and lyophilize the supernatant for 12 h to remove the NH_4HCO_3.

5. Equilibrate an RP-318 wide-pore reverse-phase HPLC column (Bio-Rad) at a flow rate of 1.1 ml/min with water containing 0.05% trifluoroacetic acid and 5% acetonitrile.

6. Dissolve the dried material in 100 μl of water containing 0.05% trifluoroacetic acid and 5% acetonitrile; inject the phosphopeptides in 100 μl sample.

7. Elute the phosphopeptides with water containing 0.05% TFA and a gradient of acetonitrile increasing linearly from 5% to 25% during 85 min.

8. Collect 1.1 ml fractions at 1 min intervals in 1.5-ml microfuge tubes.

9. Measure the radioactivity in each tube by Cerenkov counting (40% efficiency) using a Beckman scintillation counter. The percentage of acetonitrile in each sample has no effect on the efficiency of the Cerenkov radiation. Greater than 95% of the radioactivity in the trypsin digest was routinely recovered from the reverse phase HPLC column.

5.2 Analysis of the tryptic phosphopeptide map

Many tryptic phosphopeptides are obtained from the β-subunit of the WGA-purified human insulin receptor phosphorylated to steady state with $[\gamma\text{-}^{32}P]ATP$ during insulin stimulation. The major peptides have been identified and are labelled pY1, pY1a, pY2, pY3, pY4, and pY5 (*Figure 7*, A). The structures of the tryptic peptides were deduced by a combination of techniques, and their identities are summarized in *Table 5*. The peptide maps are complicated because each fragment contains multiple Tyr(P) residues and trypsin cleavage sites. Three methods to characterize the tryptic peptide map are described below.

Table 5. The structure of the tryptic phosphopeptides of the β-subunit

Peptide	Amino acid residues	Structure
pY1	1143–1153	AspIleTyr(P)GluThrAspTyr(P)Tyr(P)ArgLys
pY1a	1143–1152	AspIleTyr(P)GluThrAspTyr(P)Tyr(P)Arg
pY2/pY3	1312–1330	*Arg*SerTyr(P)GluGluHisIleProTyr(P)ThrHisMetAsnGlyGly-Lys*Lys*
pY4	1143–1152	AspIleTyr(P)GluThrAsp[TyrTyr](P)Arg
pY5	1143–1153	AspIleTyr(P)GluThrAsp[TyrTyr](P)ArgLys

The identity of these tryptic fragments were determined previously (1). The amino-acid numbering system used is that of Ullrich *et al.* (11). The amino-acids in italics indicate variable trypsin cleavage sites that give rise to the various forms of these peptides; the exact length of the pY2 and pY3 peptide has not been determined.

5.2.1 Simplification of the peptide map by removal of the C-terminal domain with trypsin

The C-terminal domain of the β-subunit is by removed mild trypsin digestion of the WGA-purified insulin receptor (22). The truncated β-subunit loses about 40% of the Tyr(P). This decrease is due to the removal of two tryptic phosphopeptides, pY2 and pY3, both of which contain Tyr-1316 and Tyr-1322 (*Figure 7*, B).

Protocol 16 Partial trypsinolysis of the insulin receptor

1. Phosphorylate the insulin receptor as described in Section 4.2.
2. Stop the autophosphorylation reaction by adding 0.5 ml of 50 mM Hepes containing 0.1% Triton X-100 and 2 mM Na_3VO_4.
3. Digest the receptor with TPCK-treated trypsin (final concentration 5 μg/ml) for 1 min at 22°C.
4. Stop digestion by adding aprotinin to a final concentration of 10 μg/ml.
5. Add 3 μg of α-PY to immunoprecipitate the truncated phosphorylated insulin receptor as described in Section 3.4 and separate the protein by SDS-PAGE,
6. Digest the 85 kd fragment of the β-subunit as described in Section 5.1 and separate the tryptic phosphopeptides. The peaks due to the C-terminal phosphorylation sites will be missing.

5.2.2 Simplification of the tryptic phosphopeptide map by inhibition of the autophosphorylation cascade with α-PY

In vitro, 1–3 μg of α-PY added to the phosphorylation reaction (Section 4.2) inhibits insulin-stimulated autophosphorylation of the β-subunit by 70 to 80% and traps one of the first phosphotyrosine intermediates (1). Since the α-PY does

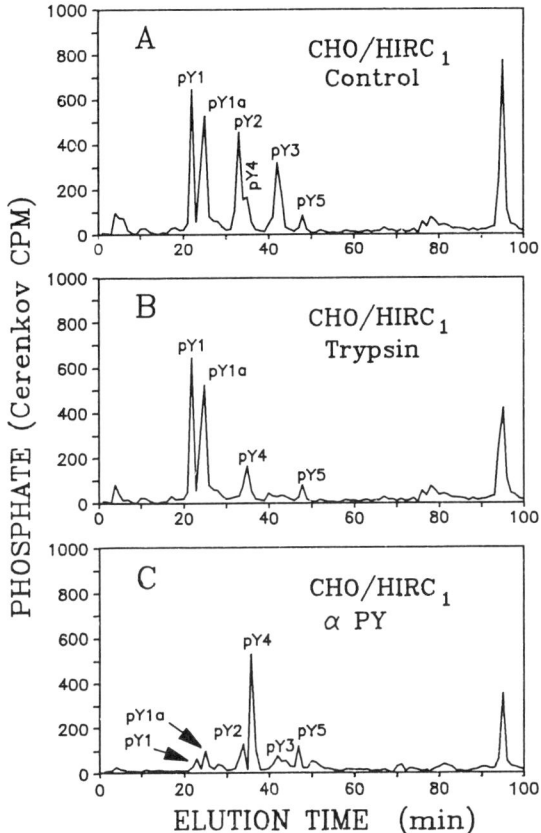

Figure 7. Separation of the tryptic phosphopeptides from the β-subunit of the human insulin receptor. The insulin receptor was purified from CHO/HIRC₁ cells by WGA-affinity chromatography. The receptor (2 μg) was incubated with 100 nM insulin for 20 min and then incubated with [γ-³²P]ATP (50 μM) for 30 min in the absence (A and B) of α-PY (Section 4.2) or presence (C) of 60 μg/ml α-PY (Section 5.2.2). Before immunoprecipitation, the phosphorylation receptor in (B) was incubated with trypsin as described in Section 4.4. The phosphorylated receptor was immunoprecipitated from each reaction with the α-PY, reduced with DTT, separated by SDS-PAGE and digested exhaustively with trypsin as described in *Protocol 14*. The tryptic phosphopeptides were separated by reverse phase HPLC as described in Section 5.1.

not bind to the insulin receptor purified from unstimulated cells, the α-PY probably binds to newly phosphorylated tyrosine residues which occur during the incubation with insulin and ATP. Tryptic peptide mapping indicates that the autophosphorylation sites in peptides pY4 (and sometimes pY5) are not inhibited by the αPY and actually increase, whereas all of the other phosphorylation sites in peptides pY1, pY1a, pY2, and pY3 are completely inhibited (*Figure 7*, C). Thus the regulatory domain of the insulin receptor is predominantly bis-phosphorylated in the presence of α-PY, containing Tyr(P)-

1146 and either Tyr(P)-1150 or Tyr(P)-1151 (*Table 5*). Since a mono-phosphotyrosyl-form of this domain was never observed, two tyrosine residues apparently undergo autophosphorylation before the α-PY binds and stops the cascade (1). Similar analysis of other insulin receptor systems may help determine the role of the autophosphorylation site in signal transmission. Moreover, autophosphorylation in the presence of α-PY provides a simple way to positively identify pY4 and pY5 on an HPLC profile.

5.2.3 Use of anti-peptide antibodies to identify tryptic phosphopeptides

Assignment of the structure of the tryptic phosphopeptides obtained from the insulin receptor can be done by immunoprecipitation with peptide specific antibodies targeting presumed regions of autophosphorylation. All of the major tryptic phosphopeptides resolved from the human insulin receptor immuno-precipitate with α-PY indicating that each peptide contains phosphotyrosine (*Figure 8*). However, the antibody to the Tyr-960 domain did not react with any of these fragments, suggesting that none of them were derived from this region of the receptor (*Figure 8*). By contrast, pY1, pY1a, and pY4 were specifically immunoprecipitated with the antibody to the Tyr-1150 domain, which verified that these peptides contained tyrosyl residues 1146, 1150, and 1151. pY2 and pY3 were specifically precipitated with the antibody to the C-terminal domain, which includes Tyr-1316 and Tyr-1322. This analysis is accomplished as in *Protocol 17*.

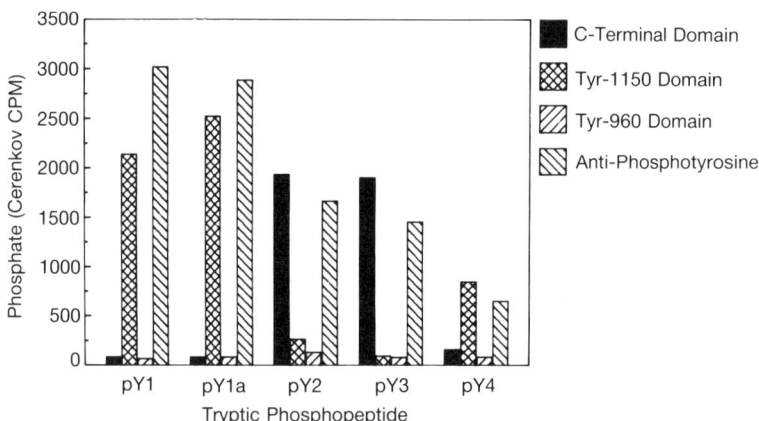

Figure 8. Identification of tryptic phosphopeptides of the human insulin receptor. The insulin receptor was phosphorylated for 30 min, and then immunoprecipitated with *a*PY, reduced with DTT, separated by SDS-PAGE and digested exhaustively with trypsin. Tryptic phosphopeptides were separated by reverse-phase HPLC and identified by the elution position. Each phosphopeptide was immunoprecipitated with the indicated domain-specific antibody as described in Section 5.2.3.

Protocol 17. Immunoprecipitation of insulin receptor tryptic peptides

1. Dry *in vacuo* HPLC fractions containing the [^{32}P]peptides and dissolve the residue in 0.5 ml of 50 mM Hepes containing 0.1% Triton X-100.

2. Add antibody (1 μg) to each sample and incubated for 12 h at 4°C.

3. Precipitate the antibody complex by adding of 50 μl of a 10% suspension of Pansorbin (Section 3.3.3).

4. Wash the precipitate three times with a 50 mM Hepes (pH 7.4) containing Triton X-100 (1%), SDS (0.1%), NaCl (150 mM), NaF (100 mM) and Na$_3$VO$_4$ (2 mM).

   Measure the radioactivity in the precipitate by Cerenkov counting.

6. Kinase activity of the purified insulin receptor

6.1 Measuring kinase activity with synthetic peptides

Phosphorylation of cellular proteins on tyrosyl residues is one of the potential mechanism of signal transmission by the insulin receptor. Thus an important assay to carry out on the receptor is the measurement of its phosphotransferase activity. Several proteins have been used as substrates for *in vitro* kinase assays. Recently we have used a synthetic peptide derived from the amino-acid sequence of the β-subunit including residues 1142–1152, called Thr-12-Lys, which is phosphorylated primarily at a single tyrosyl residue:

$$\text{Thr-Arg-Asp-Ile-Tyr-Glu-Tyr-Asp-Tyr-Tyr-Arg-Lys}$$
$$|$$
$$\text{Pi}$$

The kinase activity of the insulin receptor is measured by an initial velocity experiment described in *Protocol 18*.

Protocol 18. Insulin receptor tyrosine kinase activity assay

1. Prepare 16 reactions, by diluting in 16 microfuge tubes (8 for basal and 8 for insulin stimulated), WGA-purified receptor preparation (4 μg/tube) to 20 μl with 50 mM Hepes, pH 7.4, and a final concentration of 0.1% Triton X-100, 5 mM MnCl$_2$.

2. Measure basal kinase activity at eight different peptide concentrations by adding to 8 reaction tubes, 20 μl of Thr-12-Lys (0.2 mM, or 0.5, 1.0, 1.5, 2.0, 6.0, 8.0 or 10 mM) prepared in Hepes (pH 7.4) containing 0.1% Triton X-100 and 5 mM MnCl$_2$. Start the phosphorylation by adding 5 μl of Hepes containing 0.1% Triton X-100 and 5 μl of water containing 100 μM ATP and 20 μCi [γ-^{32}P]ATP. Incubate for 5 min.

Protocol 18. *Continued*

3. Measure insulin-stimulated activity by adding 5 μl of 10^{-6} M insulin to 8 reaction tubes. Incubate the reaction mixtures for 20 min. Start the autophosphorylation reaction by adding 5 μl of 100 μM ATP containing 20 μCi [γ-^{32}P]ATP. Incubate for 5 min at 22°C, and then add 25 μl of 2 × Thr-12-Lys (0.2 mM, or 0.5, 1.0, 1.5, 2.0, 6.0, 8.0, or 10 mM) prepared in Hepes (pH 7.4) containing 0.1% Triton X-100 and 5 mM $MnCl_2$. Incubate for 5 min.

4. Stop the reaction after 5 min by adding 20 μl of 1% bovine serum albumin followed immediately by 50 μl of 10% TCA.

5. Sediment the precipitated protein by centrifugation.

6. Apply the supernatant to a 2 × 2 cm piece of phosphocellulose paper (Whatman).

7. Wash the paper with four changes of 1 litre each 75 mM phosphoric acid.

8. Measure the radioactivity by Cerenkov counting and plot the results as pmol of phosphate incorporated against the peptide concentrations.

6.2 Taking advantage of substrate inhibition to determine the kinase activity of the receptor stimulated in the intact cell

Insulin-stimulated autophosphorylation of the β-subunit increases the phosphotransferase activity of the purified insulin receptor toward exogenous substrates such as the dodecapeptide, Thr-12-Lys (*Figure 9*). However, high concentrations of this peptide also inhibit autophosphorylation, which blocks the activation of the phosphotransferase and causes biphasic kinetic curves (*Figure 9* and reference 1). In the absence of insulin stimulation and prior autophosphorylation, the V_{max} for phosphorylation of Thr-12-Lys by the human insulin receptor is about 0.05 pmol/min (*Figure 9*). By contrast, insulin-stimulated autophosphorylation for 5 min before addition of the peptides, stimulates the phosphotransferase more than tenfold during assays at concentrations of Thr-12-Lys below 1 mM; however, Thr-12-Lys above 1 mM inhibited a portion of the stimulation, causing a biphasic kinetic curve (*Figure 9*). We have shown previously that this biphasic kinetic curve results from the inhibition of additional autophosphorylation which occurs during the *in vitro* kinase assay (1).

Inhibition of autophosphorylation *in vitro* can be used to determine the activity of the receptor kinase purified from insulin-stimulated cells. Inhibitory concentrations of Thr-12-Lys prevent *in vitro* autophosphorylation and activation of the receptor while acting as a substrate to measure kinase activity resulting from insulin stimulation of the receptor *in vivo*.

Protocol 19. Determination of receptor kinase activity through substrate inhibition

1. Stimulate two confluent dishes (15-cm) of cells with 100 nM insulin, and extract the receptor with 50 mM Hepes containing 1% Triton and protease and phosphatase inhibitors as described in Section 3.2.2.

2. Incubate the extract ($\sim$4 ml) with 0.25 ml of WGA-agarose (Vector) in the presence of phosphatase inhibitors. Transfer the suspension to a small column and remove the supernatant by centrifugation.

3. Close the column outlet, and immediately suspend the WGA-agarose in 200 μl of 50 mM Hepes containing 0.1% Triton X-100, 1 mM Na_3VO_4, and 100 mM *N*-acetylglucosamine. Incubate for 15 min at 4°C and collect the supernatant by centrifugation.

4. Carry out the phosphorylation experiment described in Section 6.1. If the receptor is fully activated *in vivo* no substrate inhibition will be observed, and the kinetic curves will be identical rectangular hyperbole in the absence and presence of insulin. If the receptor is not fully activated *in vivo*, then additional activation will occur *in vitro* and biphasic kinetic curves will emerge from the velocity verses substrate plots.

Acknowledgements

I am indebted to C. R. Kahn and my other colleagues for their suggestions and guidance throughout the years. This work has been supported in part by NIH grants DK35988 and DK38712. MFW is a scholar of the PEW Foundation, Philadelphia.

References

1. White, M. F., Shoelson, S. E., Keutmann, H., and Kahn, C. R. (1988). *J. Biol. Chem.*, **263**, 2969.
2. White, M. F., Livingston, J. N., Backer, J. M., Lauris, V., Dull, T. J., Ullrich, A., and Kahn, C. R. (1988). *Cell*, **54**, 641–49.
3. White, M. F., Maron, R., and Kahn, C. R. (1985). *Nature, Lond.*, **318**, 183.
4. Haring, H. U., White, M. F., Machicao, F., Ermel, B., Schleicher, E., and Obermaier, B. (1987). *Proc. Natl. Acad. Sci. USA*, **84**, 113.
5. Perrotti, N., Accili, D., Marcus Samuels, B., Rees Jones, R. W., and Taylor, S. I. (1987). *Proc. Natl. Acad. Sci. USA*, **84**, 3137.
6. Bernier, M., Laird, D. M., and Lane, M. D. (1987). *Proc. Natl. Acad. Sci. USA*, **84**, 1844–1848.
7. White, M. F., Stegmann, E. W., Dull, T. J., Ullrich, A., and Kahn, C. R. (1987). *J. Biol. Chem.*, **262**, 9769.

8. Izumi, T., White, M. F., Kadowaki, T., Takaku, F., Akanuma, Y., and Kasuga, M. (1987). *J. Biol. Chem.*, **262**, 1282.

9. Kadowaki, T., Koyasu, S., Nishida, E., Tobe, K., Izumi, T., Takaku, F., Sakai, H., Yahara, I., and Kasuga, M. (1987). *J. Biol. Chem.*, **262**, 7342.

10. Boonstra, J., Mummery, C. L., Feyen, A., de Hoog, W. J., van der, Saag, P. T., and de Laat, S. W. (1987). *J. Cell Physiol.*, **131**, 409.

11. Ullrich, A., Bell, J. R., Chen, E. Y., Herrera, R., Petruzzelli, L. M., Dull, T. J., Gray, A., Coussens, L., Liao, Y.-C., Tsubokawa, M., Mason, A., Seeburg, P. H., Grunfeld, C., Rosen, O. M., and Ramachandran, J. (1985). *Nature, Lond.*, **313**, 756.

12. Ebina, Y., Ellis, L., Jarnagin, K., Edery, M., Graf, L., Clauser, E., Ou, J.-H., Masiar, F., Kan, Y. W., Goldfine, I. D., Roth, R. A., and Rutter, W. J. (1985). *Cell*, **40**, 747.

13. Whittaker, J., Okamoto, A. K., Thys, R., Bell, G. I., Steiner, D. F., and Hofmann, C. A. (1987). *Proc. Natl. Acad. Sci. USA*, **84**, 5237.

14. McClain, D. A., Maegawa, H., Lee, J., Dull, T. J., Ullrich, A., and Olefsky, J. M. (1987). *J. Biol. Chem.*, **262**, 14663.

15. Cullen, B. R. (1987). *Methods in Enzymology* (ed S. L. Berger and A. R. Kimmel), Vol. 152, pp. 684–704. Academic Press, New York.

16. Kunkel, T. A., Roberts, J. D., and Zakour, R. A. (1987). *Methods in Enzymology* (ed R. Wu and L. Grossman), Vol. 154, p. 382. Academic Press, New York.

17. Tabor, S. and Richarson, C. C. (1980). *Proc. Natl. Acad. Sci. USA*, **84**, 4767.

18. Shapiro, J. (1981). *Radiation Protection. A Guide for Scientists and Physicians.* Harvard University Press, Cambridge, Mass.

19. Rodbard, D., Munson, P. J., and Thakur, A. K. (1980). *Cancer*, **46**, 2907.

20. Kahn, C. R., Baird, K. L., Flier, J. S., Grunfeld, C., Harmon, J. T., Harrison, L. C., Karlsson, F. A., Kasuga, M., King, G. L., Lang, U., Podskalny, J. M., and Van Obberghen, E. (1981). *Rec. Prog. Horm. Res.*, **37**, 447.

21. Morgan, D. O. and Roth, R. A. (1987). *Proc. Natl. Acad. Sci. USA*, **84**, 41.

22. Goren, H. J., White, M. F., and Kahn, C. R. (1987). *Biochemistry*, **26**, 2374.

23. Herrera, R., Petruzzelli, L., Thomas, N., Bramson, H. N., Kaiser, E. T., and Rosen, O. M. (1985). *Proc. Natl. Acad. Sci. USA*, **82**, 7899.

24. Laemmli, U. K. (1970). *Nature, Lond.*, **227**, 680.

25. Kasuga, M., White, M. F., and Kahn, C. R. (1985). *Methods in Enzymology* (ed L. Birnbaum and B. O'Malley), Vol. 109, pp. 609–21. Academic Press, New York.

26. Pang, D. T., Sharma, B. R., and Shafer, J. A. (1985). *Arch. Biochem. Biophys.*, **242**, 176.

27. Pang, D. T., Sharma, B., Shafer, J. A., White, M. F., and Kahn, C. R. (1985). *J. Biol. Chem.*, **260**, 7131.

28. Davis, L. G., Dibner, M. D., and Battey, J. F. (ed.) (1986). In *Basic Methods in Molecular Biology*, pp. 311–17. Elsevier, New York.

29. Takayama, S., White, M. F., and Kahn, C. R. (1988). *J. Biol. Chem.*, **263**, 3440.

30. Shoelson, S. E., White, M. F., and Kahn, C. R. (1988). *J. Biol. Chem.*, **263**, 4852.

Index

Index

protein kinase C (*cont.*)
  membrane and cytosolic forms 203
  phorbol ester binding assay 207–8
  purification 208, 214
  substrate phosphorylation in intact cells
    211–13
  translocation assay 205–6
proteinases, *see* proteases
protein kinase, *see* kinases
proteolysis, of phosphodiesterase 106
Ptdlns, *see* phosphatidyl inositol
purification, *see also* affinity purification,
  immunopurification,
  of calcium activated photoproteins 130–1
  of cAMP-dependent protein kinase 110
  of inositol lipids 161–3
  of phosphodiesterase 106–9
  of protein kinase C 208, 214
  of radioiodinated peptides 7
  of receptors 49, 54–60

quin-2 117, 127, 147

radioimmunoassay
  of cyclic nucleotides 78–81
  of prostaglandins 218
radioiodination, *see* iodination
radioisotopes
  for labelling inositol phosphates 64–171
  for labelling polypeptides 2
radiolabelling, of cells with ^{32}P-phosphate
  230–1
radioligand, *see* tracers
rate constant, for hormone binding 20–2
receptor binding
  analysis 25–32
  assay 14–19
  buffers 15
  kinetics 19–24
  of peptides 8–19
  specificity 24
  temperature 15
  theory 20
receptors, *see also* insulin receptors
  for binding assay 10
  concentration 60
  hydrodynamic properties 64–6
  recycling 34
  solubilization 13
recycling, of receptors 34
ryanodine 148

safety, with ^{32}P-phosphate 230
Scatchard plot, of insulin binding 29, 227–30
sedimentation coefficient 66

sedimentation, sucrose gradient 65
separation
  in cyclic nucleotide assay 80
  of bound and free tracer 16
silicone oil, in receptor binding assays 16
single-cell analysis, of calcium 128, 137
solubilization
  of membranes 47
  of tissue 48
specific radioactivity
  of inositol lipids 152
  of iodinated tracers 8
steady-state, receptor binding 23, 27
staurospirine, as inhibitor of protein kinase C
  212
sucrose gradient centrifugation, of receptors
  65
suramin 33

teleocidin 200
thin layer chromatography, of inositol lipids
  162
thymocytes, calcium content of 140
thyroid tissue, adenylate cyclase assay 86
toxin
  cholera 74, 93
  pertussis 74, 93
TPA, *see* phorbol ester
tracers
  immunoreactivity 7, 38
  receptor binding assay 9
  specific activity 8
  validation 7
transfection, of CHO cells 225
translocation, of protein kinase C 202–7
trichloracetic acid
  in analysis of peptide degradation 37
  for extraction of inositol phosphates 155
Triton X-100, for receptor solubilization 13,
  47
trypsin
  for insulin receptor activation 242
  for phosphopeptide mapping 243
  for protein kinase C activation 205–7
tyrosine kinase assay 47, 247–9

vasopressin, and phosphatidylcholine
  hydrolysis 188
verapamil 146

Western blotting, *see* immunoblotting
wheat germ agglutinin 14, 50, 238

zinc 127, 140